Applications of Intelligent Control to Engineering Systems

International Series on
INTELLIGENT SYSTEMS, CONTROL, AND AUTOMATION:
SCIENCE AND ENGINEERING

VOLUME 39

For other titles published in this series, go to
www.springer.com/series/6259

Kimon P. Valavanis
Editor

Applications of Intelligent Control to Engineering Systems

In Honour of Dr. G. J. Vachtsevanos

Springer

Editor
Kimon P. Valavanis
Department of Electrical & Computer Engineering
University of Denver
Denver, CO 80208
USA

ISBN 978-90-481-3017-1 e-ISBN 978-90-481-3018-4
Springer Dordrecht Heidelberg London New York

Library of Congress Control Number: 2009929445

Printed on acid-free paper

Springer is part of Springer Science+Business Media (www.springer.com)

In honour of Dr. George J. Vachtsevanos, our Dr. V.,
from his former students and colleagues

Table of Contents

Part III: Bioengineering/Neurotechnology

Part IV: Intelligent Control Systems

Preface

This edited book is published in honor of Dr. George J. Vachtsevanos, *our Dr. V*, currently Professor Emeritus, School of Electrical and Computer Engineering, Georgia Institute of Technology, on the occasion of his 70th birthday and for his more than 30 years of contribution to the discipline of Intelligent Control and its application to a wide spectrum of engineering and bioengineering systems.

The book is nothing but a very small token of appreciation from Dr. V's former graduate students, his peers and colleagues in the profession – and not only – to the Scientist, the Engineer, the Professor, the mentor, but most important of all, to the friend and human being. All those who have met Dr. V over the years and have interacted with him in some professional and/or social capacity understand this statement: George never made anybody feel inferior to him, he helped and supported everybody, and he was there when anybody needed him!

I was not Dr. V's student. I first met him and his wife Athena more than 26 years ago during one of their visits to RPI, in the house of my late advisor, Dr. George N. Saridis. Since then, I have been very fortunate to have had and continue to have interactions with him. It is not an exaggeration if I say that we all learned a lot from him. We understood that theory and applications go together and they do not oppose each other; we acquired a wide and well rounded perspective of what engineering is; we witnessed firsthand how diverse concepts and ideas are brought under the same framework and how they are applied successfully; we were fortunate to see how conventional control techniques and soft-computing techniques work in unison to complete tasks and missions in complex multi-level systems.

During his tenure at GaTech, Dr. V and his group established a unique 'school of thought' in *systems*, where the term *system* may be interpreted as being rather abstract or very specific. Without sacrificing theoretical developments and contributions, he and his group have demonstrated repeatedly, how complex systems function in real-time.

It is probably safe to claim that Dr. V and his group have made and continue to make seminal and significant contributions to four areas: Intelligent Control Systems; Unmanned Aircraft Systems; Diagnostics, Prognostics and Condition-Based Maintenance; Bioengineering and Neurotechnology. The impact of their research findings and accomplishments is evidenced by publications, citations, prototypes and final products, to say the least. What distinguishes *that* group from any other is 'sustainability and continuity'! They, and all of us, keep on working together over the years, so the *torch* as far as the school of thought is concerned is passed to from one student generation to another.

When we first toyed with the idea to have a Workshop in honor of Dr. V, in a nice Greek Island so that we can mix business and pleasure, Frank (Dr. Frank Lewis) and I jumped on the opportunity to publish this book, and Frank allowed me to take the lead. The 15 contributed chapters are just a small sample of projects in which Dr. V and his group are either involved directly, or *influenced* because of their work.

The book is divided into four parts. Part I refers to diagnostics, prognostics and condition-based maintenance, an area which is the natural outgrowth of integrated control and diagnostics. It includes four contributed chapters, however, if one wants to obtain expert knowledge on the subject, should read the book *Intelligent Fault Diagnosis and Prognosis for Engineering Systems*, by G. Vachtsevanos et al.

Part II refers to Unmanned Aerial Systems, an area pioneered and dominated for years by the GaTech group. The GTmax is most likely the most widely known autonomous unmanned helicopter, capable of functioning under failures; the Software-Enabled Control approach and the Open Control Platform (OCP) Architecture are two unique, seminal contributions that have changed the way systems are modeled and controlled. The hardware-in-the-loop and software-in-the-loop verification and validation are two approaches almost everybody uses when dealing with complex systems, not necessarily unmanned. This part includes four chapters.

Part III refers to Bioengineering and Neurotechnology. This is, to my surprise, kind of ‘a well kept secret’, because Dr. V’s group has indeed contributed greatly to studying, modeling and predicting (this is the key word, *predicting*) epileptic seizures in patients! This, coupled with contributions to neurostimulation using intelligent control techniques will advance the field of Bioengineering making the most impact in improving the quality of life of patients.

Part IV refers to Intelligent Control Systems, chronologically the first and everlasting area in which Dr. V and his group are contributing to. This part is composed of five chapters, ‘disconnected’ so to speak, giving a flavor of the diversity of disciplines Intelligent Control may be applied to.

It is unfortunate that we could not include more contributions to this edited volume. Regardless, the response of people to joining and participating in the Workshop on June 27–29 in the Island of Lemnos, Athena’s birthplace, has been beyond expectations.

The Springer group, led by Ms. Nathalie Jacobs, and Karada Publishing Services led by our point of contact, Ms. Jolanda, have been very supportive throughout this project. Both Nathalie and Jolanda have worked very hard so that the book is available during the Workshop.

Last, on a personal note, I want to state that Dr. V has supported me over the years, has guided me professionally, has been a true mentor and friend, and has backed up every crazy idea (!!!) I have come up with. I am honored by his friendship, and I owe a lot to him.

Kimon P. Valavanis
Denver, May 6, 2009

List of Contributors

Chaouki T. Abdallah
Department of Electrical & Computer Engineering
University of New Mexico
Albuquerque, NM 87131, USA
chaouki@ece.unm.edu

Konstantinos Alexis
Department of Electrical & Computer Engineering
University of Patras
Rio, Achaia 26500, Greece
kostalexis@ece.upatras.gr

H. Bae
School of Electrical & Computer Engineering
Georgia Institute of Technology
Atlanta, Georgia 30332-0250, USA

Roy Bakay
Department of Neurology
Emory University School of Medicine
Atlanta, GA 30332, USA

Laura Barnes
Automation and Robotics Research Institute
The University of Texas at Arlington
Fort Worth, TX 76118-7115, USA
lbarnes@arri.uta.edu

Yannis S. Boutalis
Laboratory of Automatic Control Systems & Robotics
Department of Electrical & Computer Engineering
Democritus University of Thrace
67100 Xanthi, Greece
ybout@ee.duth.gr

D. Brown
School of Electrical & Computer Engineering
Georgia Institute of Technology
Atlanta, Georgia 30332-0250, USA

Carl S. Byington
Impact Technologies
200 Canal View Blvd
Rochester, NY 14623, USA
carl.byington@impact-tek.com

R. Chen
Northrop Grumman Corporation
El Segundo, CA 90245, USA
robert.chen@ngc.com

Manolis A. Christodoulou
Department of Electrical & Electronic Engineering
Technical University of Crete
73100 Chania, Crete, Greece
manolis@ece.tuc.gr

Maryann D'Alessandro
School of Electrical & Computer Engineering
Georgia Institute of Technology
Atlanta, GA 30332, USA
and
Army Research Labs
Atlanta, GA 30332, USA

Marc Dichter
Department of Neurology & Bioengineering
Hospital of the University of Pennsylvania
Philadelphia, PA 19104, USA

Leonidas Dritsas
Department of Electrical & Computer Engineering
University of Patras
Rio, Achaia 26500, Greece
dritsas@ece.upatras.gr

Javier Echauz
Senior Research Scientist and President
JE Research, Inc.

Alpharetta, GA, USA
echauz@ieee.org

Charles Epstein
Department of Neurology
Emory University School of Medicine
Atlanta, GA 30332, USA

Rosana Esteller
School of Electrical & Computer Engineering
Georgia Institute of Technology
Atlanta, GA 30332, USA
and
Departamento Tecnología Industrial
Universidad Simón Bolívar
89000 Caracas, Venezuela

Hiram Firpi
Department of Internal Medicine
2294 CBRB, 285 Newton Road
Iowa City, IA 52242, USA
hiram-firpi@uiowa.edu

Richard D. Garcia
US Army Research Laboratory
Aberdeen Proving Ground, MD 21005, USA
richard.d.garcia@arl.army.mil

George Georgoulas
TEI of the Ionian Islands
Computer Technology Applications in Management & Economics
31100 Lefkada, Greece
georgoul@teiion.gr

Kai Goebel
NASA Ames Research Center
MS 269-4, Moffett Field, CA 94035, USA
kai.goebel@nasa.gov

Bonnie S. Heck
School of Electrical & Computer Engineering
Georgia Institute of Technology
Atlanta, GA 30332-0250, USA

Thomas Henry
Department of Neurology
Emory University School of Medicine
Atlanta, GA 30332, USA

Y.H. Ho
Northrop Grumman Corporation
El Segundo, CA 90245, USA

Gregory Kacprzynski
Impact Technologies, LLC
200 Canal View Blvd.
Rochester, NY 14623, USA
greg.kacprzynski@impact-tek.com

Jie Huang
Department of Mechanical and Automation Engineering
The Chinese University of Hong Kong
Hong Kong
hwzhang@mae.cuhk.edu.hk

Taimoor Khawaja
School of Electrical & Computer Engineering
Georgia Institute of Technology
Atlanta, GA 30332, USA
taimoor@ece.gatech.edu

Joud Khoury
Department of Electrical & Computer Engineering
University of New Mexico
Albuquerque, NM 87131, USA
jkhoury@ece.unm.edu

Michael Kontitsis
Computer Science and Engineering
University of South Florida
Tampa, FL 33620, USA
mkontits@cse.usf.edu

Thodoris L. Kottas
Laboratory of Automatic Control Systems & Robotics
Department of Electrical & Computer Engineering
Democritus University of Thrace
67100 Xanthi, Greece
tkottas@ee.duth.gr

Frank L. Lewis
Automation and Robotics Research Institute
Department of Electrical Engineering
The University of Texas at Arlington
Fort Worth, TX 76118-7115, USA
lewis@uta.edu

Brian Litt
Department of Neurology & Bioengineering
Hospital of the University of Pennsylvania
Philadelphia, PA 19104, USA

Linda M. Wills
School of Electrical & Computer Engineering
Georgia Institute of Technology
Atlanta, GA 30332-0250, USA

George Nikolakopoulos
Department of Electrical & Computer Engineering
University of Patras
Rio, Achaia 26500, Greece
gnikolak@ece.upatras.gr

Marcos Orchard
School of Electrical & Computer Engineering
Georgia Institute of Technology
Atlanta, GA 30332, USA
and
Department of Electrical Engineering
University of Chile
Santiago, Chile
marcos.orchard@ece.gatech.edu

Romano Patrick
School of Electrical & Computer Engineering
Georgia Institute of Technology
Atlanta, GA 30332, USA
romano.patrick@ece.gatech.edu

Page Pennell
Department of Neurology
Emory University School of Medicine
Atlanta, GA 30332, USA

Michael J. Roemer
Impact Technologies
200 Canal View Blvd
Rochester, NY 14623, USA
mike.roemer@impact-tek.com

Bhaskar Saha
MCT, NASA Ames Research Center
MS 269-4, Moffett Field, CA 94035, USA
bsaha@email.arc.nasa.gov

Abhinav Saxena
School of Electrical & Computer Engineering
Georgia Institute of Technology
Atlanta, GA 30332, USA
asaxena@ece.gatech.edu

J.B. Schroeder
AFRL/RBCC
Wright Patterson AFB, OH, USA
john.schroeder@wpafb.af.mil

Rachel Shor
School of Electrical & Computer Engineering
Georgia Institute of Technology
Atlanta, GA 30332, USA

Dimitris G. Stavrakoudis
Aristotle University of Thessaloniki
Department of Electrical & Computer Engineering
Division of Electronics and Computer Engineering
54124 Thessaloniki, Greece
jstavrak@auth.gr

G. Tannenbaum
Moog Inc.
East Aurora, New York 14052, USA
gtannenbaum@moog.com

John B. Theocharis
Aristotle University of Thessaloniki
Department of Electrical & Computer Engineering
Division of Electronics and Computer Engineering
54124 Thessaloniki, Greece
theochar@eng.auth.gr

Athanasios Tsalatsanis
Department of Industrial and Management Systems Engineering
University of South Florida
Tampa, FL 33620, USA
atsalats@mail.usf.edu

Anthony Tzes
Department of Electrical & Computer Engineering
University of Patras
Rio, Achaia 26500, Greece
tzes@ece.upatras.gr

George J. Vachtsevanos
School of Electrical & Computer Engineering
Georgia Institute of Technology
Atlanta, GA 30332, USA
gjv@ece.gatech.edu

Kimon P. Valavanis
Department of Electrical & Computer Engineering
University of Denver
Denver, CO 80208, USA
kimon.valavanis@du.edu

Ali Yalcin
Department of Industrial and Management Systems Engineering
University of South Florida
Tampa, FL 33620, USA
ayalcin@eng.usf.edu

George C. Zalidis
Lab of Applied Soil Science
Faculty of Agronomy
Aristotle University of Thessaloniki
54124 Thessaloniki, Greece
zalidis@agro.auth.gr

Bing Zhang
School of Electrical & Computer Engineering
Georgia Institute of Technology
Atlanta, GA 30332, USA
bin.zhang@ece.gatech.edu

Hongwei Zhang
Department of Mechanical and Automation Engineering
The Chinese University of Hong Kong
Hong Kong
hwzhang@mae.cuhk.edu.hk

PART I

DIAGNOSTICS, PROGNOSTICS, CONDITION-BASED MAINTENANCE

Chapter 1
Selected Prognostic Methods with Application to an Integrated Health Management System

Carl S. Byington and Michael J. Roemer

Abstract Due to the increasing desire for having more autonomous vehicle platforms and life cycle support mechanisms, there is a great need for the development of prognostic health management technologies that can detect, isolate and assess remaining useful life of critical subsystems. To meet these needs for next generation systems, dedicated prognostic algorithms must be developed that are capable of operating in an autonomous and real-time vehicle health management system that is distributed in nature and can assess overall vehicle health and its ability to complete a desired mission. This envisioned prognostic and health management system should allow vehicle-level reasoners to have visibility and insight into the results of local diagnostic and prognostic technologies implemented down at the LRU and subsystem levels. To accomplish this effectively requires an integrated suite of prognostic technologies that can be applied to critical systems and can capture fault/failure mode propagation and interactions that occur in these systems, all the way up through the vehicle level. In the chapter, the authors will present a generic set of selected prognostic algorithm approaches, as well as provide an overview of the required vehicle-level reasoning architecture needed to integrate the prognostic information across systems.

1.1 Introduction

Various health monitoring technologies have been developed for aerospace applications that aid in the detection and classification of developing system faults [16, 20, 23]. However, these technologies have traditionally focused on fault detection and isolation within an individual subsystem or system. Health management system developers are just beginning to address the concepts of prognostics and the integration of anomaly, diagnostic and prognostic technologies across subsys-

Carl S. Byington · Michael J. Roemer
Impact Technologies, LLC, 200 Canal View Blvd., Rochester, NY 14623, USA

K.P. Valavanis (ed.), Applications of Intelligent Control to Engineering Systems, 3–21.

tems and systems. Hence, the ability to first detect and isolate impending faults and then predict their future progression based on its current diagnostic state and available operating data is currently a high priority in various defense and commercial vehicle applications. Also, the capability for updating these failure predictions based on either measured or inferred features related to the progression of the fault over time is also desirable.

However, there are inherent uncertainties in any prognosis system; thus achieving the best possible prediction on a LRU/subsystem's health is often implemented using various algorithmic techniques and data fusion concepts that can optimally combine sensor data, empirical/physics-based models and historical information. By utilizing a combination of health monitoring data and model-based techniques, a comprehensive prognostic capability can be achieved throughout a component's or LRU's life, using model-based estimates when no diagnostic indicators are present and monitored features at later stages when failure indications are detectable.

Finally, these technologies must be capable of communicating the root cause of a problem across subsystems and propagating the up/downstream effects across the health management system architecture. This paper will introduce some generic prognostic and health management (PHM) system algorithmic approaches that have been previously demonstrated within various aircraft subsystem components with the ability to predict the time to conditional or mechanical failure (on a real-time basis). Prognostic and health management systems that can effectively implement the capabilities presented herein offer a great opportunity in terms of reducing the overall Life Cycle Costs (LCC) of operating systems as well as decreasing the operations/maintenance logistics footprint.

1.2 Prognostic Algorithm Approaches

In the engineering disciplines, fault prognosis has been approached via a variety of techniques ranging from Bayesian estimation and other probabilistic/statistical methods to artificial intelligence tools and methodologies based on notions from the computational intelligence arena. Specific enabling technologies include multistep adaptive Kalman filtering [12], auto-regressive moving average models [13], stochastic auto-regressive integrated moving average models [6], Weibull models [8] forecasting by pattern and cluster search [7], and parameter estimation methods [15]. From the artificial intelligence domain, case-based reasoning [1], intelligent decision-based models and min-max graphs have been considered as potential candidates from prognostic algorithms. Other methodologies, such as Petri nets, neural networks, fuzzy systems and neuro-fuzzy systems [25] have found ample utility as prognostic tools as well. Physics-based fatigue models [18, 19] have been extensively employed to represent the initiation and propagation of structural anomalies.

Next, we will provide a brief overview of a representative sample of the multitude of enabling technologies. Figure 1.1 summarizes the range of possible prognostic

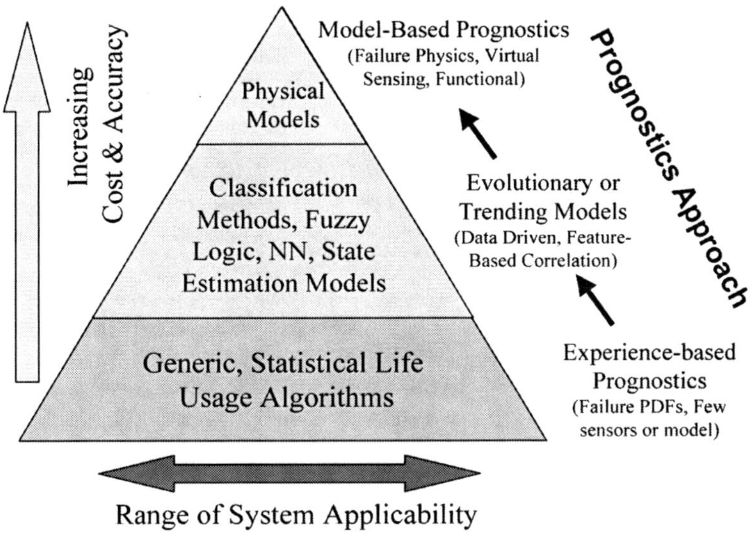

Fig. 1.1 Prognosis technical approaches.

approaches as a function of the applicability to various systems and their relative implementation cost. Prognostic technologies typically utilize measured or inferred features, as well as data-driven and/or physics-based models to predict the condition of the system at some future time. Inherently probabilistic or uncertain in nature, prognostics can be applied to failure modes governed by material condition or by functional loss. Prognostic algorithms can be generic in design but specific in terms of application. Prognostic system developers have implemented various approaches and associated algorithmic libraries for customizing applications that range in fidelity from simple historical/usage models to approaches that utilize advanced feature analysis or physics-of-failure models.

Depending on the criticality of the LRU or subsystem being monitored, various levels of data, models and historical information will be needed to develop and implement the desired prognostic approach. Table 1.1 provides an overview of the recommended models and information necessary for implementing specific approaches. Of course, the resolution of this table only illustrates three levels of algorithms, from the simplest experienced-based (reliability) methods to the most advanced physics of failure approaches that are calibrated by sensor data.

1.3 Statistical Reliability and Usage-Based Approaches

In situations where sophisticated prognostic models are not warranted due to the lower level of criticality or low failure occurrence rates and/or there is an insuf-

Table 1.1 Prognostic accuracy.

	Experience-based	Evolutionary	Physics-based
Engineering model	Not required	Beneficial	Required
Failure history	Required	Not required	Beneficial
Past operating conditions	Beneficial	Not required	Required
Current conditions	Beneficial	Required	Required
Identified fault patterns	Not required	Required	Required
Maintenance history	Beneficial	Not required	Beneficial
In general	No Sensors/No Model	Sensors/No Model	Sensors and Model

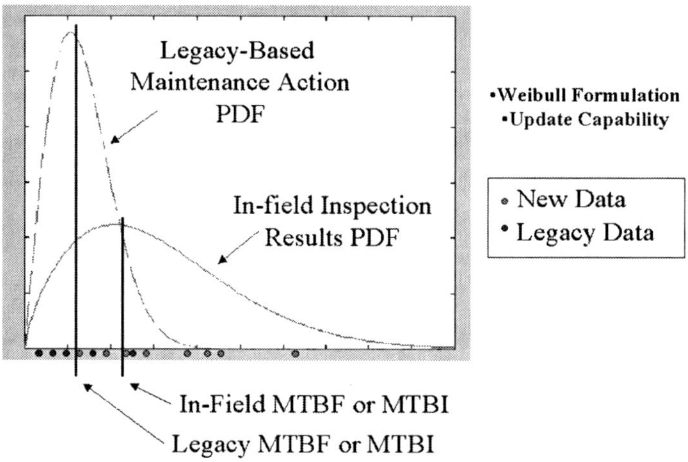

Fig. 1.2 Statistical reliability-based approach.

ficient sensor network to assess condition, a statistical reliability or usage-based prognostic approach may be the only alternative. This form of prognostic algorithm is the least complex and requires the component/LRU failure history data and/or operational usage profile data. Typically, failure and/or inspection data is compiled from legacy systems and a Weibull distribution or other statistical failure distribution can be fitted to the data [8, 21]. An example of these types of distributions is given in Figure 1.2. Although simplistic, a statistical reliability-based prognostic distribution can be used to drive interval-based maintenance practices that can then be updated on regular intervals. An example may be the maintenance scheduling for an electrical component or airframe component that has few or no sensed parameters and is not critical enough to warrant a physical model. In this case, the prognosis of when the component will fail or degrade to an unacceptable condition must be based solely on analysis of past experience or reliability. Depending on the maintenance complexity and criticality associated with the component, the prognostics system may be set up for a maintenance interval (i.e. replace every 1000 ± 20 Engine Flight Hours) then updated as more data becomes available. The benefit to having a regu-

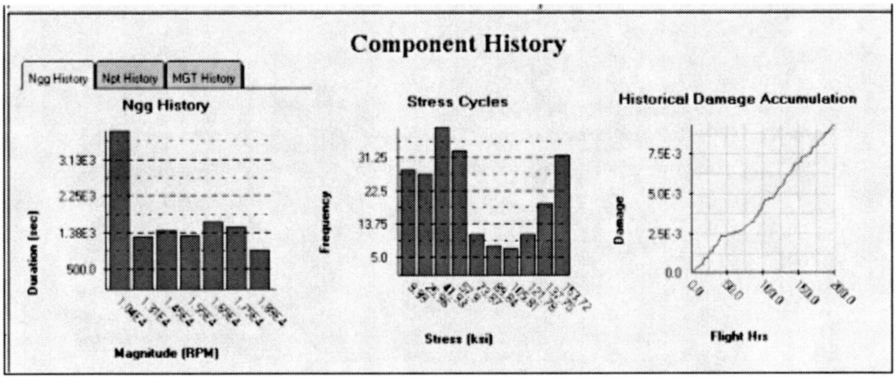

Fig. 1.3 Usage-based damage accumulation approach.

larly updated maintenance database as happens in autonomic logistics applications is significant for this application.

The next logical extension to this type of reliability-based statistical model is to correlate the failure rate data with specific operational usage profiles that are more directly related to the way a specific vehicle is used. In this manner, statistical damage accumulation models or usage models for specific components/LRUs can be directly tied to the loading profiles inferred from the high-level operations data sets, for example, fatigue cycles that are a function operating conditions such as speed or maneuvering conditions. An example of this is shown in Figure 1.3, where a usage model (in this case damage accumulation model) was developed based on the operating speed of an engine. This type of usage models are often referred to as regime recognition in the helicopter community.

It is important to recognize that this is not another form of reliability-centered maintenance, in which we replace components based upon a conservative safe-life operational time. It is a method to include the operational profile information and up-to-date reliability/inspection data in an automated algorithm that will augment existing fault detection conclusions or provide a prediction when more accurate means are not justified. More accurate prognostic methods are described further.

1.4 Signal Integrity and Anomaly Detection

One of the most important areas to consider when implementing an integrated health management system is to ensure the reliability of all measured parameters. When automated algorithms are used to identify vehicle subsystem anomalies, the diagnostic and prognostic algorithms must be confident that the anomalies are indeed occurring within the system and are not the result of normal transients or faulty sensors. Therefore, a generic signal integrity and anomaly detection method is

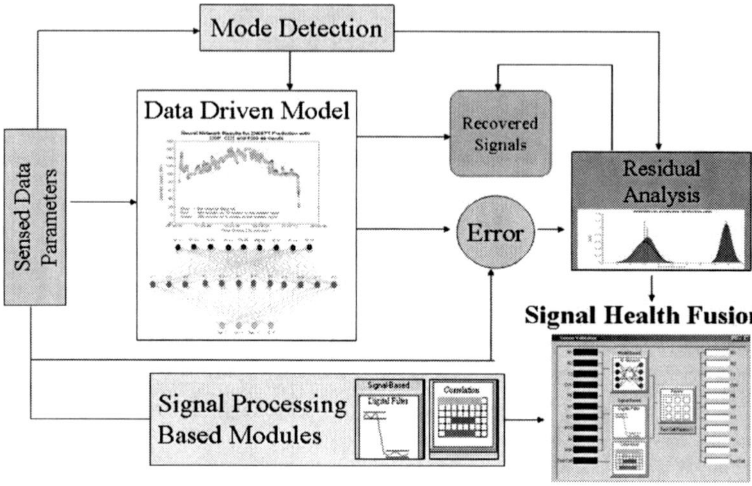

Fig. 1.4 Signal validation and anomaly detection process.

needed to act as a front-end to validate and call out health status anomalies from the sensed signals prior to further analysis.

One of the AI technologies that can be applied to address this issue is based on a probabilistic neural network (PNN) modeling technique that can use normal system operating data to detect off-nominal behavior. In this approach, a trained PNN data model is used to predict the normal behavior of the system signal data acquired, which can then be used to continuously assess the difference between the actual and predicted data. The overall implementation approach is shown in Figure 1.4. The benefits of this type of data-driven modeling approach include the ability to detect subtle or abrupt changes in system health signals over a short period of time. Mode detection is also recommended for use with the PNN algorithm to aide in determining the width of the "normal" bands utilized. At the output of the PNN, a residual analysis algorithm evaluates the errors of the current health parameters relative to "expected" values for any given level of operation based on these "normal" bands. The results of the mode detection algorithm will scale the magnitude of the residual being assessed in determining the integrity of the underlying signal.

The PNN is trained to predict a signal, with inputs that are correlated to it in some manner over an appropriate dynamic range. The operation of the PNN is simple and can be implemented in real-time without the need for supervised training that makes many neural network applications often difficult to implement and update. As shown in Figure 1.5, the PNN first compares the inputs to a set of "normal" training vectors contained in database denoted $\mathbf{IW}_{(1,1)}$. The "training data" is simply the data that represent normal behavior for a particular person and is easily substituted based on each persons "normal" condition. Once the data is stored, the second layer of the algorithm sums the contribution of each class of inputs to produce a vector of probabilities. Finally, the prediction is based on the weighting associated with

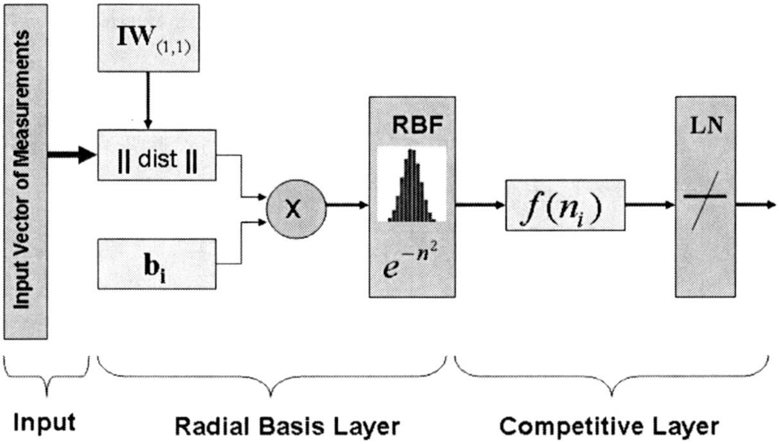

Fig. 1.5 Probabilistic Neural Network (PNN) model.

each of the probabilities and the "similarity" associated with the inputs and known "normal" data. Because of the information compression and regeneration, if an anomaly occurs, the output residual will quickly identify it and pass the information to the residual analysis module for further analysis. The PNN has been successfully applied to various anomaly detection implementations for critical systems on the Space Shuttle main engine, gas turbine engines, chemical processing plants, and machinery vibration monitoring applications, just to name a few.

The residual analysis module assesses the "normal" bands associated with each sensor signal at the current operating condition. When a signal goes outside these bands, while others remain within, an anomaly is detected associated with those specific sensors/data sets. A decision level fusion output determines the final confidence levels that a particular sensor or health feature has anomalies or is corrupted in some way.

1.5 Trend-Based Evolutionary Approaches

A trend-based or evolutionary prognostic approach relies on the ability to track and trend deviations and associated rates of change of these deviations of specific features or measurements from their normal operating condition. Figure 1.6 is an illustration of such a technique. Evolutionary prognostics may be implemented on systems or subsystems that experience conditional or slow degradation type faults such as an efficiency loss in a turbo machine. Generally, trend-based prognostics works well for system level degradation because conditional loss is typically the result of interaction of multiple components functioning improperly as a whole. This approach requires that sufficient sensor information is available to assess the current condition of the system or subsystem and relative level of uncertainty in this

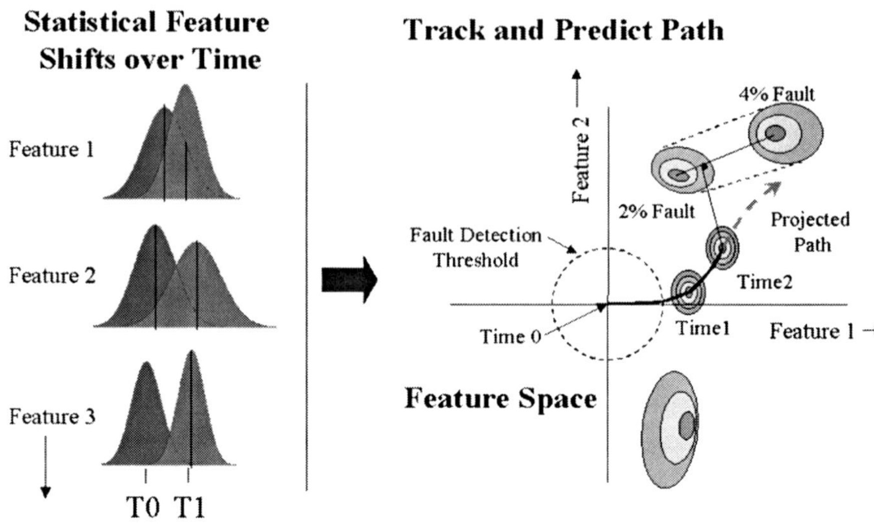

Fig. 1.6 Trend-based or evolutionary approach.

measurement. Furthermore, the parametric conditions that signify known performance related fault must be identifiable. While a physical or statistical model that can help classify a specific fault is beneficial, it is not a requirement for this technical approach. An alternative to the physical model is built in "expert" knowledge of the fault condition and how it manifests itself in the measured and extracted features.

Incipient faults and performance degradations in electrical and mechanical systems exhibit detectable features that provide a means to diagnose and predict the future progression of that fault under known operating conditions. Feature-based prognostics can be implemented for electronic systems based on changes in a variety of measurable quantities including temperature, current, and voltage at various locations in the system. Features such as heat generation, EMI, and power consumption that correlate with known faults can be extracted from the sensed data. Once these features are obtained, they can be tracked and trended over the component's life and compared with remaining useful life estimates to provide corroborative evidence of a degrading or failing condition.

1.6 Data-Driven Model-Based Approaches

In many instances, one has historical fault/failure data in terms of time domain plots of various signals leading up to the failure, or statistical data sets. In many of these cases, it is either difficult or impractical to determine a physics-based model for prediction purposes. In such situations, one may use nonlinear network approximators that can be tuned using well-established formal algorithms to provide desired outputs directly in terms of the data. Nonlinear networks include the *neural network*, which is based on signal processing techniques in biological nervous systems, and *fuzzy logic systems*, which are based on the linguistic and reasoning abilities of humans. These are similar in that they provide structured nonlinear function mappings with very desirable properties between the available data and the desired outputs.

In prediction, Artificial Neural Networks (ANNs), fuzzy systems and other computational intelligence methods, have provided an alternative tool for both forecasting researchers and practitioners [22]. Werbos [31] reported that ANNs trained with the backpropagation algorithm outperform traditional statistical methods such as regression and Box–Jenkins approaches. In a recent forecasting competition organized by Weigend and Gershenfeld [30] through the Santa Fe Institute, all winners of each set of data used ANNs. Unlike the traditional model-based methods, ANNs are data-driven and self-adaptive and they make very few assumptions about the models for problems under study. ANNs learn from examples and attempt to capture the subtle functional relationship among the data. Thus, ANNs are well suited for practical problems where it is easier to have data than knowledge governing the underlying system being studied. Generally, they can be viewed as one of many multivariate nonlinear and nonparametric statistical methods [3]. The main problem of ANNs is that the reasoning behind their decisions is not always evident. Nevertheless, they provide a feasible tool for practical prediction problems.

Hence, with an understanding of how the fault/failure signature is related to specific measurable or inferred features from the system being monitored, a data-driven modeling approach is a commonly utilized approach. Based on the selected input features that correlate with the failure progression, a desired output prediction of the time to failure is produced based on a training process in which the network will automatically adjusts its weights and thresholds based on the relationships it sees between the time to failure and the correlated feature magnitudes. Figure 1.7 shows an example of a neural network after being trained by some vibration feature data sets for predicting a gear failure. The difference between the neural network output and the "ground truth" probability of failure curve is due to error that still exists after the network parameters have optimized to minimize this error. Once trained, the neural network architecture can be used to predict the same features progressions for a different test under similar operating conditions.

1.7 State-Estimator-Based Prognostics

State estimation techniques such as Kalman filters or various other tracking filters can also be implemented as a prognostic technique. The Kalman filter [4, 12] is a dynamical systems analysis tool for estimating unknown states by combining current measurements with the most recent state estimate. It can be considered as a virtual sensor in that it takes current available sensor measurements and provides optimal estimates (or predictions) of quantities of interest that may in themselves not be directly be measurable. Knowledge of noise processes is used to minimize the es-

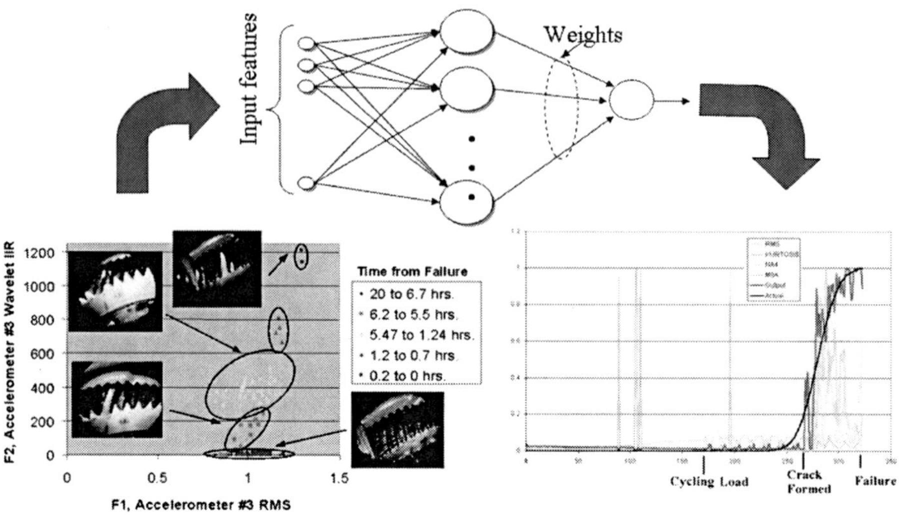

Fig. 1.7 Data-driven model-based approach

timation error covariance, via the optimal determination of the so-called Kalman gain.

It is typically implemented with the use of a linear system model, but can also be extended to non-linear systems through the use of the extended Kalman filter algorithm that linearizes the system about an operating point. The discrete-time system with internal state x_k and sensor measurements z_k may be described in terms of the recursive difference equation

$$\begin{aligned} x_{k+1} &= A_k x_k + B_k u_k - G_k w_k, \\ z_k &= H_k x_k + v_k, \end{aligned} \tag{1.1}$$

where u_k is a control input, w_k is a *process noise* that captures uncertainties in the process dynamics, such as modeling errors and unknown disturbances (e.g. wind gusts in aircraft), and v_k is a *measurement noise*. This is depicted in Figure 1.8.

In this type of application, the minimization of error between a model and measurement can be used to predict future feature states and hence the behavior of the modeled system. Either fixed or adaptable filter gains can be utilized (Kalman is typically adapted, while Alpha-Beta-Gamma is fixed) within an nth-order state variable vector.

In a slightly different application of the Kalman filter, measured or extracted features f, can be used to develop a state vector as shown below.

$$x = [f \quad \dot{f} \quad \ddot{f}]^T. \tag{1.2}$$

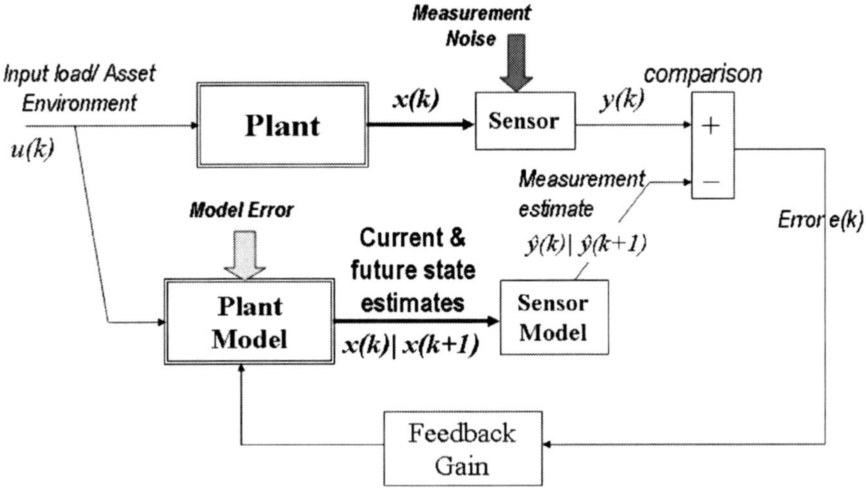

Fig. 1.8 State estimation (Kalman filter implementation) approach.

Then, the state transition equation can be used to update these states based upon a model. A simple Newtonian model of the relationship between the feature position, velocity and acceleration can be used if constant acceleration is assumed. This simple kinematic equation can be expressed as follows:

$$f(n+1) = f(n) + \dot{f}(n)t + \frac{1}{2}\ddot{f}(n)t^2, \tag{1.3}$$

where f is again the feature and t is the time period between updates. There is an assumed noise level on the measurements and model related to typical signal-to-noise problems and unmodeled physics. The error covariance associated with the measurement noise vectors is typically developed based on actual noise variances, while the process noise is assumed based on the kinematic model. In the end, the tracking filter approach is used to track and smooth the features related to predicting a failure.

1.8 Physics-Based Modeling Approaches

A physics-based model is a technically comprehensive modeling approach that has been traditionally used to understand component failure mode progression. Physics-based models provide a means to calculate the damage to critical components as a function of operating conditions and assess the cumulative effects in terms of component life usage. By integrating physical and stochastic modeling techniques, the model can be used to evaluate the distribution of remaining useful component life as a function of uncertainties in component strength/stress properties, loading or

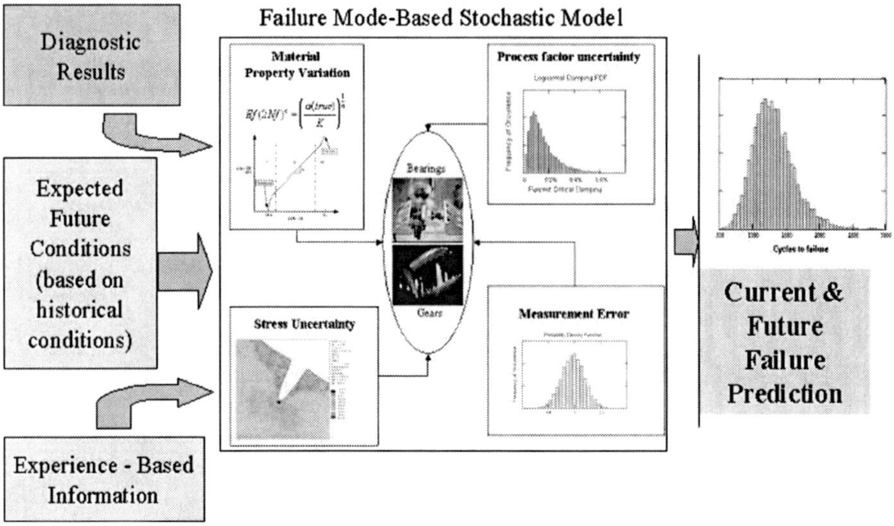

Fig. 1.9 Physics-based modeling approach.

lubrication conditions for a particular fault. Statistical representations of historical operational profiles serve as the basis for calculating future damage accumulation. The results from such a model can then be used for real-time failure prognostic predictions with specified confidence bounds. A block diagram of this prognostic modeling approach is given in Figure 1.9. As illustrated at the core of this figure, the physics-based model utilizes the critical, life-dependent uncertainties so that current health assessment and future remaining useful life (RUL) projections can be examined with respect to a risk level.

Model-based approaches to prognostics differ from feature-based approaches in that they can make RUL estimates in the absence of any measurable events, but when related diagnostic information is present (such as the feature described previously) the model can often be calibrated based on this new information. Therefore, a combination or fusion of the feature-based and model-based approaches provides full prognostic ability over the entire life of the component, thus providing valuable information for planning which components to inspect during specific overhauls periods. While failure modes may be unique from component to component, this combined model-based and feature-based methodology can remain consistent across different types of critical components or LRUs.

To perform a prognosis with a physics-based model, an operational profile prediction must be developed using the steady state and transient loads, temperatures or other on-line measurements. With this capability, probabilistic critical component models can then be "run into the future" by creating statistical simulations of future operating profiles from the statistics of past operational profiles or expected future operating profiles.

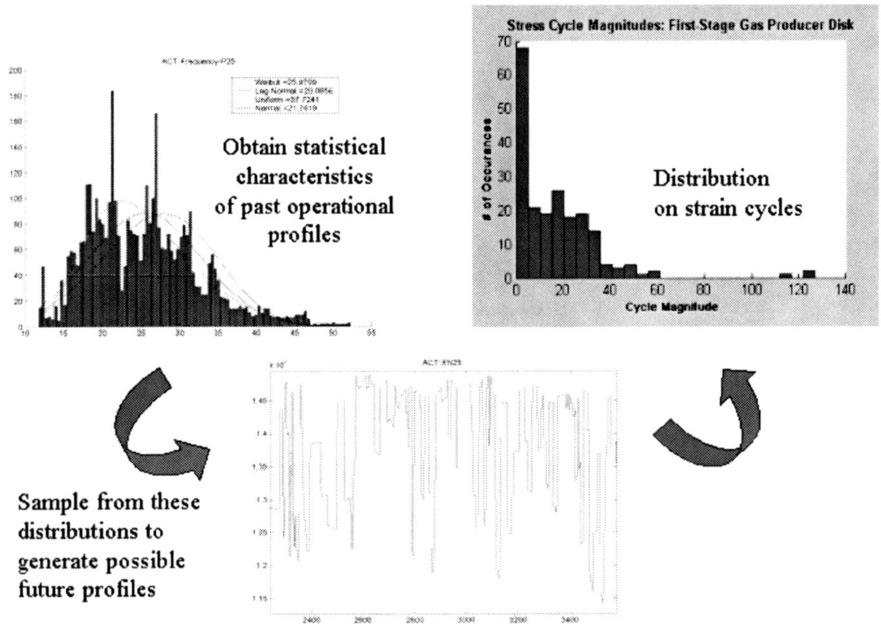

Fig. 1.10 Operation profile and loading model for prognosis.

Based on the authors' past experience correlating operational profile statistics and component or LRU life usage, the non-linear nature associated with many damage mechanisms is dependent on both the inherent characteristics of the profiles and operational mix types. Significant component damage resulting from the large variability in the operating environment and severity of the missions directly affects the vehicle component life-times. Very often, component lives driven by fatigue failure modes are dominated by unique operational usage profiles or a few, rare, severe, randomly occurring events, including abnormal operating conditions, random damage occurrences, etc. For this reason, the authors recommend a statistical characterization of loads, speeds, and conditions for the projected future usage in the prognostic models as shown in Figure 1.10.

1.9 Prognosis Remaining Useful Life Probability Density Function

A comprehensive description on the probabilistic techniques in prognostics as related to predicting the remaining useful life (RUL) is given by Engel et al. [5]. The seminal notions presented in this paper serve to clarify our thinking about remaining useful life prediction. A key concept in this framework is the *remaining useful life*

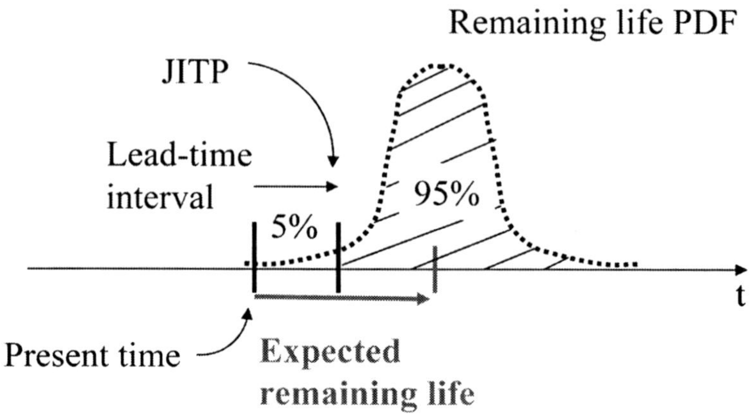

Fig. 1.11 A probability density function for prognosis.

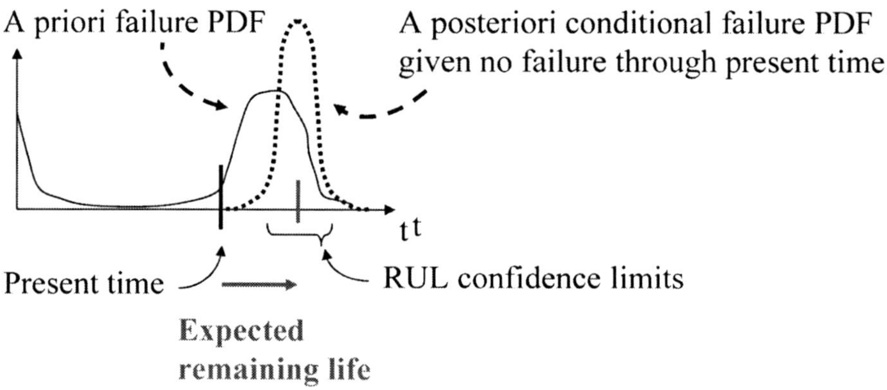

Fig. 1.12 Updated prognosis probability density function .

failure probability density function (PDF). In this representation, a component or LRU is recommended to be removed from service prior to attaining a high probability of failure, set based on the criticality. This concept is depicted in Figure 1.11, in terms of the RUL PDF, where a *just in time point* is defined for removal from service that corresponds to a 95% probability that the component has not yet failed.

A key issue, unfortunately, is that the RUL PDF is actually a *conditional PDF* that changes as time advances. In fact, one must recompute the RUL PDF at each time *t* based on the new information that *the component has not yet failed at that time*. This concept is shown in Figure 1.12. One starts with an a priori PDF similar to the hazard function. Then, as time passes, one must recompute the a posteriori RUL PDF based on the fact that the failure has not yet occurred. This involves renormalizing the PDF at each time so that its area is equal to one. As time passes, the variance of the RUL PDF decreases; that is the PDF becomes narrower. This

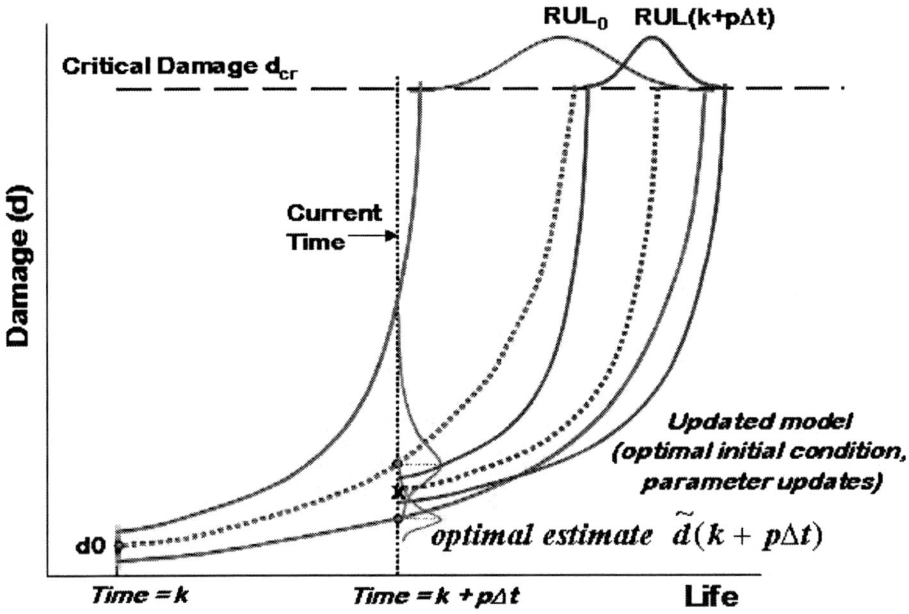

Fig. 1.13 Adaptive prognosis concept.

corresponds to the fact that, as time passes and one approaches the failure point, one becomes more and more certain about the time of failure and its predicted value becomes more accurate.

1.10 Adaptive Prognosis

As a direct extension to the concept presented above, the idea of updating the prognosis PDF based on additional state awareness (fault detection and diagnostic) information that can become available over time is also desirable. The adaptive prognosis concept entails that information available at the current time (which may or may not be diagnostic in nature) be used to modify future predictions, hence updating the prognosis PDF. This idea is illustrated in Figure 1.13 [5] and briefly described next.

Consider point d_0 in Figure 1.13 to be the mean initial damage condition for a prognostic model. A prognosis of life, from time k to predetermined damage level is found to be represented by RUL_0 or Remaining Useful Life. Suppose that some imperfect measurement $z(k)$ regarding the damage state becomes available at time $k = k + p \cdot T$. The challenge is to find optimal current damage state to re-initialize the model and/or adjust model parameters so that a calibrated and more accurate prognosis can be established.

Through utilization of a new initial condition, $\tilde{d}(k)$ at time $k = k + p \cdot T$ as shown in Figure 1.11, it is apparent that the prediction mean has shifted and the confidence bounds on the resulting RUL has less variance than the original (blue prediction). The prediction accuracy improvement would generally mean that a decision to take action based on failure probability will likely reduce lost operational availability over a run-to-failure maintenance plan.

1.11 Distributed Prognosis System Architecture

The cornerstone of an effective Prognostic and Health Management (PHM) system is the information/data architecture and the ability for understanding and managing the anomaly, diagnostic, and prognostic (A/D/P) information from the LRU level all the way up through to the subsystem and vehicle level reasoners. This concept is briefly illustrated in Figure 1.14, where faults detected and predicted at the LRU level are assessed through the hierarchy of reasoners in order to determine the root causes of vehicle malfunctions and contingency option for impending failures.

In general, the A/D/P technologies implemented at the lower levels (LRUs) are used to detect and predict off-nominal conditions or damage accumulating at an accelerated rate. In the distributed PHM architecture, this information is analyzed through the hierarchy of reasoners to make informed decisions on the health of the vehicle subsystems/systems and how they affect total vehicle capability. This integration across LRUs, subsystems and systems is vital to correctly isolating the root-cause of failures and understanding the propagation of up/downstream effects of the faults. Integration of the individual subsystem PHM results is eventually accomplished with the vehicle level reasoner, which will assess the intra-system A/D/P results in order to prioritize the recommended maintenance action(s) to perform in order to correct the problem.

A distributed PHM architecture, such as that shown below, has many benefits including:

1. Optimal computational resource management (i.e. placing high bandwidth processing at the lowest level and only passing up critical features).
2. Supports the concept of "Smart LRU/Subsystem", where the most detailed "intelligence" about the system exists (i.e. supplier/designer responsibility).
3. Provides the ability to isolate and assess the extent of multiple faults and battle damage, hence improving survivability of the vehicle.
4. Hierarchical reasoners have a "built-in" data management capability for containing erroneous information and utilizing multiple data and information sources.
5. Ability to capture and localize system degradations (as opposed to only hard failures), based on increased health awareness of the lowest-level LRUs, hence providing a more accurate vehicle availability assessment.

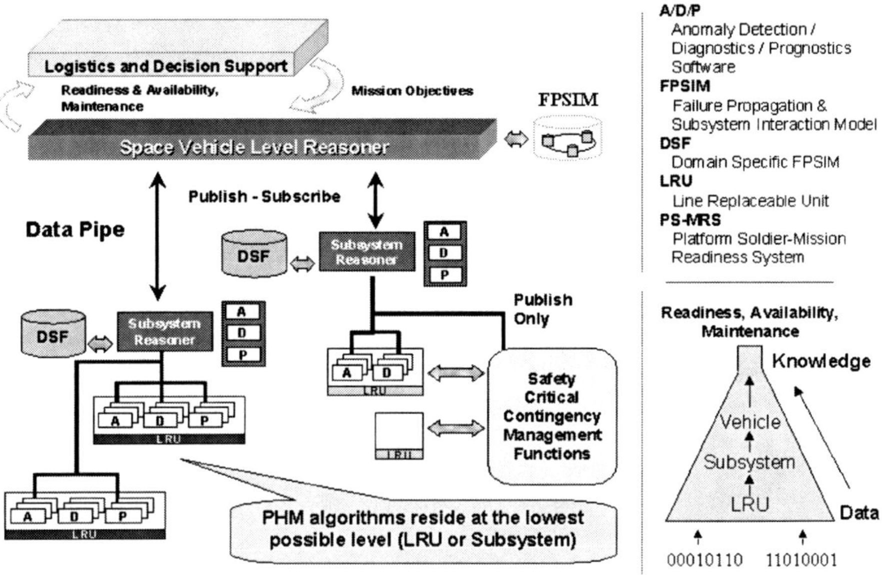

Fig. 1.14 Distributed prognostic system architecture.

1.12 Conclusions

This chapter has briefly reviewed some generic prognosis algorithmic approaches and introduced the theoretical and notional foundations associated with probabilistic predictions and required software architectures for performing prognostics on critical aerospace systems.

Prognosis is certainly the Achilles' heel of the Prognostics and Health Management (PHM) system and presents major challenges to the CBM/PHM system designer primarily because it entails large-grain uncertainty. Long-term prediction of a fault evolution to the point that may result in a failure requires means to represent and manage the inherent uncertainty. Moreover, accurate and precise prognosis demands good probabilistic models of the fault growth and statistically sufficient samples of failure data to assist in training, validating and fine tuning prognostic algorithms. Prognosis performance metrics, robust algorithms and test platforms that may provide needed data have been the target of CBM/PHM researchers over the recent past. Many accomplishments have been reported but major challenges still remain to be addressed.

In addition, due to the inherent uncertainties in prognosis approaches, which are the aggregate of many unknowns and can result in considerable prediction variability, the concept of adaptive prognosis was also introduced. In that case, available, albeit imperfect, information is used to update elements of the prognostic model. Only one of many approaches for accomplishing this was briefly introduced, i.e.

the Kalman Filter. Other statistical update techniques include Bayesian updating, constrained optimization and particle filtering.

The prognosis process by which features and models are integrated to obtain the best possible prediction on remaining useful life still has many remaining challenges. It is a significant challenge to design systems so that measured data can be fused and used in conjunction with physics-based models to estimate current and future damage states. Furthermore, multiple models will often be required that may or may not use various feature inputs. Finally, the feedback mechanism in a prognosis system design cannot be ignored. Specifically, the prognosis system must be capable of intelligently calibrating a priori initial conditions (i.e. humidity, strain and temperature have changed), random variable characteristics or switching prognostic models in an automated yet lucid process to empower better operational and logistical decisions for vehicle platforms.

References

1. D.W. Aha, Special Issue on Lazy Learning, *Artificial Intelligence Review* **11**(1–5), 1–6, 1997.
2. C.G. Atkeson, A.W. Moore, and S. Schaal, Locally weighted learning, *Artificial Intelligence Review* **11**(1–5), 11–73, 1997.
3. J. Cheng and D.M. Titerington, Neural networks: A review from a statistical perspective, *Statistical Science* **9**(1), 2–54, 1994.
4. M. Drexel, and J.H. Ginsberg, Mode isolation: A new algorithm for modal parameter identification, *Journal of the Acoustical Society of America* **110**(3), 1371–1378, 2001.
5. S.J. Engel, B.J. Gilmartin, K. Bongort and A. Hess, Prognostics, The real issues involved with predicting life remaining, *Aerospace Conference Proceedings* **6**, 457–469, March 2000.
6. Jardim-Goncalves, 1996.
7. C. Frelicot, A fuzzy-based prognostic adaptive system, RAIRO-APII-JESA, *Journal Européen des Systèmes Automatisés* **30**(2–3), 281–299, 1996.
8. P.G. Groer, Analysis of time-to-failure with a Weibull model, in *Proceedings of the Maintenance and Reliability Conference*, Marcon, 2000.
9. M.T. Hagan and M. Menhaj, Training feedforward networks with the Marquard algorithm, *IEEE Transactions on Neural Networks* **5**(6), 1994.
10. N. Khiripet, An architecture for intelligent time series prediction with causal information, Ph.D. Thesis, Georgia Institute of Technology, May 2001.
11. J.A. Leonard, M.A. Kramer and L.H. Ungar, A neural network architecture that computes its own reliability, *Computers & Chemical Engineering* **16**(9), 819–835, 1992.
12. F.L. Lewis, *Optimal Estimation: With an Introduction to Stochastic Control Theory*, John Wiley & Sons, New York, April 1986.
13. Lewis, 1992.
14. J.S. Liu and R. Chen, Sequential Monte Carlo methods for dynamical systems, *Journal for American Statistical Association* **93**, 1032–1044, 1998.
15. L. Ljung, *System Identification: Theory for the User*, 2nd ed., Prentice-Hall, New Jersey, 1999.
16. K.A. Marko, J.V. James, T.M. Feldkamp, C.V. Puskorius, J.A. Feldkamp and D. Roller, Applications of neural networks to the construction of virtual sensors and model-based diagnostics, in *Proceedings of ISATA 29th International Symposium on Automotive Technology and Automation*, 3–6 June, pp. 133–138, 1996.
17. M.L. Minsky, Step toward artificial intelligence, *Proceedings IRE* **49**, 8–30, 1961.
18. D. Muench, G. Kacprzynski, A. Liberson, A. Sarlashkar and M. Roemer, Model and sensor fusion for prognosis, Example: Kalman filtering as applied to corrosion-fatigue and FE models, *SIPS Quarterly*, Review Presentation, 2004.

19. Ray, 1996.
20. J.R. Schauz, Wavelet neural networks for EEG modeling and classification, PhD Thesis, Georgia Institute of Technology, 1996.
21. A. Schömig and O. Rose, On the suitability of the Weibull distribution for the approximation of machine failures, in *Proceedings of the 2003 Industrial Engineering Research Conference*, May 18–20, Portland, OR, 2003.
22. R. Sharda, Neural network for the MS/OR analyst: An application bibliography, *Interfaces* **24**(2), 116–130, 1994.
23. J. Shiroishi, Y. Li, S. Liang, T. Kurfess and S. Danyluk, Bearing condition diagnostics via vibration and acoustic emission measurements, *Mechanical Systems and Signal Processing* **11**(5), 693–705, September 1997.
24. D.F. Specht, A general regression neural network, *IEEE Transactions on Neural Networks* **2**(6), 568–576, November 1991.
25. L. Studer and F. Masulli, On the structure of a neuro-fuzzy system to forecast chaotic time series, in *Proceedings of the International Symposium on Neuro-Fuzzy Systems*, 29–31 August, pp. 103–110, 1996.
26. R.S. Sutton, Introduction: The challenge of reinforcement learning, *Machine Learning* **8**, 225–227, 1992.
27. D.S.J. Veaux, J. Schweinsberg and J. Ungar, Prediction intervals for neural networks via nonlinear regression, *Technometrics* **40**(4), 273–282, November 1998.
28. P. Wang and G. Vachtsevanos, Fault prognostics using dynamic wavelet neural networks, *Artificial Intelligence for Engineering Design, Analysis and Manufacturing* **15**(4), 349–365, September 2001.
29. W. Weibull, A statistical distribution function of wide applicability, *Journal of Applied Mechanics* **18**, 293, 1951.
30. A.S. Weigend and N.A. Gershenfeld, *Time Series Prediction: Forecasting the Future and Understanding the Past*, Addison-Wesley, MA, 1993.
31. P.J. Werbos, Generalization of back propagation with application to recurrent gas market model, *Neural Networks* **1**, 339–356, 1988.

Chapter 2
Advances in Uncertainty Representation and Management for Particle Filtering Applied to Prognostics*

Marcos Orchard, Gregory Kacprzynski, Kai Goebel, Bhaskar Saha and George Vachtsevanos

Abstract Particle filters (PF) have been established as the de facto state of the art in failure prognosis. They combine advantages of the rigors of Bayesian estimation to nonlinear prediction while also providing uncertainty estimates with a given solution. Within the context of particle filters, this paper introduces several novel methods for uncertainty representations and uncertainty management. The prediction uncertainty is modeled via a rescaled Epanechnikov kernel and is assisted with resampling techniques and regularization algorithms. Uncertainty management is accomplished through parametric adjustments in a feedback correction loop of the state model and its noise distributions. The correction loop provides the mechanism to incorporate information that can improve solution accuracy and reduce uncertainty bounds. In addition, this approach results in reduction in computational burden. The scheme is illustrated with real vibration feature data from a fatigue-driven fault in a critical aircraft component.

Marcos Orchard
Electrical Engineering Department, University of Chile, Santiago, Chile;
e-mail: morchard@ing.uchile.cl

Gregory Kacprzynski
Impact Technologies, LLC, 200 Canal View Blvd., Rochester, NY 14623, USA;
e-mail: greg.kacprzynski@impact-tek.com

Kai Goebel
NASA Ames Research Center, MS 269-4, Moffett Field, CA 94035, USA;
e-mail: kai.goebel@nasa.gov

Bhaskar Saha
MCT, NASA Ames Research Center, MS 269-4, Moffett Field, CA 94035, USA;
e-mail: bsaha@email.arc.nasa.gov

George Vachtsevanos
School of Electrical and Computer Engineering, Georgia Institute of Technology, Atlanta, GA 30332, USA; e-mail: gjv@ece.gatech.edu

K.P. Valavanis (ed.), Applications of Intelligent Control to Engineering Systems, 23–35.

2.1 Introduction

Uncertainty management of prognostics holds the key for a successful penetration of prognostics as a key enabler to health management in industrial applications. While techniques to manage the uncertainty in the many factors contributing to current health state estimation – such as signal-to-noise ratio (SNR) on diagnostic features, optimal features with respect to detection statistics and ambiguity set minimization – have received a fair amount of attention due to the maturity of the diagnostics domain, uncertainty management for prognostics is an area which still awaits significant advances.

Those shortcomings non-withstanding, a number of approaches have been successfully suggested for uncertainty representation and management in prediction. In particular, probabilistic, soft computing methods and tools derived from evidential theory or Dempster–Shafer theory [1] have been explored for this purpose. Probabilistic methods are mathematically rigorous assuming that a statistically sufficient database is available to estimate the required distributions. Possibility theory (fuzzy logic) offers an alternative when scarce data and even incomplete or contradictory data are available. Dempster's rule of combination and such concepts from evidential theory as belief on plausibility (as upper and lower bounds of probability) based on mass function calculations can support uncertainty representation and management tasks. The authors in [2] introduced a Neural Network construct called Confidence Prediction Neural Network to represent uncertainty in the form of a confidence distribution while managing uncertainty via learning during the prediction process. The scheme employs Parzen windows as the kernel and the network is based on Specht's General Regression Neural Network [3]. Radial Basis Function Neural Nets (RBFNN), Probabilistic Neural Nets (PNN) and other similar constructs from the neural net and neuro-fuzzy arena have been deployed as candidates for uncertainty representation and management. For example, Leonard et al. [4] used an RBFNN to obtain confidence limits for a prognosticator. Probabilistic reliability analysis tools employing an inner-outer loop Bayesian update scheme [5, 6] have also been used to "tune" model hyperparameters given observations. However, the scalability of this rigorous approach for more than a few parameters is unproven and relies on the assumption that all distributions are unimodal.

This paper introduces a generic and systematic methodology to the uncertainty representation and management problem in failure prognosis by capitalizing on notions from Bayesian estimation theory and, specifically, particle filtering (PF) [7, 9–10] for long-term prognosis in non-linear dynamic systems with non-Gaussian noise, appropriate kernels to reduce the impact of model errors and feedback correction loops to improve the accuracy and precision of the remaining useful life estimates. Prediction uncertainty is modeled via rescaled Epanechnikov kernels, considering the current state pdf estimate as initial condition of stochastic dynamic models, and is assisted with regularization algorithms. Uncertainty management is accomplished through parametric adjustments in a feedback correction loop of the state model and its noise distributions. It is assumed that for a specific application domain, the sources of uncertainty have been identified, raw data are available (for example, vibration data, load profiles, etc.), key fault indicators or features are extracted on-line from such sensor data that are characteristic of the health of critical components/subsystems, and fault detection, isolation and identification routines exploit these features to classify with prescribed confidence and false alarm rates the presence of a fault. In a probabilistic fault diagnosis framework, these features are expressed as probability density functions and are used to initialize the prognostic algorithms.

2.2 Uncertainty Representation and Management in Long-Term Prediction: A Particle Filtering-Based Approach

2.2.1 Failure Prognosis and Uncertainty Representation

Nonlinear filtering is defined as the process of using noisy observation data to estimate at least the first two moments of a state vector governed by a dynamic nonlinear, non-Gaussian state-space model. From a Bayesian standpoint, a nonlinear filtering procedure intends to generate an estimate (of the posterior probability density function $p(x_t \mid y_{1:t})$ for the state, based on the set of received measurements. Particle Filtering (PF) is an algorithm that intends to solve this estimation problem by efficiently selecting a set of N particles $\{x^{(i)}\}_{i=1\ldots N}$ and weights $\{w_t^{(i)}\}_{i=1\ldots N}$, such that the state pdf may be approximated by [7]

$$\tilde{\pi}_t^N(x_t) = \sum_{i=1}^{N} w_t^{(i)} \delta(x_t - x_t^{(i)}). \tag{2.1}$$

Prognosis, and thus the generation of long-term prediction, is a problem that goes beyond the scope of filtering applications since it involves future time horizons. Hence, if PF-based algorithms are to be used, it is necessary to propose a procedure with the capability to project the current particle population in time in the absence of new observations [7].

Any adaptive prognosis scheme requires the existence of at least one feature providing a measure of the severity of the fault condition under analysis (fault dimension). If many features are available, they can always be combined to generate a single signal. In this sense, it is always possible to describe the evolution in time of the fault dimension through the nonlinear state equation:

$$\begin{cases} x_1(t+1) = x_1(t) + x_2(t) \cdot F(x(t), t, U) + \omega_1(t), \\ x_2(t+1) = x_2(t) + \omega_2(t), \end{cases}$$
$$y(t) = x_1(t) + \upsilon(t), \tag{2.2}$$

where $x_1(t)$ is a state representing the fault dimension under analysis, $x_2(t)$ is a state associated with an unknown model parameter, U are external inputs to the system (load profile, etc.), $F(x(t), t, U)$ is a general time-varying nonlinear function, $y(t)$ represents feature measurements, and $\omega_1(t)$, $\omega_2(t)$ and $\upsilon(t)$ are white noises (not necessarily Gaussian). The nonlinear function $F(x(t), t, U)$ may represent a model based on first principles, a neural network, or even a fuzzy system.

By using the aforementioned state equation to represent the evolution of the fault dimension in time, it is possible to generate long term predictions for the state pdf in a recursive manner using the current pdf estimate for the state [7]:

$$\begin{aligned}\tilde{p}(x_{t+p} \mid y_{1:t}) &= \int \tilde{p}(x_t \mid y_{1:t}) \prod_{j=t+1}^{t+p} p(x_j \mid x_{j-1})\,\mathrm{d}x_{t:t+p-1} \\ &\approx \sum_{i=1}^{N} w_t^{(i)} \int \cdots \int p(x_{t+1} \mid x_t^{(i)}) \prod_{j=t+2}^{t+p} p(x_j \mid x_{j-1})\,\mathrm{d}x_{t+1:t+p-1}.\end{aligned} \tag{2.3}$$

The evaluation of these integrals, though, may be difficult and/or may require significant computational effort. This paper proposes a methodology to solve this problem using a PF algorithm to estimate the current state pdf (from feature measurements associated to the fault dimension) and a combination of rescaled kernel functions and resampling schemes to reconstruct the estimate of the state pdf for future time instants.

Consider, in this sense, a discrete approximation for the predicted state pdf

$$\hat{p}(x_{t+k} \mid \hat{x}_{1:t+k-1}) \approx \sum_{i=1}^{N} w_{t+k-1}^{(i)} K\left(x_{t+k} - E[x_{t+k}^{(i)} \mid \hat{x}_{t+k-1}^{(i)}]\right), \tag{2.4}$$

where $K(\cdot)$ is a kernel density function, which may correspond to the process noise pdf, a Gaussian kernel or a rescaled version of the Epanechnikov kernel:

$$K_{\text{opt}}(x) = \begin{cases} \dfrac{n_x + 2}{2c_{n_x}}\left(1 - \|x\|^2\right) & \text{if } \|x\| < 1, \\ 0 & \text{otherwise,} \end{cases} \tag{2.5}$$

where c_{n_x} is the volume of the unit sphere in $\mathbb{R}^{n_x}$.

Furthermore, if the density is Gaussian with unit covariance matrix, the optimal bandwidth is given by

$$\begin{aligned} h_{\text{opt}} &= A \cdot N^{-1/n_x+4}, \\ A &= \left(8c_{n_x}^{-1} \cdot (n_x + 4) \cdot \left(2\sqrt{\pi}\right)^{n_x}\right)^{1/n_x+4}. \end{aligned} \tag{2.6}$$

The Epanechnikov kernel is particularly recommended in the special case of equally weighted samples [8], and thus it is well suited for uncertainty representation in long term predictions where no future measurements are available for a weight update procedure.

Given $\{x_t^{(i)}\}_{i=1\ldots N}$ and $\{w_t^{(i)}\}_{i=1\ldots N}$ as initial conditions, it is possible to represent the uncertainty inherent to the predicted state pdf by performing an inverse transform resampling procedure for the particle population [7]. This method obtains a fixed number of samples for each future time instant, avoiding problems of excessive computational effort. In fact, after the resampling scheme is performed, the weights may be expressed as: $\{w_{t+k}^{(i)}\}_{i=1\ldots N} = N^{-1}$. Furthermore, if only Epanechnikov kernels are used, it is ensured that the representation of the uncertainty will be bounded.

To avoid loss of diversity among particles and minimize the effect of model errors in the long term predictions, an additional step inspired by the Regularized Particle Filter [7] is performed for $k > 1$. In this step, it is assumed that the state covariance matrix $\hat{S}_{t+k}$ is equal to the empirical covariance matrix of $\hat{x}_{t+k}$:

Long Term Predictions: Regularization of Predicted State PDF

- Apply modified inverse transform resampling procedure.
  For $i = 1, \ldots, N$, $w_{t+k}^{(i)} = N^{-1}$
- Calculate $\hat{S}_{t+k}$, the empirical covariance matrix of
  $\{E[x_{t+k}^{(i)} \mid |\hat{x}_{t+k-1}^{(i)}], w_{t+k}^{(i)}\}_{i=1}^{N}$
- Compute $\hat{D}_{t+k}$ such that $\hat{D}_{t+k}\hat{D}_{t+k}^{T} = \hat{S}_{t+k}$
- For $i = 1, \ldots, N$, draw $\varepsilon^i \sim K$, an Epanechnikov kernel and assign
  $\hat{x}_{t+k}^{(i)*} = \hat{x}_{t+k}^{(i)} + h_{t+k}^{\text{opt}}\hat{D}_{t+k}\varepsilon^i$

It must be noted that the proposed method for uncertainty representation allows considering information for on-line measurements to estimate the uncertainty that is present in the system at the moment that the long-term predictions are generated. Moreover, the use of kernel functions and resampling techniques (with a limited number of particles) naturally permits to represent uncertainty in future time instants, just as other approaches, but using less computational resources. Figure 2.1 shows a representation of the particle filtering-based uncertainty representation scheme.

After the completion of the algorithm, it is possible to isolate the particles that define the bounds for the predicted pdf at future time instants. The collection of all these particles results in minimum and maximum bounds for the predicted state in time. Moreover, these bounds intrinsically incorporate, measure, and represent model uncertainty (through the estimation of unknown parameters) and measurement noise (since the initial condition for long-term predictions corresponds to the output of the Particle Filtering procedure).

For the test case of a fatigue failure in a critical aircraft component discussed in the sequel (treated in Section 2.4 of this paper), the uncertainty representation results are shown in Figure 2.2.

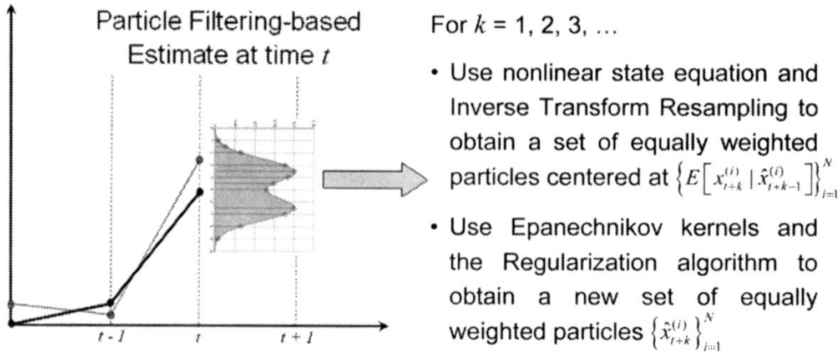

Fig. 2.1 Particle filtering-based uncertainty representation.

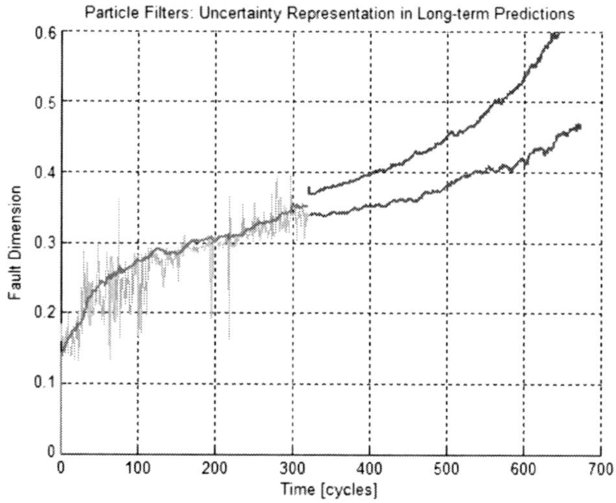

Fig. 2.2 Particle filtering-based uncertainty representation; results for the case of a fatigue fault progression in a critical aircraft component.

2.3 Uncertainty Management in Long-Term Predictions

The issue of uncertainty management, in a Particle Filtering-based prognosis framework, is basically related to a set of techniques aimed to improve the estimate at the current time instant, since both the expectation of the predicted trajectories for particles and bandwidth of Epanechnikov kernels depend on that pdf estimate.

In this sense, it is important to distinguish between two main types of adjustments that may be implemented to improve the current representation of uncertainty for future time instants:

- Adjustments in unknown parameters in the state equation.
- Adjustments in the parameters that define the noise pdf's embedded in the state equation. These parameters will be referred to as "hyperparameters".

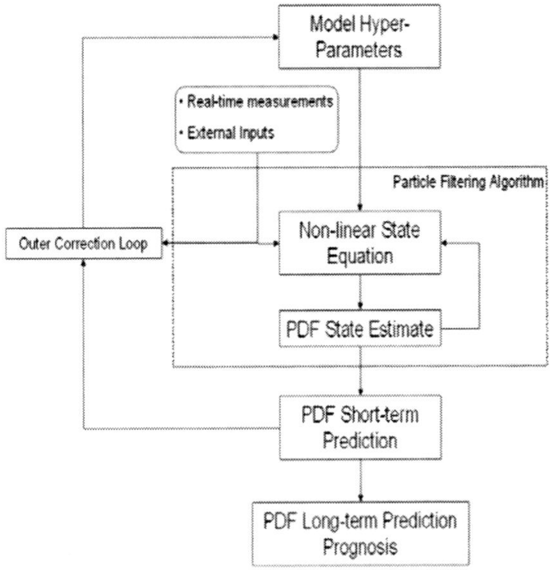

Fig. 2.3 Particle filtering-based uncertainty management system.

Accuracy of long-term predictions is directly related to the estimates of x_t and the model hyper-parameters that affect $E[x_t \mid x_{t-1}]$. Precision in long-term predictions, on the other hand, is directly related to the hyper-parameters that describe the variance of the noise structures considered in the state equation. In this sense, any uncertainty management system for future time instants, within a Particle Filtering-based framework, should follow the general structure presented in Figure 2.3, where the performance of the algorithm is evaluated in terms of the short-term prediction error (which depends on the PF-based pdf estimate). Whenever the performance criteria for the short-term prediction error are not met, an *outer correction loop* directly modifies both model parameters and hyper-parameters.

By modifying the hyper-parameters via an *outer correction loop*, short-term predictions may be used to improve the efficiency of the particle filtering-based estimate, and thus the subsequent generation of long-term predictions. There is a large variety of *outer correction loops* that may be applied for this purpose. One of them, aimed to modify the variance of the noise term in the state equation, is proposed and analyzed in [7]. In this case, let $\omega_2(t)$ represent the model uncertainty. Then for the example detailed in the next section of the paper, the recommended parameters are:

$$\begin{cases} \mathrm{var}\{\omega_2(t+1)\} := p\,\mathrm{var}\{\omega_2(t)\}, \text{ if } \dfrac{\|\mathrm{Pred\_error}(t)\|}{\|\mathrm{Feature}(t)\|} < Th, \\ \mathrm{var}\{\omega_2(t+1)\} := q\,\mathrm{var}\{\omega_2(t)\}, \text{ if } \dfrac{\|\mathrm{Pred\_error}(t)\|}{\|\mathrm{Feature}(t)\|} > Th, \end{cases} \tag{2.7}$$

where Pred_error(t) is the short-term prediction error computed at time t, $\|\cdot\|$ is any well-defined norm (usually L_2-norm), $0 < p < 1$, $q > 1$, and $0 < Th < 1$ are scalars. In particular, $p \in [0.925, 0.975]$, $q \in [1.10, 1.20]$, and $Th = 0.1$. These values have been determined through exhaustive analysis of simulations considering scenarios with different combinations of values for the parameters p, q, and Th. The range for short-term predictions depends on the system under analysis, although a 5-step is recommended to ensure rapid adaptation of the scheme.

Outer correction loops may be also implemented using neural networks, fuzzy expert systems, PID controllers, among others. Additional correction loops include the modification of the number of particles used for 1-step or long-term prediction purposes and the reduction of the threshold for the use of the importance resampling algorithm.

2.4 An Illustrative Example

As an illustrative example, consider the case of propagating fatigue crack on a critical component in a rotorcraft transmission system. The objective in this seeded fault test is to analyze how a cyclic load profile affects the growth of an axial crack. Although the physics-based model for a system of these characteristics may be complex, it is possible to represent the growth of the crack (fault dimension) using the much simpler population-growth-based model [9, 10]:

$$\begin{cases} x_1(t+1) = x_1(t) + C \cdot x_2(t) \cdot (a - b \cdot t + t^2)^m + \omega_1(t), \\ x_2(t+1) = x_2(t) + \omega_2(t), \end{cases}$$
$$y(t) = x_1(t) + \upsilon(t), \tag{2.8}$$

where $x_1(t)$ is a state representing the fault dimension, $x_2(t)$ is a state associated with an unknown model parameter, $y(t)$ are vibration-based feature measurements, C and m are constants associated with the fatigue properties of the material. The constants a and b depend on the maximum load and duration of the load cycle (external input U). Given a set of vibration feature data such as described in [11], it is possible to use this model to obtain an approximate (and noisy) estimate of the crack length via the use of a PF-based algorithm [7]. Once the estimate of the state pdf is available, it can be used as initial condition of the model to generate long-term predictions, if the integrals in the aforementioned recursive expression are evaluated.

Figures 2.4 and 2.5 show the results for the application of (the regularization procedure to $\hat{p}(x_{t+1} \mid \hat{x}_{1:t})$, i.e. the predicted state pdf for time $t+1$, in the example case. First, the PF-based algorithm is used to obtain a pdf state estimate for the state model (2.8) at time $t = 320$. The analysis of feature data at that point indicates that the length of the crack is approximately 0.35 units. It is of interest to prognosticate the time instant when the fault dimension reaches 0.45 units, which is the expected

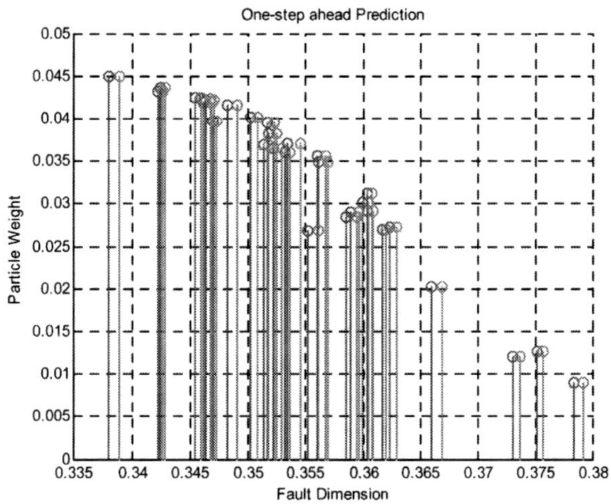

Fig. 2.4 One-step ahead prediction from a Particle Filtering standpoint. The black (blue) samples represent the state pdf at time $t = 320$, while the light-gray (magenta) samples illustrate $\hat{p}(x_{321} \mid \hat{x}_{1:320})$.

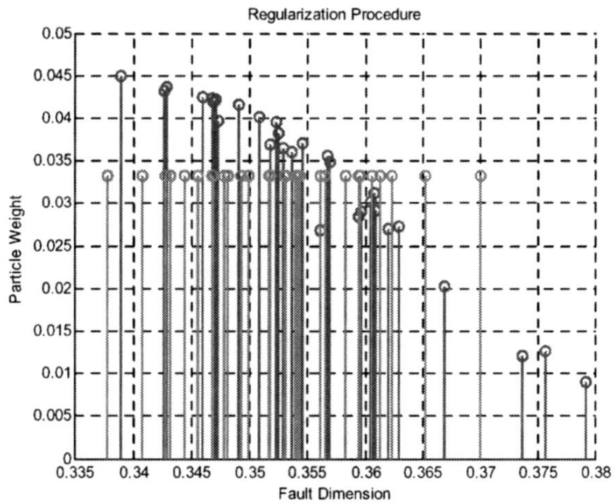

Fig. 2.5 Illustration of regularization procedure.

value of the hazard zone of the component under consideration [7]. Figure 2.4 illustrates how the use of model (2.8) helps to propagate the particle population in time: $E[x_{t+1}^{(i)} \mid \hat{x}_t^{(i)}]$ is computed, and the particle weights are kept constant. As a result, the expected value of the crack length should increase in time.

The second step of the proposed algorithm is the regularization procedure; see Figure 2.5. This step modifies the whole particle population to improve the uncertainty representation for the predicted state pdf at time $t = 321$. As a result, a

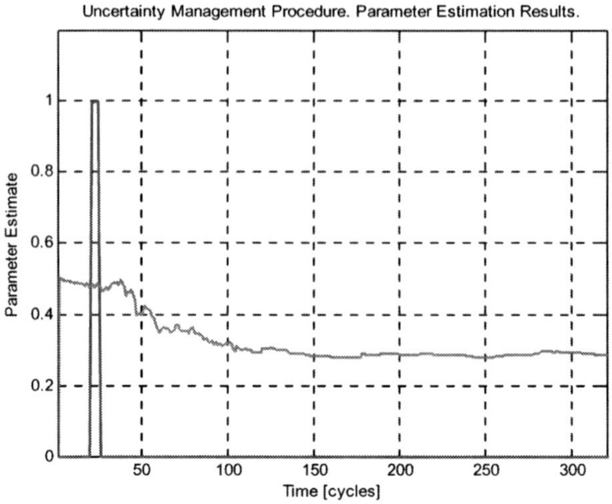

Fig. 2.6 Uncertainty management procedure; results of hyper-parameters update via an outer correction loop.

new population of equally weighted samples is obtained representing the probability density function at that particular time. The procedure may be repeated as needed until the expected value of the population reaches 0.45 units.

For the uncertainty management scheme, a 5-step prediction error has been used in the design of the *outer correction loop*. As expected, the longer the period used to calculate the prediction error, the larger the delay in the feedback loop. Several aspects must be considered in a proper selection of this parameter – as well as for p, q and Th in Equation (2.7) – including time constants of the system and the variability of model parameters.

The final implementation of the *correction loop* is shown in (2.9), while Figure 2.6 shows the results.

$$\begin{cases} \text{var}\{\omega_2(t+1)\} := 0.95\,\text{var}\{\omega_2(t)\}, \text{ if } \dfrac{\|\text{Pred\_error}(t)\|}{\|y(t)\|} < 0.1, \\ \text{var}\{\omega_2(t+1)\} := 1.20\,\text{var}\{\omega_2(t)\}, \text{ if } \dfrac{\|\text{Pred\_error}(t)\|}{\|y(t)\|} > 0.1, \end{cases} \tag{2.9}$$

Figure 2.6 clearly shows a period of time where the *outer correction loop* actually increments the variance of the noise profile $\omega_2(t)$ in the dynamic model (2.8); see black (blue) lines in Figure 2.6. After this period, it is observed that the state estimate (light-gray (magenta) line) rapidly converges. After that condition is reached, the prediction error decreases considerably and therefore the variance of the noise used for the "artificial evolution" estimation method decreases exponentially.

Figure 2.7 depicts the results obtained when computing the pdf of the Remaining Useful Life (RUL) for the same test case under study, considering that the critical fault dimension corresponds to 0.45 units. A procedure to obtain the RUL pdf from

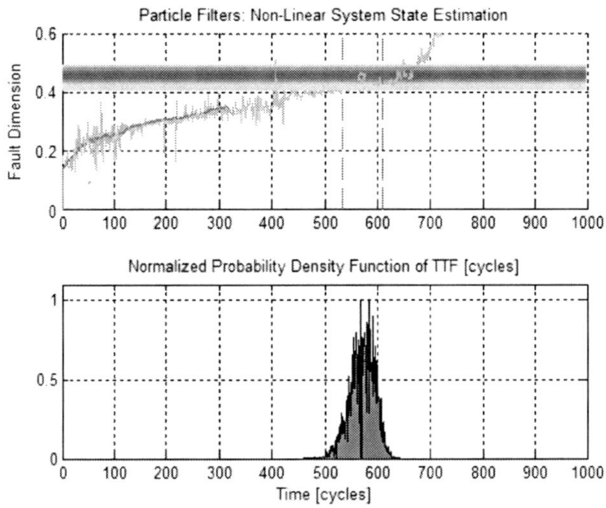

Fig. 2.7 Particle filtering-based uncertainty management system. Prognosis results for the case of axial crack in a gearbox plate.

the predicted path of the state pdf is detailed and discussed in [7]. Basically, the RUL pdf is the probability of failure at future time instants. This probability can be obtained from long-term predictions, when the empirical knowledge about critical conditions for the system is included in the form of thresholds for main fault indicators, also referred to as the hazard zones. The hazard zone in this case is represented as a horizontal band around 0.45 units in Figure 2.7. Once the RUL pdf estimate is generated, it is possible to obtain any necessary statistics about the evolution of the fault in time, either in the form of expectations or 95% confidence intervals.

It is desired for the expected value of this RUL pdf (computed at time $t = 320$) to be close to the actual time-to-failure. In this sense, the efficacy of the approach can be evaluated using two performance metrics: accuracy and precision. Accuracy measures how close the RUL expectation is to the actual time-to-failure. Precision, on the other hand, indicates the variance associated to RUL estimates. Precise RUL estimates imply a pdf with small variance.

The performance of the proposed Particle filtering-based approach has been compared with another implementation that uses the Extended Kalman filter to generate an estimate of the current state pdf. Results indicate that the Particle filteringbased approach, in combination with the proposed *outer correction loop*, provides better results in terms of accuracy and precision of the RUL pdf estimates; see Figure 2.8. Moreover, both applications require a similar amount of time to perform all the calculations, given that the Particle filterbased prognosis algorithm does not necessitate an extremely large number of particles to achieve its objective [7].

More accurate and more precise pdf estimates provide the basis for timely corrective actions and thus, help to avoid catastrophic failure events during the operation of the unit/process under supervision. In this sense, the proposed methodology

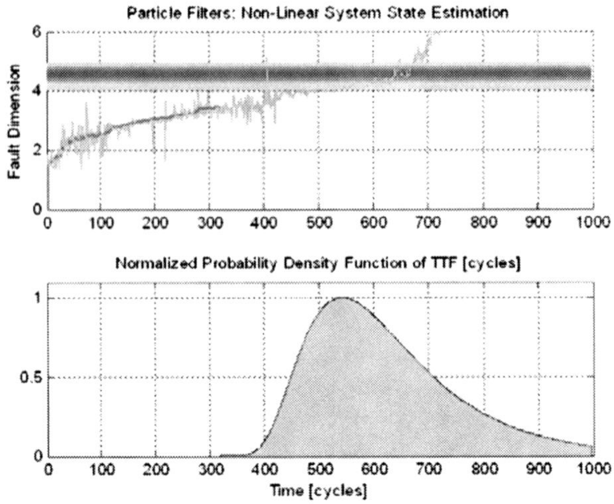

Fig. 2.8 Prognosis results for the case of axial crack in a gearbox plate, using an Extended Kalman filter-based approach.

has proven to be a valuable tool in Prognostics and Health Management (PHM) systems.

2.5 Conclusion

Uncertainty representation and management in failure prognosis present major challenges to the PHM user community. The stochastic nature of long-term predictions necessitates appropriate and effective means that can represent and manage uncertainty in almost real time. The on-platform utility of health and usage monitoring systems must be accompanied by robust prognostic algorithms if these technologies are to provide useful health information impacting safety of operation and cost of ownership. This paper proposes advances to uncertainty representation and management that will help in faster, more accurate, and more precise prognostics. Specifically, an architecture has been proposed that features feedback correction loops that in turn can reduce the impact of model errors and thus improve the accuracy and precision of the remaining useful life estimates. In addition, the use of appropriate kernels allows for an elegant and fast updating mechanism within the particle filter paradigm.

Results were obtained using data from a cracked gearbox plate. The performance (accuracy and precision) of the particle filter was superior when compared to a more traditional Extended Kalman Filter approach. Future work should explore the impact of injecting additional information into the correction loop such as domain expert information, maintenance information, etc. In addition, learning algorithms should be investigated that can automatically establish optimal or near-optimal values for parameters p, q, and Th and for the range for short-term prediction (here arbitrarily set to 5).

Acknowledgment

The authors would like to acknowledge and thank USRA/RIACS and NASA Ames for supporting research in the area of prognostic uncertainty management and for Dr. Orchard's financial support from Conicyt via Fondecyt #11070022.

References

1. G. Shafer, *A Mathematical Theory of Evidence*, Princeton University Press, Princeton, NJ, 1976.
2. N. Khiripet, G. Vachtsevanos, A. Thakker and T. Galie, A new confidence prediction neural network for machine failure prognosis, in *Proceedings of Intelligent Ships Symposium IV*, Philadelphia, PA, April 2–3, 2001.
3. Specht, D.F., A general regression neural network, *IEEE Transactions on Neural Networks* **2**(6), 568–576, November 1991.
4. J.A. Leonard and M.A. Kramer, Radial basis function networks for classifying process faults, *IEEE Control Systems* **11**, 31–38, 1991.
5. T.A. Cruse, *Probabilistic Systems Modeling and Validation*, HCF 2004, March 16–18, 2004.
6. J.L. Beck and S.K. Au, Bayesian updating of structural models and reliability using Marcov chain Monte Carlo simulation, *Journal of Engineering Mechanics* **128**(4), 380–391, April, 2002.
7. M. Orchard, A particle filtering-based framework for on-line fault diagnosis and failure prognosis, Ph.D. Thesis, Department of Electrical and Computer Engineering, Georgia Institute of Technology, 2007.
8. C. Musso, N. Oudjane and F. Le Gland, Improving regularized particle filters, in *Sequential Monte Carlo Methods in Practice*, A. Doucet, N. De Frietas and N. Gordon (Eds.), Springer-Verlag, New York, 2001.
9. M. Orchard, B. Wu and G. Vachtsevanos, A particle filter framework for failure prognosis, in *Proceedings of World Tribology Congress III*, Washington DC, September 12–16, 2005.
10. R. Patrick, M. Orchard, B. Zhang, M. Koelemay, G. Kacprzynski, A. Ferri and G. Vachtsevanos, An integrated approach to helicopter planetary gear fault diagnosis and failure prognosis, in *Proceedings of 42nd Annual Systems Readiness Technology Conference, AUTOTESTCON 2007*, Baltimore, MD, September 2007.
11. R. Patrick-Aldaco, A model based framework for fault diagnosis and prognosis of dynamical systems with an application to helicopter transmissions, Ph.D. Thesis, Electrical and Computer Engineering, Georgia Institute of Technology, Atlanta, 2007.

Chapter 3
A Novel Blind Deconvolution De-Noising Scheme in Failure Prognosis

Bing Zhang, Taimoor Khawaja, Romano Patrick, George Vachtsevanos,*
Marcos Orchard and Abhinav Saxena[1]

Abstract With increased system complexity, Condition-Based Maintenance (CBM) becomes a promising solution to system safety by detecting faults and scheduling maintenance procedures before faults become severe failures resulting in catastrophic events. For CBM of many mechanical systems, fault diagnosis and failure prognosis based on vibration signal analysis are essential techniques. Noise originating from various sources, however, often corrupts vibration signals and degrades the performance of diagnostic and prognostic routines, and consequently, the performance of CBM. In this paper, a new de-noising structure is proposed and applied to vibration signals collected from a testbed of the main gearbox of a helicopter subjected to a seeded fault. The proposed structure integrates a blind deconvolution algorithm, feature extraction, failure prognosis, and vibration modeling into a synergistic system, in which the blind deconvolution algorithm attempts to arrive at the true vibration signal through an iterative optimization process. Performance indexes associated with quality of the extracted features and failure prognosis are addressed, before and after de-noising, for validation purposes.

List of Symbols

f^s	Planetary carrier rotation frequency
N_t	Number of teeth in the annular gear

Bing Zhang · Taimoor Khawaja · Romano Patrick · George Vachtsevanos · Abhinav Saxena
School of Electrical and Computer Engineering, Georgia Institute of Technology, Atlanta, GA 30332 USA; e-mail: {bin.zhang, taimoor, romano.patrick, gjv, asaxena}@ece.gatech.edu

Marcos Orchard
School of Electrical and Computer Engineering, Georgia Institute of Technology, Atlanta, GA 30332 USA and Electrical Engineering Department, University of Chile, Av. Tupper 2007, Santiago, Chile; e-mail: marcos.orchard@ece.gatech.edu

* Author for correspondence.

K.P. Valavanis (ed.), Applications of Intelligent Control to Engineering Systems, 37–63.

p	Index of gear in consideration
N_p	Number of gears
m	Index of harmonics of tooth meshing frequency
M	Total number of tooth meshing harmonics in consideration
n	Index of harmonics of carrier rotation frequency
N	Total number of carrier rotation harmonics in consideration
δ	Angular phase shift caused by a crack on the plate
α_n	Magnitude of modulating signal at its harmonic nf^s
β_m	Magnitude of vibration signal from a single gear at its harmonic $mN_t f^s$
$\varphi_{p,m,n}$	Magnitude of frequency components from gear p at sideband $mN_t + n$
η	Remainder of $(mN_t + n)/N_p$
$\lambda_{m,n}$	Magnitude of combined vibration signal (superposition of vibration signals from different gears) at sideband $mN_t + n$
$\bar{\theta}$	Angular phase evenly separated by N_p gears
$W_{m,n}$	Weighting factor of nonlinear projection at sideband $mN_t + n$
θ_p	Angular position of gear p

3.1 Introduction

The increasing demand for system safety and reliability requires that faults in complex dynamic systems be detected and isolated as early as possible so that maintenance practices can be scheduled before faults become severe. Traditional breakdown and scheduled maintenance practices are, therefore, replaced by Condition-Based Maintenance (CBM) to meet this need [1]. CBM is an integration of signal processing, feature extraction, fault detection and isolation, failure prognosis, and decision making. In order to implement CBM, the health of critical components and subsystems must be monitored and reliable diagnostic/prognostic strategies developed [1]. Then, the system health can be assessed and maintenance practices can be scheduled based on the remaining useful life of components/systems to avoid catastrophic events.

Performance of failure prognostic routines, however, is closely related to the features (also known as condition indicators), derived from sensor data, which reveal the evolution and propagation of a failure in the system [2–5]. For many mechanical systems, features are typically extracted from vibration data [6]. During the operation of such a system, if a fault occurs, it is expected that the vibration signals will exhibit a characteristic signature which reveals the severity and location of the fault. Noise in the system, however, often corrupts the vibration signals and masks the indication of faults, especially in their early stages, thus curtailing the ability to accurately diagnose and predict failures. Therefore, it is important to develop a good and reliable de-noising scheme to improve the signal-to-noise ratio and make the characteristics of the fault perceptible in the vibration data. This process will improve the quality of the obtained features, potentially lower the fault detection threshold which increasing the accuracy of the diagnostic and prognostic algorithms.

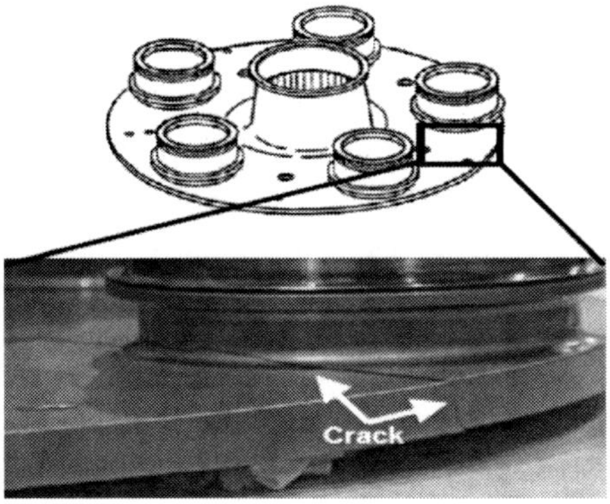

Fig. 3.1 The crack of planetary gear carrier plate of the UH-60A helicopter.

The main transmission of Blackhawk and Seahawk helicopters employs a five-planet epicyclic gear system, which is a critical component directly related to the availability and safety of the vehicle [6, 7]. Recently, a crack in the planetary carrier plate was discovered during regular maintenance, as shown in Figure 3.1. This resulted in major overhaul, re-design and replacement of gear plates with a high cost associated with these activities. Manual inspection of all transmissions is not only costly, but also time prohibitive [2]. CBM could provide a cost-effective solution to reduce the work burden and enhance vehicle safety [2]. Many research efforts towards designing and implementing CBM on helicopters, such as vibration signal preprocessing [3, 8], vibration signal modeling [9], features extraction [3, 4], detection of cracks, and prediction of crack length [10], have been carried out. The objective of this paper is to propose a new vibration pre-processing structure, which synergizes vibration de-noising, vibration modeling, feature extraction, and failure prognosis, to enhance the performance of CBM. The planetary gear system with a seeded crack on the gear carrier plate will be used to verify the proposed method.

For an epicyclic gear system, the widely used de-noising technique is Time Synchronous Averaging (TSA), which can be implemented in the time or frequency domain [3–6, 8]. This operation enhances the components at the frequencies that are multiples of the shaft frequency, which are often related to the meshing of gear teeth [2, 3]. At the same time, it tends to average out external random disturbances and noise that are asynchronous with the rotation of the gear.

Other de-noising algorithms include Blind Source Separation (BSS) [11–13], stochastic resonance [14], and adaptive schemes [15, 16]. BSS aims to extract individual, but physically different, excitation sources from the combined output measurement [11]. Due to the complex environment and the large number of noise sources in mechanical systems, the application of BSS is severely hindered [11]. A practical solution is to focus on the main vibration source that contributes mostly to the vibration, while it treats all other sources as a combined noise. Then, the objective is reduced to separating the vibration source from noise, which, in this sense, is the cumulative contribution of many different sources. This leads to a blind deconvolution de-noising algorithm [17–19].

Previous research work reported in [9] has provided a good understanding of the true vibration signals, originating from the epicyclic gearbox under both healthy and faulty operational conditions. However, little knowledge about the noise profile is available. To remove noise and recover the actual vibration signal, a blind deconvolution algorithm developed for a similarly formulated image processing problem [17] will be modified and employed. The paper addresses in detail the structure of the overall de-noising scheme, the analysis of vibration mechanisms, the blind deconvolution algorithm, as well as its experimental verification. The results show that the proposed de-noising scheme can substantially improve the signal-to-noise ratio, feature performance, and the precision of the failure prognostic algorithm.

3.2 The De-Noising Scheme Architecture

The proposed overall de-noising scheme is illustrated in Figure 3.2. An accelerometer mounted on the gearbox frame collects vibration signals and the TSA signal $s(t)$ is calculated. The blind deconvolution de-noising algorithm is carried out in the frequency domain, and hence $s(t)$ is Fourier transformed to arrive at $S(f)$. Then, the de-noising algorithm is applied to $S(f)$ which outputs the de-noised vibration data in the frequency domain $B(f)$. If the time domain signal is required, $B(f)$ can be inverse Fourier transformed to obtain $b(t)$. From $B(f)$ and $b(t)$, features can be extracted and fused to be used subsequently for fault diagnosis and failure prognosis. With the main objective being remaining useful life prediction, the failure prognosis algorithm also provides an estimate of crack length on the planet gear carrier plate as a function of time [10]. Both the estimated crack length and load profile of the helicopter serve as inputs to the vibration model [9], which generates the noise-free modeled vibration signal $m(t)$. This modeled vibration signal is Fourier transformed into the frequency domain and its frequency spectra are normalized to obtain the weighting factor vector $W(f)$, which is used in a nonlinear projection of the blind deconvolution de-noising algorithm.

The architecture of the proposed de-noising algorithm is decomposed and shown in Figure 3.3. In this scheme, a nonlinear projection, which is based on vibration analysis in the frequency domain, and a cost function minimization are critical components, which are described in the sequel. Initially, an inverse filter $\bar{Z}(f)$ must be defined. This inverse filter is an initial estimate of the modulating signal in the frequency domain, and converges to a filter through an optimization routine that recovers the vibration signal from the noisy measured data $S(f)$. The initial inverse filter $\bar{Z}(f)$ is convoluted with $S(f)$ to obtain a rough estimate of the noise-free vibration signal $\bar{B}(f)$. The signal $\bar{B}(f)$ passes through the nonlinear projection, which maps to a subspace that contains only known characteristics of the vibration signal, to yield $B_{\text{nl}}(f)$. The difference between $\bar{B}(f)$ and $B_{\text{nl}}(f)$ is denoted as $E(f)$. By adjusting $\bar{Z}(f)$ iteratively to minimize $E(f)$, and when $E(f)$ reaches a minimal

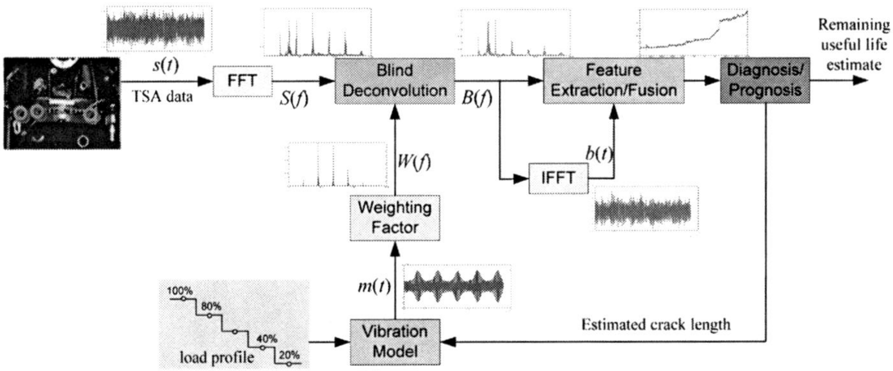

Fig. 3.2 The overall structure of the de-noising scheme.

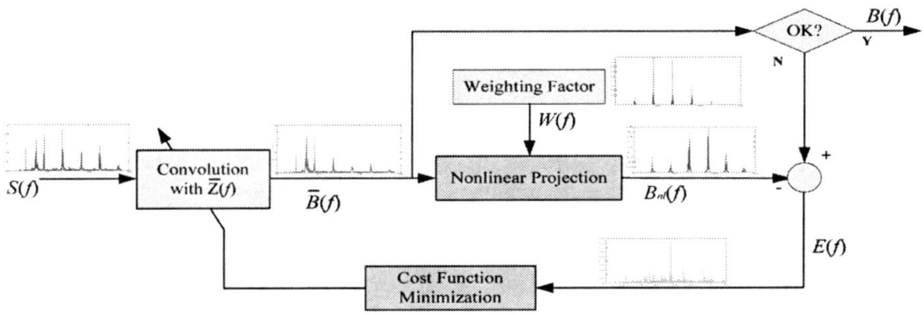

Fig. 3.3 Blind deconvolution de-noising scheme.

value, the signal $\bar{B}(f) \to B(f)$ can be regarded as the de-noised vibration signal. At the same time, $\bar{Z}(f)$ converges to $Z(f)$. Through an inverse Fourier transform, the de-noised vibration signal in the time domain can be obtained as well.

3.3 Vibration Data Analysis

The vibration signals are derived from the main transmission gearbox of Blackhawk and Seahawk Helicopters. The gearbox is an epicyclic gear system with five planet gears, whose configuration is illustrated in Figure 3.4. The gearbox is mounted on a test cell and with a seeded crack fault on the planetary gear carrier. The following sections intend to describe the expected vibration data in the healthy and faulty gear carrier plate, respectively.

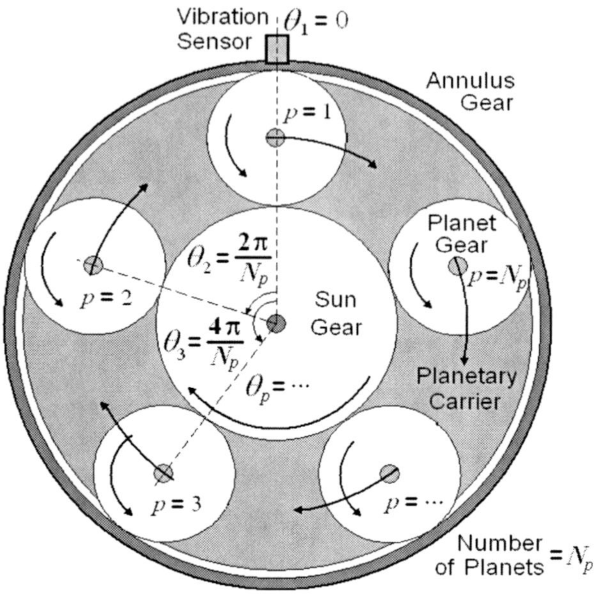

Fig. 3.4 The configuration of an epicyclic gear system.

3.3.1 Healthy Gearbox

Let us assume that the gearbox is an ideal healthy system without cracks. The accelerometer is mounted at a fixed point at position $\theta = 0$. Since the vibration signal is generated from the meshing of gear teeth and the planetary gears are rotating inside the angular gear, the vibration signal is amplitude-modulated to the static accelerometer. That is, the observed vibration amplitude will be large when the planetary gear is close to the accelerometer and it will be small when the planetary gear is far. Suppose that there is only one planetary gear, then the vibration observed by the transducer should have the largest amplitude when the planetary gear is at $\theta = 0, 2\pi, 4\pi, \ldots$. Similarly, the vibration should have the smallest amplitude at $\theta = \pi, 3\pi, 5\pi, \ldots$. Suppose that the planetary carrier has a rotation frequency f^s. The vibration amplitude modulating signal for this single planet gear and its spectra, which show components at harmonics of f^s, are illustrated in Figures 3.5a and 3.5b, respectively. In Figure 3.5b, n is the index of harmonics of f^s and α_n is the amplitude of the component of the modulating signal at frequency nf^s.

In the ideal case, the $N_p = 5$ planetary gears are evenly spaced. Then, the planetary gear p at time instant t has a phase

$$\theta_p = 2\pi \left(f^s t + \frac{p-1}{N_p} \right). \tag{3.1}$$

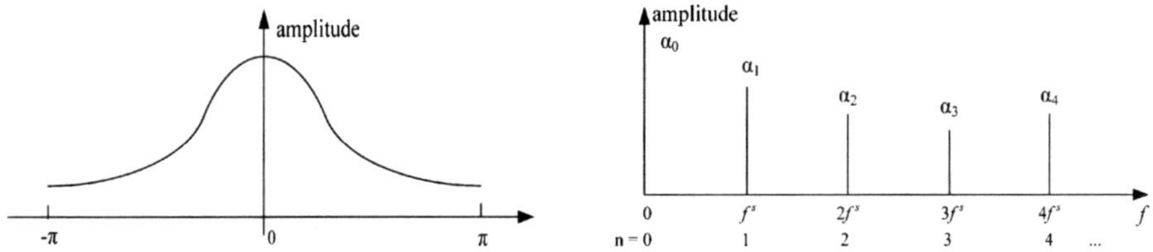

(a) Time domain modulating signal (b) Frequency domain modulating signal

Fig. 3.5 Vibration amplitude modulation signal.

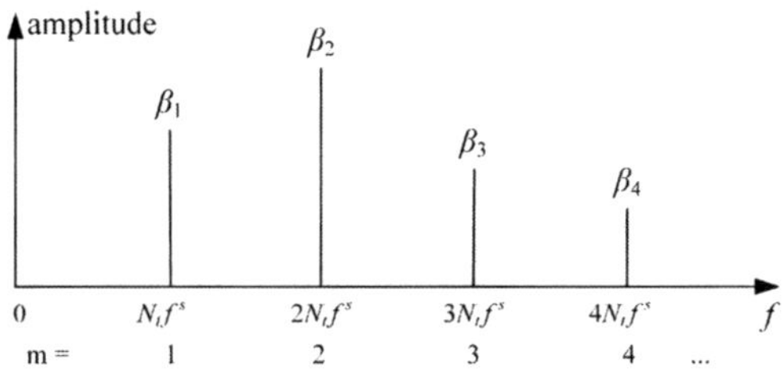

Fig. 3.6 Spectrum of tooth meshing vibration of a single planet gear.

The amplitude modulating signal for planetary gear p can be written in the time domain as

$$a_p(t) = \sum_{n=-N}^{N} \cos_n \cos(n\theta_p), \tag{3.2}$$

where N is the number of sidebands about the harmonics under consideration.

In this case, it is natural to assume that all mesh vibrations generated from different planetary gears are of the same amplitude but of different phase shifts. The annulus gear has $Nt = 228$ teeth. Since the speed at which teeth meshing is proportional to the angular velocity of the planetary carrier, the meshing vibration appears at frequencies $Nt\ f^s$ [7, 9]. In addition, suppose that, in the frequency domain, the meshing vibration signal has amplitudes of β_m at its harmonics $mNtf^s$, the spectra are illustrated in Figure 3.6. Then, the vibration signal generated from planet gear p can be written in the form of

$$b_p(t) = \sum_{m=1}^{M} \beta_m \sin(mN_t\theta_p), \tag{3.3}$$

where M is the number of harmonics under consideration.

The observed vibration signal of planetary gear p, with respect to the static accelerometer, is given as the product of the meshing vibration signal and the amplitude modulating signal. It is denoted as $y_p(t)$, with frequency spectra as shown in Figure 3.7, and is given by

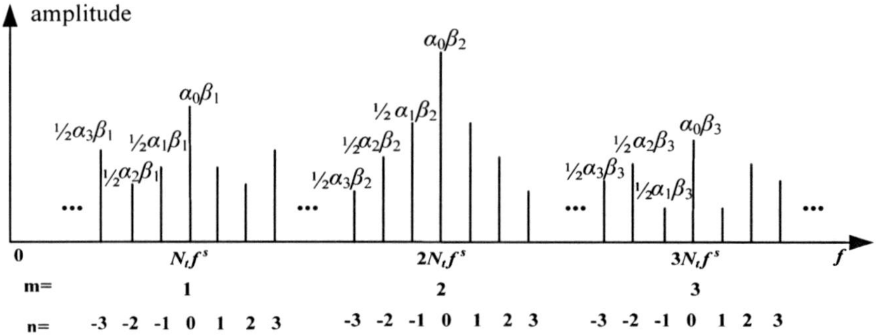

Fig. 3.7 Vibration spectrum of a single planet.

$$y_p(t) = a_p(t)b_p(t) = \frac{1}{2}\sum_{m=1}^{M}\sum_{n=-N}^{N} \alpha_n \beta_m \sin(mN_t + n)\theta_p). \tag{3.4}$$

Note the sidebands around the meshing vibration harmonics, as compared to Figure 3.6. The "sidebands" are defined as the frequency components that appear as a harmonically spaced series [7, 9]. The position of the sidebands can be located by $mN_t + n$ or (m, n) with m being the index of meshing vibration harmonics and n the index of the modulating signal harmonics.

When there are more than one, say N_p, Np planetary gears, the vibration signal observed by the accelerometer is the superposition of the Np vibration signals generated from Np different planetary gears. This superposition vibration signal has the form:

$$\begin{aligned} y(t) &= = \frac{1}{2}\sum_{p=1}^{N_p}\sum_{m=1}^{M}\sum_{n=-N}^{N} \alpha_n \beta_m \sin(mN_t + n)\theta_p) \\ &= \frac{1}{2}\sum_{p=1}^{N_p}\sum_{m=1}^{M}\sum_{n=-N}^{N} \alpha_n \beta_m \sin\left(2\pi(p-1)\frac{mN_t + n}{N_p}\right), \end{aligned} \tag{3.5}$$

where Equation (3.1) and the fact that $\sin(2k\pi + \theta) = \sin(\theta)$ for any integer k are used.

Since the planetary gears are evenly spaced, the phase angle of the sidebands will be evenly spaced along 2π [7]. From Equation (3.5), it is obvious that if sideband $mN_t + n$ is not a multiple of N_p and $(mN_t + n)/N_p$ has a remainder of η, the vibration components from different gears are evenly spaced by an angle $2\eta\pi/N_p$. In this case, when the vibrations generated from different planetary gears are combined, these sidebands add destructively and become zero as illustrated in Figure 3.8a, in which $\varphi_{p,m,n}$ with $1 \le p \le 5$ indicates the frequency components of gear p. Those frequency components appear at sidebands where $mN_t + n \ne kN_p$ are termed as Non-Regular Meshing Components (NonRMC).

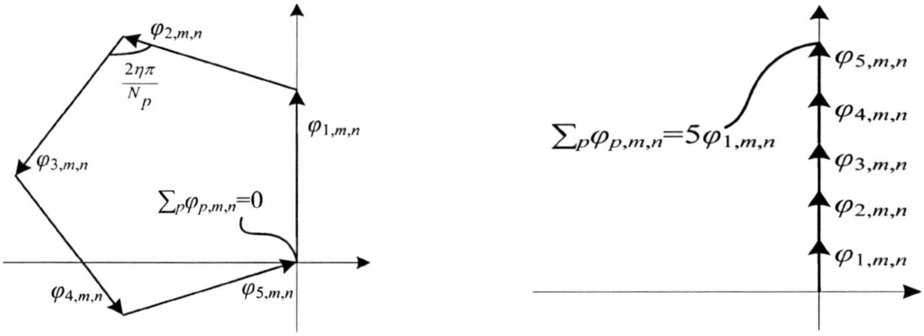

(a) $mN_t+n \neq kN_p$ (NonRMC) (b) $mN_t+n = kN_p$ (RMC)

Fig. 3.8 Examples of superposition of vibration signal for a healthy gear plate with $N_p = 5$.

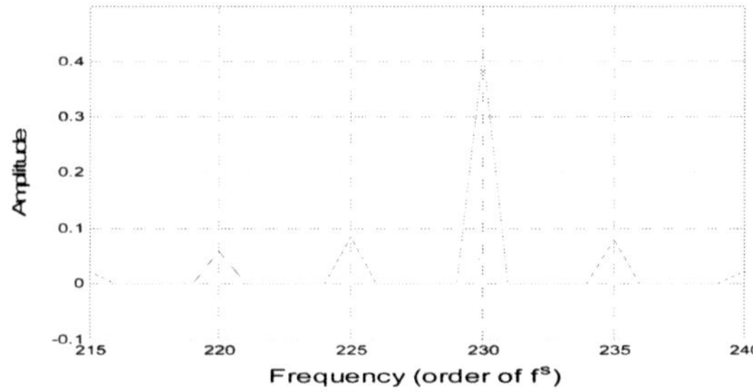

Fig. 3.9 Vibration spectrum of combined signal with $N_t = 228$ and $N_p = 5$.

On the contrary, if sideband $mN_t + n$ is a multiple of N_p, the remainder of $(mN_t + n)/N_p$ will be zero. In this case, the vibration components from different gears do not have a phase difference. When the vibration signals from different planetary gears are combined, these sidebands add constructively and are reinforced as illustrated in Figure 3.8b. These frequency components that appear at sidebands where $mN_t + n = kN_p$ are referred to as Regular Meshing Components (RMC) or apparent sidebands.

This process of frequency components adding destructively/constructively finally generates asymmetrical sidebands. Partial frequency spectra of the sidebands around the first order harmonic that illustrates this asymmetry are shown in Figure 3.9, where $N_p = 5$ and $N_t = 228$. Note that the peak of the spectrum does not appear at $n = 0$ (order 228) but at $n = -3$ (order 225), $n = 2$ (order 230), etc. The largest spectral amplitude (also known as dominant sideband) appears at the frequency closest to the gear meshing frequency [6].

According to the above vibration analysis in the frequency domain and previous research results [6, 7, 9], for an ideal system, only terms at frequencies that are multiples of the number of planetary gears (i.e. RMC) survive while the terms at other frequencies (i.e. NonRMC) vanish.

Then, the Fourier transform of the vibration data can be written as

$$Y(f) = f_{\text{hea}}(\lambda_{m,n}((mN_t + n)f^s)), \tag{3.6}$$

where $\lambda_{m,n}$ is the magnitude of the spectral amplitude at $(mN_t + n)f^s$ and f_{hea} is the nonlinear projection for an ideal healthy gearbox given by:

$$f_{\text{hea}} = \begin{cases} 1 & \text{if } mN_t + n \text{ is a multiple of } N_p, \\ 0 & \text{otherwise.} \end{cases} \tag{3.7}$$

3.3.2 Faulty Gearbox

When there is a crack on the planetary gear carrier, as in Figure 3.1, the five gears will not be evenly separated along 2π. Suppose that, at time instant t, one of the five planetary gears has an angle shift δ caused by the crack. Then, this planetary gear has a phase of $\theta_p + \delta$ at time instant t with θ_p being given in Equation (3.1). For this gear, the modulating signal becomes:

$$a'_p = \sum_{n=-N}^{N} \alpha_n \cos(n(\theta_p + \delta)). \tag{3.8}$$

The vibration signal generated from this gear is:

$$b'_p = \sum_{m=1}^{M} \beta_m \sin(mN_t(\theta_p + \delta)). \tag{3.9}$$

Accordingly, the modulated vibration signal from this gear is written as

$$y'_p(t) = a'_p(t)b'_p(t) = \frac{1}{2}\sum_{m=1}^{M}\sum_{n=-N}^{N} \alpha_n\beta_m \sin((mN_t + n)(\theta_p + \delta)). \tag{3.10}$$

If we denote by $\bar{\theta} = 2\pi(p-1)/N_p$, then when the vibration signals from the five planetary gears are superposed (note that the other four gears do not have a phase shift), the observed vibration signal should have the form of (suppose the phase shift happens on the last gear):

$$y'(t) = \frac{1}{2}\sum_{p=1}^{N_p-1}\sum_{m=1}^{M}\sum_{n=-N}^{N} \alpha_n\beta_m \sin((mN_t + n)\theta_p)$$

$$+ \frac{1}{2}\sum_{m=1}^{M}\sum_{n=-N}^{N} \alpha_n\beta_m \sin((mN_t + n)(\theta_p + \delta))$$

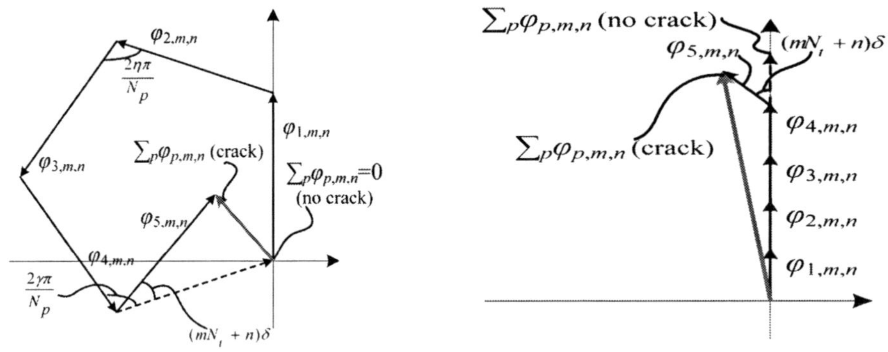

(a) $mN_t + n \neq kN_p$ (NonRMC) (b) $mN_t + n = kN_p$ (RMC)

Fig. 3.10 Examples of superposition of vibration signal for a faulty gear plate with $N_p = 5$.

$$= \frac{1}{2} \sum_{m=1}^{M} \sum_{n=-N}^{N} \alpha_n \beta_m \left(\sum_{p=1}^{N_p - 1} \alpha_n \beta_m \sin((mN_t + n)\bar{\theta}) + \sin((mN_t + n)(\bar{\theta} + \delta)) \right). \tag{3.11}$$

Due to this phase shift, when $mN_t + n$ is not a multiple of N_p, the vibration components from different gears are not evenly spaced. This can be illustrated in Figure 3.10a. It is obvious that the vibration components are not canceled in this case. This results in higher NonRMC frequency components. On the other hand, when $mN_t + n$ is a multiple of N_p, the vibration components from different gears are not exactly in phase. The vibration component from the last gear has a phase difference, which results in lower RMC frequency components, as shown in Figure 3.10b.

To see this effect in real vibration signals, the frequency spectra around the first harmonic for two different crack sizes at 1.35 and 4.4 inches are shown in Figure 3.11. The components in the circles are the NonRMC. This is consistent with the analysis in Figure 3.10a. When $(mN_t + n)\delta$ changes from 0 to π, the NonRMC changes from a minimum to a maximum amplitude while the RMC from a maximum to a minimum. On the other hand, when $(mN_t + n)\delta$ changes from π to 2π, the NonRMC becomes from a maximum amplitude to a minimum while the RMC from a minimum to a maximum.

Then, it is clear that the nonlinear projection, given in (3.7), under the assumption of a healthy gearbox is not suitable for a faulty one. A reasonable modification of the nonlinear projection is given as follows. From previous research in [9], a vibration model in the frequency domain is established with the load profile and the crack size being two of the inputs. Note that the load profile is known and the crack size can be estimated from the prognostic algorithm [10]. Then, the modeled noise-free vibration signal $m(t)$ generated from the vibration model is Fourier transformed in

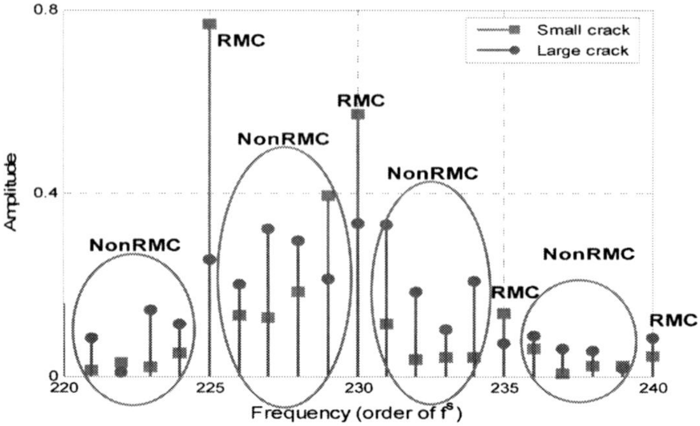

Fig. 3.11 The influence of crack size on frequency spectra.

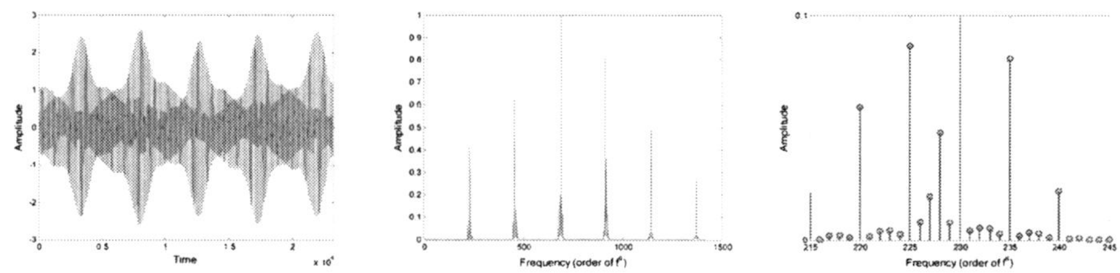

(a) (b) (c)

Fig. 3.12 The model vibration signal and weighting factor. (a) The model vibration signal with crack; (b) The weighting factor, normalized model vibration spectra; (c) The weighting factor at the 1st harmonic.

the frequency domain to arrive at $M(f)$. The magnitude of $M(f)$ is normalized to obtain weighting factors $W(f)$. For illustration purposes, a model signal $m(t)$ and the weighting factor $W(f)$ derived from its frequency spectra, are shown in Figures 3.12a and 3.12b. The weighting factors around the first harmonic are zoomed in Figure 3.12c.

When a frequency domain signal $\bar{B}(f)$ is fed into the nonlinear projection, its frequency components are multiplied by the weighting factor $W(f)$ to arrive at the output of the nonlinear projection $B_{\mathrm{nl}}(f)$. Suppose that, for a sideband located at $(mN_t + n)f^s$, $\bar{B}(f)$ has a magnitude of $\lambda_{m,n}$ then, after the nonlinear projection, the magnitude of $B_{\mathrm{nl}}(f)$ at this sideband is given by $W_{m,n}(f)\lambda_{m,n}$, where $W_{m,n}(f)$ is the weighting factor at sideband $mN_t + n$. It is clear that the nonlinear projection in this case is:

$$f_{\mathrm{nl}} = W_{m,n}(f) \quad \forall (mN_t + n) \in D_{\mathrm{sup}} \,. \tag{3.12}$$

Note that the nonlinear projection under the healthy gearbox case in Equation (3.7) is a special case of (3.12).

From the above analysis of the system vibration behavior, the following assumptions are made and used in the blind deconvolution de-noising scheme:

1. The majority of vibration information is contained in limited number of sidebands around the harmonics, which form the D_{sup}. Note that the frequency spectra fade quickly on both sides of the harmonics, this assumption is reasonable considering subsequent results.
2. The nonlinear projection that maps a signal into a subspace contains only known characteristics of the vibration signal is given in (3.12).
3. The amplitude of the modulating signal $a(t)$ decreases monotonically on either side of its maximum value until it reaches the minimum. Note that this assumption is somewhat restrictive.

3.4 Blind Deconvolution De-Noising Scheme

From the vibration analysis of the gearbox, we know that the vibration signals collected from the transducer are amplitude modulated [7, 9]. Multiple sources of noise may further corrupt the signal. A simplified model for such a complex signal may be defined as

$$s(t) = a(t)b(t) + n(t), \tag{3.13}$$

where $s(t)$ is the noisy vibration signal, $b(t)$ is the noise-free un-modulated vibration signal, $a(t)$ is the modulating signal, and $n(t)$ is the cumulative additive noise.

Note that the modulating signal $a(t)$ is itself affected by noise in the system. Let denote the ideal or noise-free modulating signal and $n_a(t)$ the noise introduced in this signal. Consequently, $a(t)$ can be represented as

$$a(t) = \hat{a}(t) + n_a(t). \tag{3.14}$$

Thus, Equation (13) can be rewritten as follows:

$$\begin{aligned} s(t) &= (\hat{a}(t) + n_a(t)b(t) + n(t) \\ &= \hat{a}(t)b(t) + \hat{n}(t), \end{aligned} \tag{3.15}$$

where $\hat{n} = n_a(t)b(t) + n(t)$ contains the total additive noise in the system. On the other hand, the factor $\hat{a}(t)$ describes the multiplicative noise in the system. The goal for a de-noising scheme, such as the one described here, is to recover the unknown vibration signal $b(t)$ from the observed signal $s(t)$ given partial information about the noise sources and characteristics of the vibration signal.

A typical approach would be to find the inverse of $\hat{a}(t)$,

$$\hat{z}(t) = 1/\hat{a}(t) \tag{3.16}$$

such that

$$\begin{aligned} b(t) &= (s(t) - \hat{n}(t)) \cdot \hat{z}(t) \\ &= s(t)\hat{z}(t) - \hat{n}(t)\hat{z}(t). \end{aligned} \tag{3.17}$$

Note however that little can be assumed about $\hat{n}(t)$ and $\hat{a}(t)$ is not available, is not applicable. To solve this problem, rather than using $\hat{z}(t)$, we propose an iterative denoising scheme that starts with $\bar{z}(t)$, a very rough initial estimate of the inverse of the modulating signal, which demodulates the observed signal $s(t)$ to give a rough noise-free vibration signal:

$$\bar{b}(t) = s(t) \cdot \bar{z}(t). \tag{3.18}$$

If partial knowledge about how the plate system is influenced by the modulating signal $\hat{a}(t)$ and a reasonable understanding of the true vibration signal is available, then the ideal characteristics of the vibration signal can be obtained by projecting this estimated signal $\bar{b}(t)$ into a subspace with only the known ideal characteristics of the vibration signal to yield a refined signal $b_{\text{nl}}(t)$. Since this non-linear projection, as the subscript signifies, removes all uncharacterisitic components that exist in the rough estimate $\bar{b}(t)$, it is necessary to stress the importance of a good understanding of the underlying process. An iterative scheme then refines these results by minimizing the error between the two signals $\bar{b}(t)$ and $b_{\text{nl}}(t)$, i.e.

$$\min \|e(t)\| = \min \|\bar{b}(t) - b_{\text{nl}}(t)\|. \tag{3.19}$$

Previous research results detail the spectral characteristics of vibration signals for rotating equipment [6, 7, 9]. It is appropriate, therefore, to investigate the measured noisy vibration in the frequency domain. The convolution theorem states that the product of two signals in the time domain is equivalent to their convolution in the frequency domain. Thus, model (15) can be written in the frequency domain as

$$S(f) = \hat{A}(f) * B(f) + \hat{N}(f), \tag{3.20}$$

with $*$ being the convolution operator and $S(f)$, $\hat{A}(f)$, $B(f)$ and $\hat{N}(f)$ are the Fourier transforms of $s(t)$, $\hat{a}(t)$, $b(t)$ and $\hat{n}(t)$, respectively. Then, the goal in the frequency domain is to recover $B(f)$.

Writing Equation (3.18) in the frequency domain, we have

$$\bar{B}(f) = S(f) * \bar{Z}(f) \tag{3.21}$$

with $\bar{B}(f)$ and $\bar{Z}(f)$ being the Fourier transforms of $\bar{b}(t)$ and $\bar{z}(t)$, respectively. Passing $\bar{B}(f)$ through the nonlinear projection, it yields $B_{\text{nl}}(f)$. Then, in the frequency domain, we will minimize the difference between $B_{\text{nl}}(f)$ and $\bar{B}(f)$:

$$\min \|E(f)\| = \min \|\bar{B}(f) - B_{\text{nl}}(f)\|. \tag{3.22}$$

The iterative process refines $\bar{Z}(f)$ to minimize Equation (3.22). When it reaches the minimal value, $\bar{Z}(f)$ converges to $Z(f)$. Then, with this $Z(f)$ replacing $\bar{Z}(f)$ in Equation (3.21), a good estimate for $B(f)$ is obtained as

$$B(f) = S(f) * Z(f). \tag{3.23}$$

Lastly, the estimate is transformed back into the time domain to recover the noise-free vibration signal $b(t)$.

$$b(t) = F^{-1}\{B(f)\} = \int_{-\infty}^{\infty} e^{i2\pi ft} B(f) df. \tag{3.24}$$

To solve this problem, the following additional assumptions are made:

4. $Z(f)$ exists and is absolutely summable, i.e. $\sum Z(f) < \infty$.
5. Since the modulating signal $\hat{a}(t)$ is always positive, its inverse should be positive too.

Hence, the Fourier transform of its inverse $Z(f)$ contains a dc component.

With the above two assumptions being taken into consideration, the cost function is defined as

$$J = \sum_{f \in D_{\text{sup}}} [\bar{B}(f) - B_{\text{nl}}(f)]^2 + \left(\sum Z(f) - 1\right)^2, \tag{3.25}$$

where D_{sup} is the frequency range that contains the main vibration information. Because of the periodic fade of signal spectra between harmonics, a window centered at harmonic frequencies is used to define critical frequencies. All these windows form the support D_{sup}. In Equation (3.25), assumptions 4 and 5 are used to arrive at the second term to avoid an all-zero inverse filter $Z(f)$, which leads to the trivial solution for error minimization. Moreover, an iterative optimization routine is required to implement this scheme. The iterative conjugate gradient method is called upon to address the optimization problem. This method has faster convergence rate in general as compared to the steepest descent method [20].

3.5 Experimental Studies

The initial length of the seeded crack on the carrier is 1.344 inches and it grows with the evolving operation of the gearbox. The gearbox operates over a large number of Ground-Air-Ground (GAG) cycles at different torque levels. This way, vibration data are acquired at different torque levels and different crack lengths.

3.5.1 Actual Crack Growth

The ground truth crack length data at discrete GAG cycles are available and tabulated in Table 3.1. With these data, the crack length for GAG cycles from 1 to 1000

Table 3.1 The ground truth data of crack length (inches).

GAG	1	36	100	230	400	550	650	714	750	988
Crack	1.34	2.00	2.50	3.02	3.54	4.07	4.52	6.21	6.78	7.63

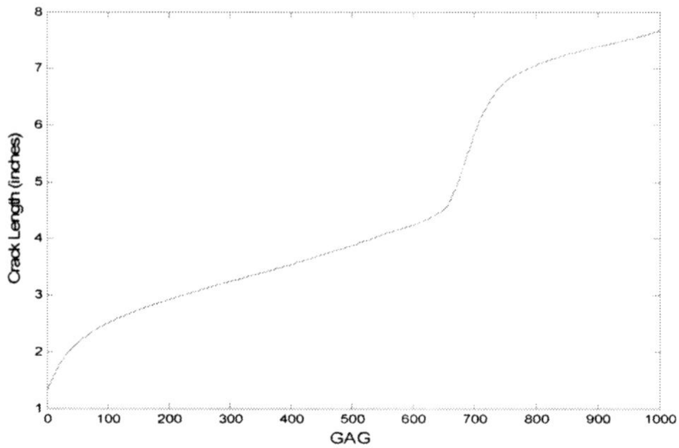

Fig. 3.13 The growth of crack length versus GAG cycles.

can be obtained via interpolation, which results in the crack length growth curve shown in Figure 3.13.

3.5.2 Load Profile

In each GAG cycle over the first 320 GAG cycles, the torque increases from 20 to 40%, then to 100%, and finally to 120% and then decreases to 20% for the next cycle. From the 321st GAG cycle on, the torque in each cycle increases from 20 to 40%, then to 93%, and then decreases to 20% for the next cycle. The torque profiles in these two cases are shown in Figure 3.14. Because no consistent high torque level data throughout the entire spectrum of GAG cycles is available and there is no substantial difference between the 93 and 100100% torque in the first 320 GAG cycles and that at 93% torque in the later GAG cycles are considered as consisting one experiment. In this case, torque levels at 20, 40, and 100% are investigated.

3.5.3 On-Line Real-Time De-Noising

It is important to implement the de-noising scheme on-line. Therefore, the reduction of the computational burden is critical. The real-time implementation of the de-noising scheme involves the structure of the inverse filter $Z(f)$, simplification of

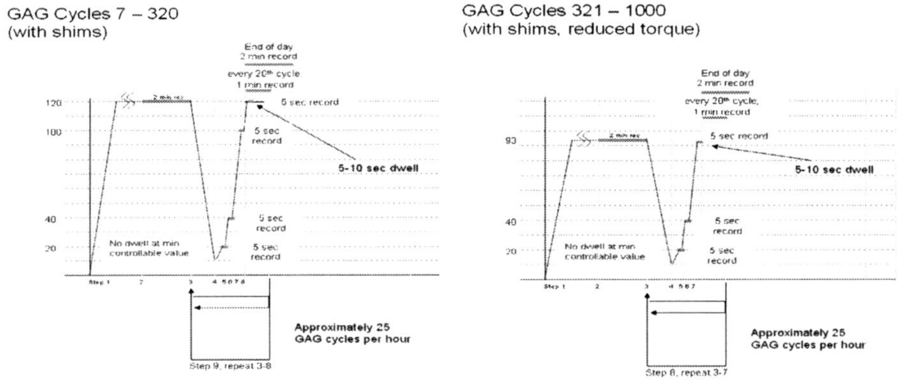

(a) GAG cycles 7-320

(b) GAG cycles 321-1000

Fig. 3.14 The load profile of the gearbox.

the convolution operation, the initial $Z(f)$, and simplification of the optimization routine.

3.5.3.1 The Structure of the Inverse Filter

For the gearbox with five gears as in Figure 3.4, the vibration signal observed by the transducer should have its largest amplitude at phase $\theta_i = 0, 2\pi/N_p, 4\pi/N_p, \ldots$. The smallest amplitude appears at phase $\theta_i = \pi/N_p, 3\pi/N_p, 5\pi/N_p, \ldots$. This indicates that the frequency of the modulating signal $a(t)$ is much lower than that of the vibration signal $b(t)$. The same is true for the inverse of $a(t)$, $1/a(t)$, whose frequency spectra fade quickly at high frequencies. Therefore, at the implementation phase, the inverse filter $\bar{Z}(f)$ can be truncated to a short signal containing only limited coefficients to save computation time and resources. This also helps to improve the efficiency of the algorithm.

3.5.3.2 Good Initial Inverse Filter

A good initial guess for $\bar{Z}(f)$ will also improve the convergence rate of the blind deconvolution approach. In the process of de-noising, some $Z(f)$ values are collected and saved from previous results. When the system operates under similar operating conditions, the previously saved $Z(f)$ can be retrieved and serves as the initial estimate of $\bar{Z}(f)$ under the current operating condition. It is anticipated that this step will speed up convergence.

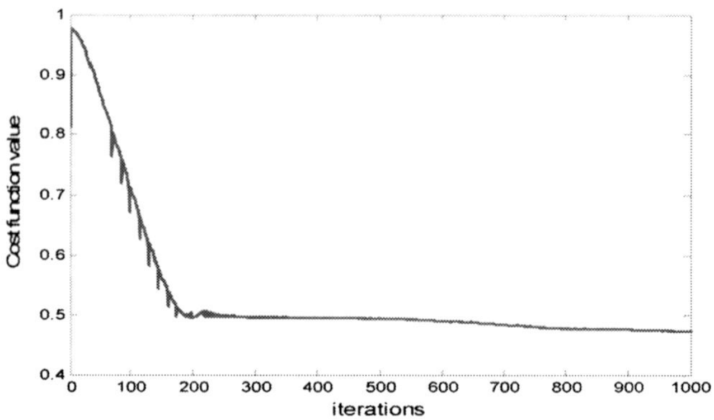

Fig. 3.15 Convergence of cost function along iterations.

3.5.3.3 Simplification of the Optimization Routine

Optimization of the de-noising algorithm is a recursive process. The number of iteration is closely related to the required computation time. To reach a minimal value of the cost function, a large number of iterations are needed. However, at the later stages of optimization, the improvement in the cost function value is rather minimal. This suggests that we can terminate the optimization routine before it reaches the minimal value and this will not degrade performance significantly.

For example, the convergence curve of the cost function versus iteration number from a specific example is illustrated in Figure 3.15. It can be seen that the cost function becomes stable from about the 200th iteration on. Hence, for online applications, the optimization routine can be terminated after a small number of iterations resulting in considerable computational savings.

3.5.3.4 Simplification of the Frequency Domain Signal Convolution

Earlier research results [9] suggest that the main vibration characteristic signature resides in frequency range below $1500f^s$ Hz, which covers up to six harmonics. In addition, rapid fade of the signal spectra between the harmonics enable us to define D_{sup} as a series of windows on each side of the harmonics. Therefore, D_{sup} is a collection of six windows with each window being centered at the harmonic frequency and having a certain order of sidebands on both sides. Then, the convolution operation only considers frequency spectra on D_{sup}. To achieve this goal, the convolution operation is divided into three steps and this process can reduce the computational burden significantly.

Step 1: Picking up frequency spectra on D_{sup} and separating them into six segments as shown in Figure 3.16. The original signal spectra are shown in Figure 3.16a. The

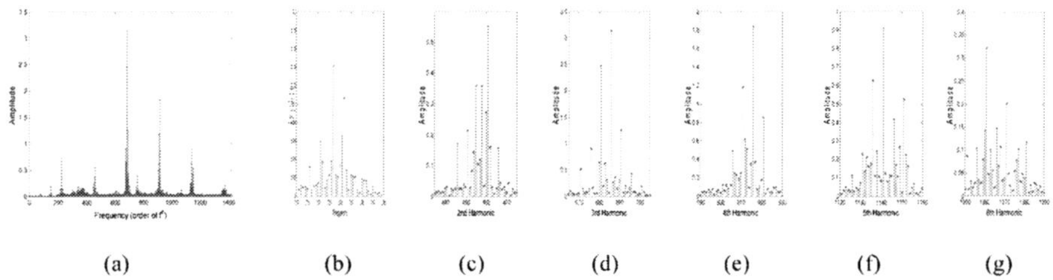

Fig. 3.16 The separation of frequency spectra (with a resolution of f^s). (a) TSA signal spectra; (b) 1st harmonic spectra; (c) 2nd harmonic spectra; (d) 3rd harmonic spectra; (e) 4th harmonic spectra; (f) 5th harmonic spectra; (g) 6th harmonic spectra.

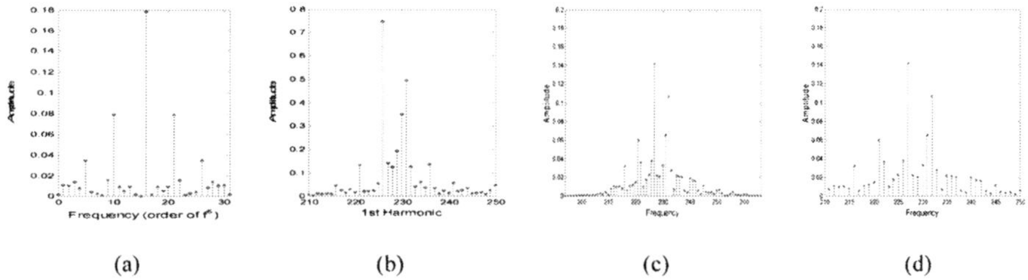

Fig. 3.17 The convolution of the segment of frequency spectra (taking the 1st harmonic as an example). (a) Inverse filter; (b) measured vibration signal spectra around the 1st harmonic; (c) convolution result; (d) truncated convolution result.

separated spectra at different harmonics are shown in Figures 3.16b–3.16g.

Step 2: Convolving each segment with the inverse filter and truncating the convolution results to the length of each segment as illustrated in Figure 3.17. Figures 3.17a and 3.17b show the inverse filer and segment of the measured vibration signal (the sidebands at the 1st harmonic as shown in Figure 3.16b), respectively. Figures 3.17c and 3.17d show the convolution results, before and after the truncation, respectively.

Step 3: Putting the truncated convolution results back to the frequency spectra on D_{sup} to obtain the convolution result.

3.5.4 Experiments and Performance Metrics

The signals are normalized and limited within $[-1, 1]$ so that all of them in the proposed scheme can be treated similarly. Since the inverse filter $\bar{Z}(f)$ is a low frequency signal and we do not assume any prior knowledge, it is truncated to contain only 15 spectral lines and assumed to be a simple discrete impulse located at frequency f^s, as shown in Figure 3.18a. The optimal inverse filter $Z(f)$ resulting from the blind deconvolution scheme is shown in Figure 3.18b for comparison purposes.

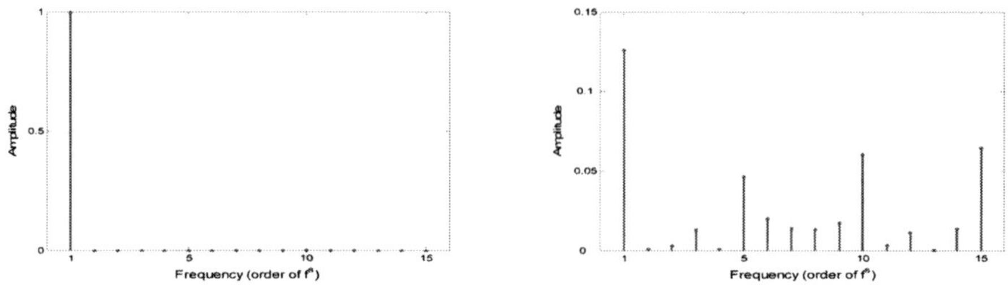

(a) the initial inverse filter (b) The optimal inverse filter

Fig. 3.18 The spectra of the inverse filter.

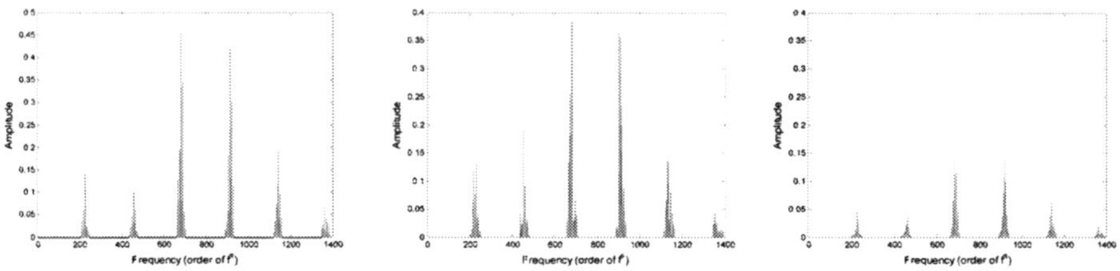

(a) $\bar{B}(f)$ (b) $B_{nl}(f)$ (c) $E(f)$

Fig. 3.19 The results of nonlinear projection.

The large frequency components at 5, 10, and 15 are consistent with the modulating signal $a(t)$ for a five-planet-gear plate, whose inverse should show frequency components at 5, 10, and 15, as predicted from the theoretical analysis.

The nonlinear projection is given in Equation (12) and is updated in the closed-loop de-noising scheme, shown in Figure 3.2, with the varying load profile and crack size. As an example, $\bar{B}(f)$, $B_{\text{nl}}(f)$, and their difference $E(f)$ are shown in Figure 3.19. $E(f)$ is the first term in the cost function (25) that must be minimized through the optimization routine. Finally, in the time domain, the measured noisy vibration signal $s(t)$, the noise-free vibration signal recovered from blind deconvolution $b(t)$, and the noise signal $n(t)$ are shown in Figure 3.20.

3.5.4.1 V-4-1 Signal-to-Noise Ratio

The signal-to-noise ratio (SNR), before and after de-noising, is investigated and the results at 40 and 100% torque levels are shown in Figures 3.21a and 3.21b, respectively. The improvement in the SNR value is significant.

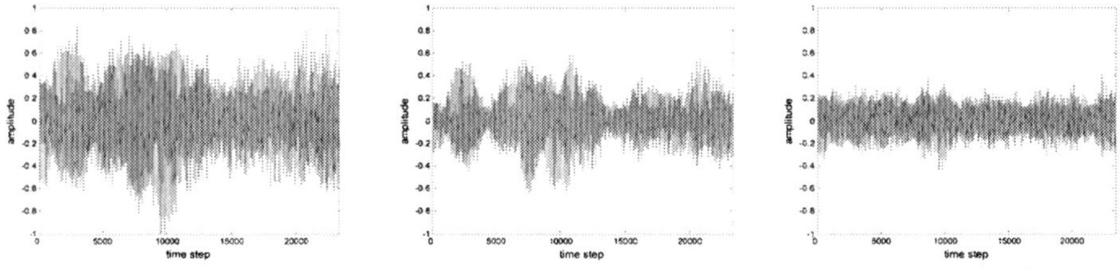

(a) (b) (c)

Fig. 3.20 The estimated vibration signal before and after nonlinear projection and their difference. (a) Measured vibration signal $s(t)$; (b) recovered noise-free vibration signal $b(t)$; (c) noise signal $n(t)$.

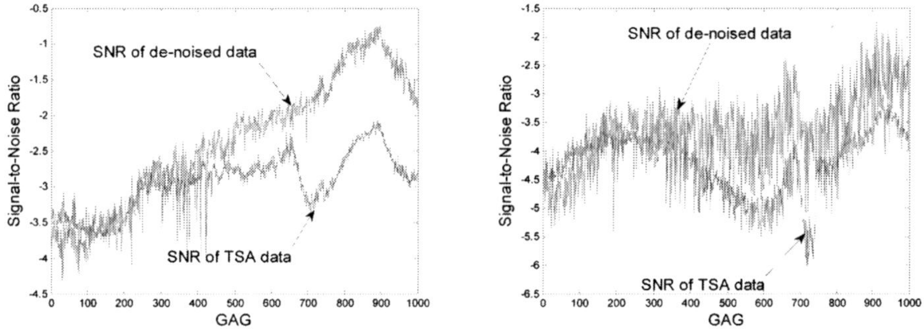

(a) Signal-to-noise ratio at 40% torque (b) Signal-to-noise ratio at 100% torque

Fig. 3.21 Signal-to-noise ratio at 40% torque.

3.5.4.2 V-4-2 Feature Performance Evaluation

Although the blind deconvolution routine shows a significant improvement in the SNR, it is desirable that it improves also the quality of the features or condition indicators. The accuracy and precision of mappings between the evolution of features and the actual crack growth have an important impact on the performance of diagnostic and prognostic algorithms and the CBM system overall. To evaluate the quality of the features, several performance indexes or metrics are introduced.

The first performance index is an overall accuracy measure defined as the linear correlation coefficient between the raw feature values and the crack length growth along the GAG cycle axis (CCR) [9]. Suppose x is the feature vector and y the crack growth curve with and their means, respectively. The correlation coefficient between them is

$$CCR(x, y) = \sqrt{\frac{ss_{xy}^2}{ss_{xx}\, ss_{yy}}}, \tag{3.26}$$

where $ss_{xy} = \sum(x_i - \bar{x}_i)(y_i - \bar{y}_i)$, $ss_{xx} = \sum(x_i - \bar{x}_i)^2$ and $ss_{yy} = \sum(y_i - \bar{y}_i)^2$, respectively. The extracted feature will be used to map and predict the crack growth.

Hence, a high correlation coefficient is expected to generate an accurate estimate of the actual crack length and is preferred. Thus, for a good feature, the value of the correlation coefficient should be near 1.

Because of changes in operating conditions and other disturbances, the feature values along the GAG axis are often very noisy and need to be smoothed through a low-pass filtering operation. In this case, the correlation coefficient is calculated based on the smoothed feature curve $\bar{x}$ (CCS) [9]. The calculation of CCS is the same as (3.26) with x being replaced by $\bar{x}$. In the following experiments, the smooth low-pass filter is a butterworth filter with cutoff frequency of 0.01 of the sampling frequency.

The third performance index is a precision measure corresponding to a normalized measure of the signal dispersion. It is referred to as percent mean deviation (PMD) [9] and defined by

$$\text{PMD}(x, \tilde{x}) = \frac{\sum_{i=1}^{l_x} \frac{x_l - \tilde{x}_i}{\tilde{x}_i}}{l_x} \times 100, \tag{3.27}$$

where l_x is the number of entities in the feature vector x.

In the fault detection and prognostic algorithms [10], the extracted feature values are used as a measurement input. Therefore, this performance index is closely related to the detection threshold and precision of the prognosis results (remaining useful life). From a precise feature, the fault detection and prognostic algorithms can detect the incipient failure and/or predict the fault growth with a high confidence. PMD for a precise feature should have a very small value close to 0.

The previous indexes evaluate the performance on the entire failure progression process. It is also important to evaluate how the failure progresses in a small period of time. To achieve this goal, a moving window along the time axis is applied to the feature vector so that the correlation coefficient and PMD in the moving windows can be evaluated to provide a "local performance assessment. The two adjacent windows are overlapped so that the performance indexes in this case are curves along the time axis.

The relative size of the NonRMC sidebands to the RMC sidebands in D_{sup} or part of D_{sup} may indicate the presence of a fault on the planetary carrier. As the crack grows, the operating condition deviates from the ideal one. In this case, the constructive superposition of vibration signals from different planetary gears is attenuated while the destructive superposition is reinforced. This phenomenon results in smaller RMC and larger NonRMC sidebands. This indicates that the ratio between RMC and NonRMC sidebands can be used as a characteristic feature. Based on this notion, features are successfully extracted in the frequency domain. One of these features, Sideband Ratio (SBR) [3], will be used to demonstrate the effectiveness of the de-noising routine.

Let X denote the number of the sidebands in consideration on both sides of the dominant components of the harmonics, the sideband ratio is then defined as the ratio between the energy of the NonRMC and all sidebands:

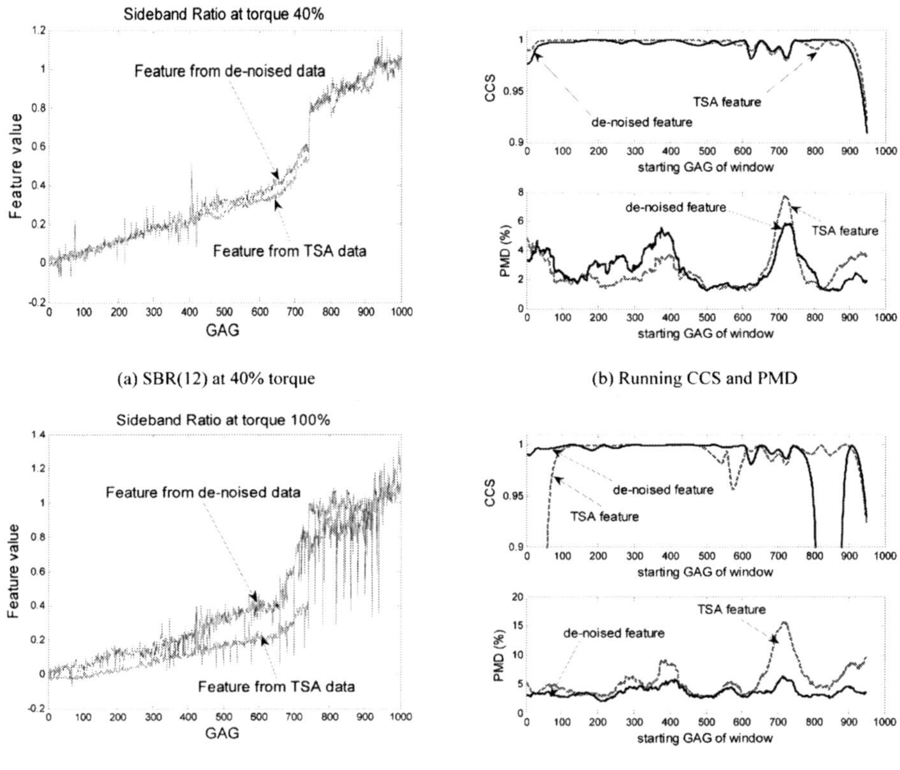

(a) SBR(12) at 40% torque

(b) Running CCS and PMD

(c) SBR(12) at 100% torque

(d) Running CCS and PMD

Fig. 3.22 Feature SBR(12) before and after de-noising.

$$\mathrm{SBR}(X) = \frac{\sum_{h=1}^{m}\sum_{g=-X}^{X}\ \mathrm{NonRMC}}{\sum_{h=1}^{m}\sum_{g=-X}^{X}\ (\mathrm{NonRMC}+\mathrm{RMC})}. \tag{3.28}$$

The results of SBR(12) at 40% torque level are shown in Figure 3.22a. The running version of the performance indexes is shown in Figure 3.22b. The results at 100% torque are shown in Figures 3.22c and 3.22d, respectively.

The correlation coefficient and PMD for this feature at different torque levels are summarized in Table 3.2. Substantial feature improvements are achieved via the application of the de-nosing routine. In the tables, D-N stands for de-noised.

3.5.4.3 Failure Prognosis

A fault detection/failure prognosis framework based on particle filtering algorithms has been successfully developed, and applied to the prediction of the remaining useful life (RUL) of the critical components [10]. The approach provides with a

Table 3.2 The performance indexes of feature Sideband Ratio.

Torque	20%		40%		100%	
	TSA	D-N	TSA	D-N	TSA	D-N
CCS	0.943	0.975	0.979	0.985	0.953	0.983
CCS	0.950	0.982	0.986	0.992	0.971	0.991
PMD	2.06%	2.01%	2.57%	2.73%	5.57%	3.57%

particle-filter-based estimate of the state probability density function (pdf), generates p-step ahead long-term predictions, and uses available empirical information about hazard thresholds to estimate the RUL pdf. The implementation of this algorithm requires a failure progression model, feature measurements to reveal the failure progression, a nonlinear mapping between failure severity and feature value, and external inputs such as the operational load profile to the system. More details about this prognosis framework can be found in [10]. In this section, results are shown to compare the algorithm performance when using the features derived in the previous section, before and after de-noising. This will illustrate the efficiency of the proposed de-noising routine, on the prediction of the planetary carrier RUL in the helicopter testbed.

Once the RUL pdf is calculated, the performance of failure prognostic algorithm can be evaluated. For this purpose, statistics such as 95% confidence intervals (CI) and RUL expectations may be used. In our experiment with a seeded crack on the planetary carrier, failure is defined when the crack reaches 6.21 inches in length, i.e. the crack reaches the edge of the carrier plate. The ground truth data in Table 3.1 show that the crack size reaches this value at the 714th GAG cycle.

Since the gearbox operates on the basis of GAG cycles, the RUL expectations and 95% CI are also given with the unit of "GAG cycle". Long-term predictions are generated at the end of the 365th GAG cycle, using the current estimate for the state pdf as initial condition. The state pdf estimate, in turn, is based on feature measurements considering two cases: before and after the de-noising routine. Results for each case are depicted in Figures 3.23a and 3.23b, respectively. It must be noted that the 95% CI in both scenarios are quite similar precision-wise. However, the RUL expectation is closer to the ground truth value when a de-noised feature is used, showing that the de-noise routine improves the accuracy of the algorithm.

As time goes on, differences between the two cases become evident. Table 3.3 summarizes the results obtained when performing the prognostic algorithm at the 400th, 450th, and 500th GAG cycles, respectively. In terms of precision of the prediction, it is clear that prognosis results with the de-noised feature offer a considerable reduction in the length of the 95% CI. For instance, when the long-term prediction is performed at the 450th GAG cycle, this reduction is in the order of about 140 minutes (46 GAG cycles). At the same time, the expected values for the RUL obtained from the de-noised feature are also more accurate that those from the TSA feature. It is important to note that, in real applications, a predicted failure time lower than the actual one is preferred since it does not involve the risk of continuing

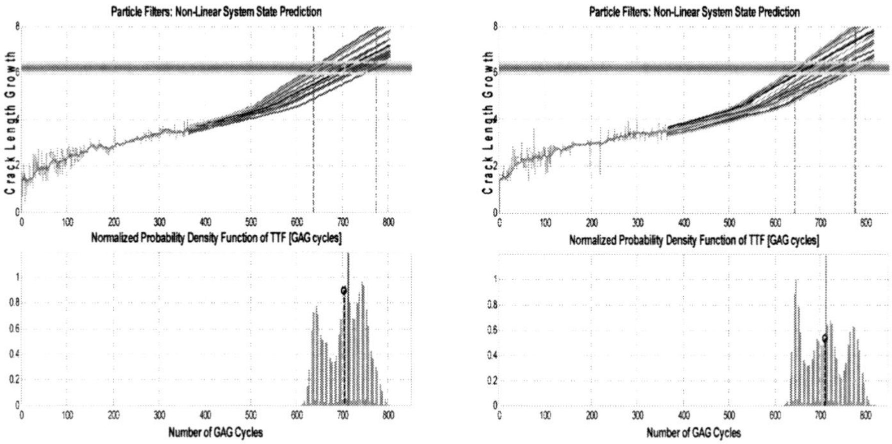

Fig. 3.23 Prognosis results started from the 365th GAG cycle.

Table 3.3 The performance indexes of failure prognosis.

Feature	Performance	Initial GAG cycle for long-term prediction			
		365	400	450	500
TSA	95% CI	[637 773]	[663 832]	[680 828]	[682 812]
	Length of CI	136	169	148	130
	Expected value	705	747	754	747
De-noised	95% CI	[638 777]	[647 775]	[618 720]	[695 784]
	Length of CI	139	128	102	89
	Expected value	710	711	670	746

the operation beyond safety. A lower failure time prediction only makes a conservative decision to schedule maintenance earlier, but a higher value of the failure time in the prediction might postpone the timely maintenance, resulting in damage or loss of vehicle.

3.6 Conclusion

The successful implementation of CBM technologies is achieved through the application of a failure prognostic algorithm, which in turn requires a high quality feature to propagate the failure in time with respect to the operating conditions of the system. Good features are often extracted from vibration signals collected from a noisy environment and noise might mask the signatures of the faults/failures. Therefore, signal de-noising and preprocessing is important in the implementation of CBM. This paper introduce the development of a new vibration signal blind deconvolution de-noising scheme in synergy with feature extraction, failure prognosis, and vibration modeling to improve the signal-to-noise ratio, the quality of features, as well as the accuracy and precision of failure prognostic algorithms. The proposed scheme is applied to the failure prognosis of a critical component of a helicopter testbed and the results demonstrate the efficiency of the proposed scheme.

Acknowledgement

The research reported in this paper was partially supported by DARPA and Northrop Grumman under the DARPA Prognosis program Contract No. HR0011-04-C-003. We gratefully acknowledge the support and assistance received from our sponsors.

References

1. G. Vachtsevanos, F. Lewis, M. Roemer, A. Hess and B. Wu, *Intelligent Fault Diagnosis and Prognosis for Engineering Systems*, Wiley, 2006.
2. A. Saxena, B. Wu and G. Vachtsevanos, A methodology for analyzing vibration data from planetary gear system using complex morelt wavelets, in *Proceedings of American Control Conference*, Portland, OR, June, Vol. 7, pp. 4730–4735, 2005.
3. B. Wu, A. Saxena, T. Khawaja, R. Patrick, G. Vachtsevanos and R. Sparis, An approach to fault diagnosis of helicopter planetary gears, in *Proceedings of IEEE AUTOTESTCON*, San Antonio, TX, September pp. 475–481, 2004.
4. B. Wu, R.P.A. Saxena and G. Vachtsevanos, Vibration monitoring for fault diagnosis of helicopter planetary gears, in *Proceedings of IFAC World Congress*, Prague, Czech Republic, July 2005.
5. M. Lebold, K. McClintic, R. Campbell, C. Byington and K. Maynard, Review of vibration analysis methods for gearbox diagnostics and prognostics, in *Proceedings of 54th Meeting of the Society for Machineary Failure Prevention Technology*, Virgina Beach, VA, May, pp. 623–634, 2000.
6. J. Keller and P. Grabill, Vibration monitoring of a UH-60A main transmission planetary carrier faul, in *Proceedings of the American Helicopter Society 59th Annual Forum*, Phoenix, AZ, May, pp. 1–11, 2003.
7. P. McFadden and J. Smith, An explanation for the asymmetry of the modulation sidebands about the tooth meshing frequency in epicyclic gear vibration, *Proceedings Institution of Mechanical Engineers, Part C: Mechanical Engineering Science* **199**(1), 65–70, 1985.
8. A. Szczepanik, Time synchronous averaging of ball mill vibrations, *Mechanical Systems and Signal Processing* **3**, 99–107, January 1989.
9. R. Patrick, A model-based framework for fault diagnosis and prognosis of dynamical system with an application to helicopter transmissions, PhD Proposal, School of Electrical and Computer Engineering, Georgia Institute of Technology, Atlanta, GA, September 2006.
10. M. Orchard, A particle filtering-based framework for on-line fault diagnosis and failure prognosis, PhD Proposal, School of Electrical and Computer Engineering, Georgia Institute of Technology, Atlanta, GA, 2006.
11. J. Antoni, Blind separation of vibration components: Principles and demonstrations, *Mechanical Systems and Signal Processing* **19**, 1166–1180, 2005.
12. G.R. Ayers and J.C. Dainty, Iterative blind deconvolution method and its applications, *Optics Letters* **13**, 547–549, July 1988.
13. G. Gelle, M. Colas and G. Delaunay, Blind sources separation applied to rotating machines monitoring by acoustical and vibrations analysis, *Mechanical Systems and Signal Processing* **14**, 427–442, July 2000.

14. B. Klamecki, Use of stochastic resonance for enhancement of low-level vibration signal components, *Mechanical Systems and Signal Processing* **19**, 223–237, 2005.
15. G. Hillerstrom, Adaptive suppression of vibrations – A repetitive control approach, *IEEE Trans. Control Systems Technology* **4**(1), 72–78, 1996.
16. J. Antoni and R. Randall, Unsupervised noise cancellation for vibration signals: Part I – Evaluation of adaptive algorithms, *Mechanical Systems and Signal Processing* **18**, 89–101, 2004.
17. D. Kundur and D. Hatzinakos, A novel blind deconvolution scheme for image restoration using recursive filtering, *IEEE Trans. Signal Processing* **26**, 375–390, February 1998.
18. M.Z.R. Peled and S. Braun, A blind deconvolution separation of multiple sources, with application to bearing diagnostics, *Mechanical Systems and Signal Processing* **19**, 1181–1195, 2005.
19. A.K. Nandi, D. Mampel and B. Roscher, Blind deconvolution of ultrasonic signals in nondestructive testing applications, *IEEE. Trans. Signal Processing* **45**, 1382–1390, May 1997.
20. R. Prost and R. Goutte, Discrete constrained iterative deconvolution with optimized rate of convergence, *Signal Processing* **7**, 209–230, December 1984.

Chapter 4
Particle Filter Based Anomaly Detection for Aircraft Actuator Systems

D. Brown, G. Georgoulas, H. Bae, G. Vachtsevanos, R. Chen, Y.H. Ho,
G. Tannenbaum and J.B. Schroeder

Abstract This paper describes the background, simulation and experimental evaluation of an anomaly detector for Brushless DC motor winding insulation faults in the context of an aircraft Electro-Mechanical Actuator (EMA) application. Results acquired from an internal Failure Modes and Effects Analysis (FMEA) study identified turn-to-turn winding faults as the primary mechanism, or mode, of failure. Physics-of-failure mechanisms used to develop a model for the identified fault are provided. The model was implemented in Simulink to simulate the dynamics of the motor with a turn-to-turn insulation winding fault. Then, an experimental test procedure was devised and executed to validate the model. Additionally, a diagnostic feature, identified by the fault model and derived using Hilbert transforms, was validated using the Simulink model and experimental data for several fault dimensions. Next, a feature extraction routine preprocesses monitoring parameters and passes the resulting features to a particle filter. The particle filter, based on Bayesian estimation theory, allows for representation and management of uncertainty in a computationally efficient manner. The resulting anomaly detection routine declares a fault only when a specified confidence level is reached at a given false alarm rate. Finally, the real-time performance of the anomaly detector is evaluated using LabVIEW.

D. Brown · G. Georgoulas · H. Bae · G. Vachtsevanos
Georgia Institute of Technology, Atlanta, GA 30332-0250, USA

R. Chen · Y.H. Ho
Northrop Grumman Corp., El Segundo, CA 90245, USA

G. Tannenbaum
Moog Inc., East Aurora, NY 14052, USA

J.B. Schroeder
AFRL/RBCC, Wright Patterson AFB, OH, USA

K.P. Valavanis (ed.), Applications of Intelligent Control to Engineering Systems, 65–88.

Nomenclature

Symbol	Units	Description
θ	rad	Motor (or rotor) position
ω	rad/s	Motor (or rotor) speed
$\boldsymbol{\Psi}_{abcs}$	Wb	Stator back-emf vector
ψ_m	Wb	Magnitude of magnetic flux linkage
$\boldsymbol{\Psi}_m$	Wb	Magnetic flux linkage vector
ψ_s	V	Series back-emf
ψ_s^f	V	Series back-emf with fault
$\mathbf{i}_{abcs}$	A	Stator winding current vector
k	–	Number of turns between fault
$\mathbf{L}_{abcs}$	H	Stator inductance matrix
L_{ii}	H	Self inductance of ith phase winding
L_{ij}	H	Mutual inductance between ith and jth phase windings
L_{ij}^*	H^{-1}	Inverse stator inductance of the ith and jth phase winding
L_s	H	Series inductance
L_s^f	H	Series inductance with fault
N	–	Number of turns per winding
P	–	Number of magnetic poles
$\mathbf{R}_{abcs}$	Ω	Stator resistance matrix
R_f	Ω	Turn-to-turn winding fault resistance
R_{ii}	Ω	Resistance of ith phase winding
R_s	Ω	Series resistance
R_s^f	Ω	Series resistance with fault
t	s	Continuous time index
T_e	$N \cdot m$	Motor torque
$\mathbf{U}_{abcs}$	V	Phase-to-neutral voltage vector
w_i^f	–	Winding fault dimension for the ith phase winding

Vector/Matrix		Scalar Representation
$\boldsymbol{\Psi}_{abcs}$	=	$[\psi_a \; \psi_b \; \psi_c]^T$
$\boldsymbol{\Psi}_m$	=	$[\psi_{am} \; \psi_{bm} \; \psi_{cm}]^T$
$\mathbf{i}_{abcs}$	=	$[i_a \; i_b \; i_c]^T$
$\mathbf{U}_{abcs}$	=	$[U_{as} \; U_{bs} \; U_{cs}]^T$
$\mathbf{L}_{abcs}$	=	$\begin{bmatrix} L_{aa} & L_{ab} & L_{ac} \\ L_{ba} & L_{bb} & L_{bc} \\ L_{ca} & L_{cb} & L_{cc} \end{bmatrix}$

$$\mathbf{R}_{abcs} = \begin{bmatrix} R_{aa} & 0 & 0 \\ 0 & R_{bb} & 0 \\ 0 & 0 & R_{cc} \end{bmatrix}$$

4.1 Introduction

The military and commercial sectors have made major commitments to install on platform Health and Usage Monitoring Systems (HUMS) to acquire on-line in real-time appropriate data and to develop models, algorithms and software that can efficiently and effectively detect faults and predict the remaining useful life (RUL) of failing components with confidence while minimizing false alarm rates. This paper proposes a comprehensive approach to develop, analyze and validate novel fault diagnosis and failure prognosis algorithms and methods for anomaly detection.

Introduced is a novel integrated framework for fault diagnosis and failure prognosis that relies on systems engineering principles and takes advantage of physics-of-failure models, Bayesian estimation methods and measurements acquired through seeded fault testing and/or on-board the aircraft. The proposed Bayesian estimation framework for diagnosis and prognosis for nonlinear, non-Gaussian systems begins with a systems engineering process to identify critical components and their failure models, sensing and monitoring requirements and processing algorithms. Fundamental to this approach is the development of physics-based failure or fatigue models and the optimum selection and extraction of features or Condition Indicators (CIs) from raw data that form the characteristic signatures of specific fault modes. The latter are selected on the basis of such criteria as sensitivity to particular fault modes and their correlation to ground truth data. The proposed framework employs a nonlinear state-space model of the plant, i.e. critical aircraft component, with unknown time-varying parameters and a Bayesian estimation algorithm, called particle filtering, to estimate the probability density function (PDF) of the state in real time [1]. The state PDF is used to predict the evolution in time of the fault indicator, obtaining as a result the PDF of the RUL for the faulty component/system. A critical fault is detected and identified by calling on the particle filter-based module that expresses the fault growth dynamics. Prognosis has been called the Achilles' heel of CBM due to major challenges arising from the inherent uncertainty in prediction. Prognosis may be understood as the result of the procedure where long-term (multi-step) predictions – describing the evolution in time of a fault indicator – are generated with the purpose of estimating the RUL of a failing component. The same particle filtering framework and nonlinear state model suggested above will be used to estimate the RUL [2].

Figure 4.1 depicts an overview of the Prognostics Health Management (PHM) architecture [3]. The proposed architecture entails generic elements applicable to a variety of aircraft systems. In this study, a specific system namely an Electro-

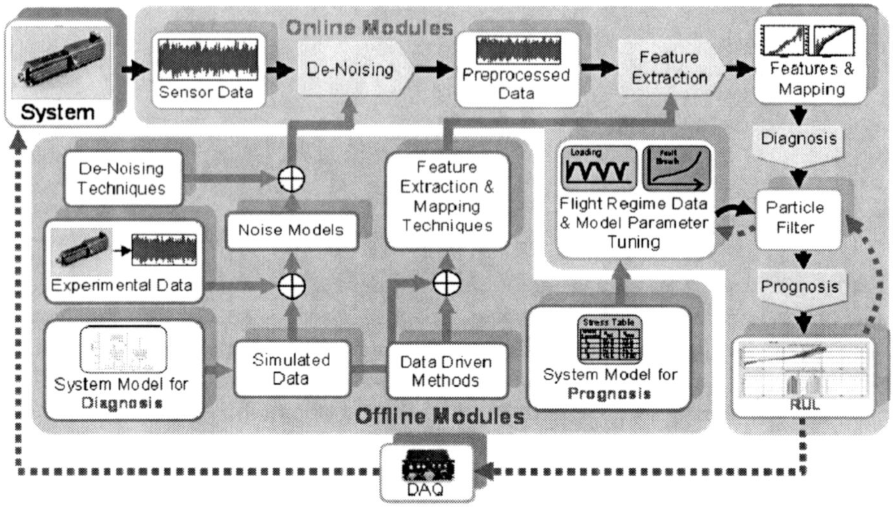

Fig. 4.1 PHM architecture.

Mechanical Actuator (EMA) was selected as a prototypical testbed for proof-of-concept.

To further expand on each element the paper has been divided into the following sections: Section 4.2 provides a description of the EMA under investigation. A FMEA study is presented to identify the primary failure mode of interest, turn-to-turn winding insulation faults for a DC brushless motor; Section 4.3 presents the theory of operation for a brushless DC motor along with a fault insertion model to mimic turn-to-turn winding insulation faults with a discussion of preliminary simulation results; Section 4.4 discusses the experimental results obtained from a test conducted with a commercial off-the-shelf (COTS) Brushless DC motor with seeded turn-to-turn winding insulation faults. A description of the fault insertion test setup is provided along with the fault insertion profile. Experimental results are presented to validate the simulated model and primary feature. Also, the real-time performance of the anomaly detector is investigated using LabVIEW; Section 4.5 summarizes the results of the study and suggests future applications for the resulting PHM architecture.

4.2 System Background

4.2.1 EMA Description

The actuator evaluated for this study is a triplex redundant rotary EMA, designed to operate with a triplex redundant Electronic Control Module (ECM). The EMA,

Fig. 4.2 Photo of the triplex redundant EMA5. Copyright © Moog Inc 2009. Photo courtesy of Moog Inc.

shown in Figure 4.2, consists of three brushless DC motors, each using a resolver for motor commutation, where all motors are driving a single string output shaft drive mechanism through a gear system.

A redundant position sensor provides output shaft position to the three controllers for loop closure of each channel as well as telemetry to the controller for redundancy management. This actuator is suitable for this study since it can operate in a single channel (simplex) mode or in fully redundant, active-active-active system mode.

The system is designed to employ active-active-active redundancy, providing a two-fault-tolerant system. In the active-active-active system, all three motors are actively driven in a torque sharing mode where the torque applied to the load is the cumulative sum of the three motor torques. A Vehicle Management Computer (VMC) determines the drive for each motor/channel and corresponding servo controllers. The VMC also monitors each motor channel for failures or degraded performance, so it can shut it down when necessary.

4.2.2 Fault Modes

Results from an internal FMEA study of the EMA, shown in Figure 4.2, are provided in Table 4.1. According to the results of the study, the leading modes of failure are associated with the bearings, position feedback sensors and brushless motors. Since the three leading modes of failure have similar failure rates, the brushless DC motors were chosen for this study. Refer to [4–5] for previous work in the area of fault diagnosis for feedback position sensors and [6–8] for related work for motor bearings.

Table 4.1 FMEA for the EMA in Figure 4.2.

Item/Component	Failure Rate (hr^{-1})	Failure Mode/Cause
Bearings	1.78E-05	Increased friction due to – bearing wear – galling
Position Sensor	1.70E-05	Loss/incorrect feedback signal from resolver: – turn-to-turn short – turn-to-ground short – open circuit
Brushless DC Motor	1.03E-05	Breakdown of stator assembly insulation due to – turn-to-turn short – phase-to-phase short – turn-to-ground short – open circuit

Note: in this study emphasis was placed on the motor/resolver assembly thereby omitting the failure rates of the power electronics. Refer to [9] for additional information regarding a FMEA study for EMA electronic components.

4.2.3 Failure Effects

The primary failure mechanisms for the brushless motor, identified in Table 4.1, are insulation turn-to-turn/turn-to-phase short circuits and open circuits. However, according to the manufacturer, turn-to-turn winding insulation faults are most likely to occur due to the geometry of the stator winding configuration, or layout, number of windings per turn, and separation distance between different phase windings. According to Malik et al. [10], Nandi and Toliyat [11], and Tavner and Penman [12], stator insulation can fail due to several reasons: high stator core or winding temperature; contamination from oil, moisture and dirt; short circuit or starting stresses; electrical discharges; and leaking in the cooling system. For the EMA under investigation, winding temperature is the dominant failure mechanism due to the excessive environmental factors. As a consequence, the principle effects of a turn-to-turn winding insulation short result in a three-phase impedance imbalance in the stator windings. This leads to asymmetries in the phase currents and phase-to-neutral voltages leading to increased harmonic generation and overall performance degradation [13, 14]. Therefore, the remaining subsections discuss methods for modeling the resulting asymmetry and a feature used to identify the corresponding imbalance.

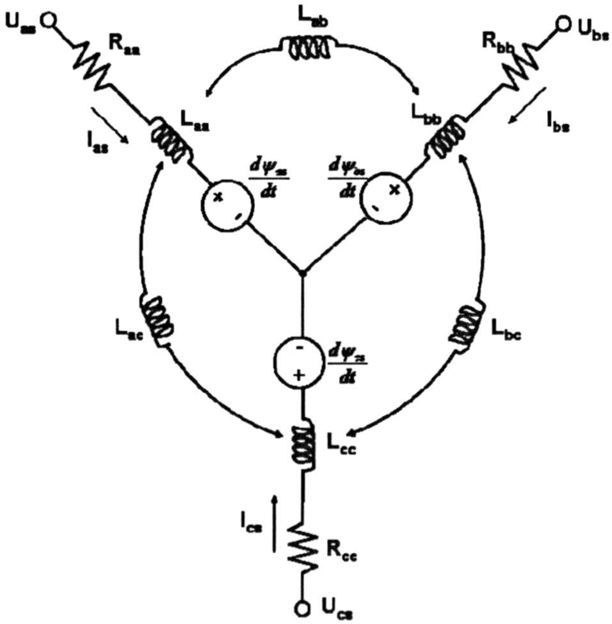

Fig. 4.3 Three-phase wye-connected electrical diagram for a brushless DC motor.

4.3 Theory

4.3.1 Brushless DC Motor Model

The electrical diagram for a brushless DC motor can be modeled as a three-phase wye-connected electrical circuit as illustrated in Figure 4.3. The diagram consists of three phase windings (a, b and c). Each phase winding consists of a self inductance, winding resistance and back-emf as represented by R_{ii}, L_{ii} and ψ_{is} where the subscript i referes to the ith phase winding accordingly. Additionally, each phase winding is coupled by a mutual inductance L_{ij} where the subscripts i and j refers to the ith and jth windings accordingly [15].

Using Kirchoff's current law, a set of differential equations can be obtained expressing the stator voltage vector $\mathbf{U}_{abcs}$, in terms of stator current vector $\mathbf{i}_{abcs}$, stator back-emf vector $\mathbf{\Psi}_{abcs}$, and stator resistance vector $\mathbf{R}_{abcs}$ [16],

$$\frac{d}{dt}\mathbf{i}_{abcs} = \mathbf{L}_{abcs}^{-1}\left[\mathbf{U}_{abcs} - \mathbf{R}_{abcs}\mathbf{i}_{abcs} - \frac{d}{dt}\mathbf{\Psi}_{abcs}\right]. \tag{4.1}$$

The last term in Equation (4.1), the stator back-emf vector, can be expanded using the winding inductance matrix $\mathbf{L}_{abcs}$ and magnetic flux linkage vector $\mathbf{\Psi}_m$ [16],

$$\mathbf{\Psi}_{abcs} = \mathbf{L}_{abcs}\mathbf{i}_{abcs} + \mathbf{\Psi}_{\mathrm{m}}. \tag{4.2}$$

Flux linkages at the stator windings due to the permanent magnets on the rotor are expressed in Equation (4.3) where the rotor position and flux linkage magnitude are represented using the symbols θ and $\mathbf{\Psi}_m$ [17].

$$\mathbf{\Psi}_m = \psi_{\mathrm{m}} \left[\sin(\theta) \sin\left(\theta - \frac{2\pi}{3}\right) \sin\left(\theta + \frac{2\pi}{3}\right) \right]^T . \tag{4.3}$$

Additionally, the inverse matrix of Equation (4.1) can be expressed in general form where each element represented as L^*_{ij} corresponds to elements in $\mathbf{L}^{-1}_{abcs}$,

$$\mathbf{L}^{-1}_{abcs} = \begin{bmatrix} L^*_{aa} & L^*_{ab} & L^*_{ac} \\ L^*_{ba} & L^*_{bb} & L^*_{bc} \\ L^*_{ca} & L^*_{cb} & L^*_{cc} \end{bmatrix} . \tag{4.4}$$

Typically, the *round rotor approximation* is invoked to simplify the computation of the inverse inductance matrix by neglecting variations in motor inductance with position [16],

$$\frac{d}{d\theta} \mathbf{L}_{abcs} \approx 0. \tag{4.5}$$

Now, let the symbol W_m represent the energy stored in the magnetic flux,

$$W_m = \frac{P}{2} \left\{ \frac{1}{2} \mathbf{i}^T_{abcs} \mathbf{L}_{abcs} \mathbf{i}_{abcs} + \mathbf{i}^T_{abcs} \mathbf{\Psi}_m \right\} . \tag{4.6}$$

The motor torque, T_e, can be computed by taking the derivative W_m with respect to the rotor position θ. The resulting expression for T_e, is given in Equation (4.7) where the symbol P denotes the number of magnetic poles [18].

$$T_e = \frac{P}{2} \left\{ \frac{1}{2} \mathbf{i}^T_{abcs} \left[\frac{d}{d\theta} \mathbf{L}_{abcs} \right] \mathbf{i}_{abcs} + \mathbf{i}^T_{abcs} \left[\frac{d}{d\theta} \mathbf{\Psi}_{abcs} \right] \right\} . \tag{4.7}$$

4.3.2 Equivalent Insulation Fault Model

A schematic representing a winding fault for any single phase winding is provided in Figure 4.4.

The symbols L_s, R_s and ψ_s represent the nominal series inductance, resistance and back-emf voltage of a winding for no fault conditions. The symbols N, k and R_f represent the number of total winding turns, number of winding turns between the fault, and the resistance of the insulation fault respectively.

The circuit network in Figure 4.4 can be reduced to a single resistor, inductor and voltage source, represented as R^f_s, L^f_s and ψ^f_s by applying the Thevenin circuit transformation. The equivalent Thevenin circuit is presented in Figure 4.5.

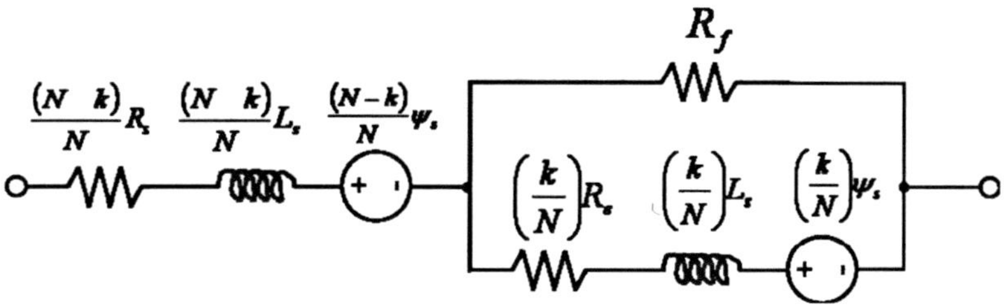

Fig. 4.4 Schematic of insulation fault model.

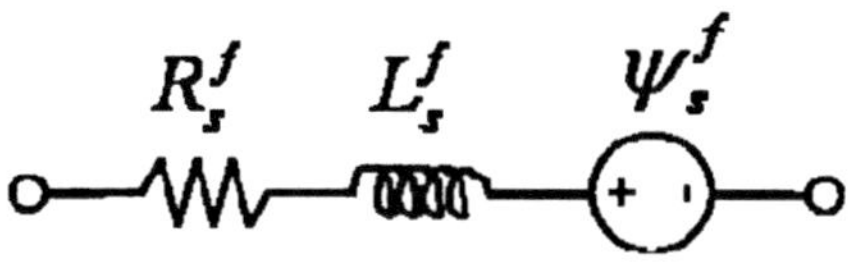

Fig. 4.5 Thevenin circuit transformation of the winding fault model.

By assuming the resistance of the insulation fault, R_f, is much larger than the series resistance of the winding, R_s, and the reactance term ωL_s, simplified expressions for R_s^f, L_s^f and ψ_s^f can be obtained in terms of L_s, R_s, ψ_s and w^f,

$$[R_s^f(t) \quad L_s^f(t) \quad \psi_s^f(t)]^T \approx w^f(t)[R_s \quad L_s \quad \psi_s]^T, \tag{4.8}$$

where the symbol w^f, expressed below in Equation (4.9), represents the dimension of the turn-to-turn winding insulation fault,

$$w^f(R_f, R_s, k, N) = 1 - \frac{k}{N}\left(1 - \left[1 + \frac{k}{N}\left(\frac{R_s}{R_f}\right)\right]^{-2}\right), \tag{4.9}$$

where $0 < w^f \leq 1$. It can also be shown if L_{ij} is the mutual inductance between two different phase windings, i and j, then the corresponding change in L_{ij} for a turn-to-turn winding insulation fault condition can be approximated as

$$L_{ij}^f(t) \approx L_{ij}\sqrt{w_i^f(t)w_j^f(t)}, \tag{4.10}$$

where the fault dimension for each phase (a, b and c) are defined as w_a^f, w_b^f and w_c^f accordingly. By replacing each electrical parameter with its equivalent expression from Equations (4.8) and (4.10) using the corresponding fault dimension for the associated phase winding, a modified three-phase wye-connected electrical diagram, shown in Figure 4.6, can be obtained.

Finally, a model for the brushless DC motor was implemented in Simulink, shown in Figure 4.7, by using Equations (4.1–4.10) presented earlier.

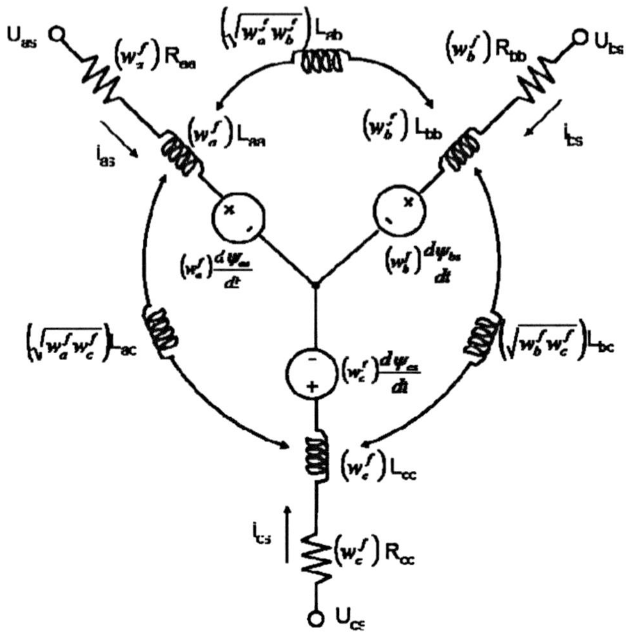

Fig. 4.6 Modified three-phase wye-connected electrical diagram for a brushless DC motor.

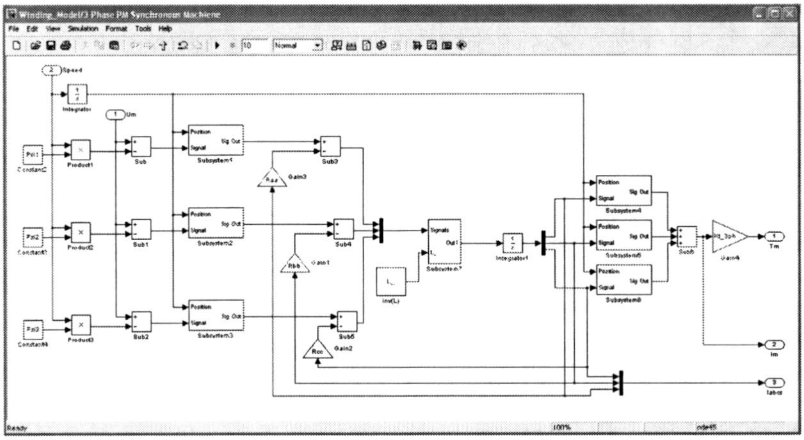

Fig. 4.7 Simulink brushless DC motor model.

4.3.3 Simulated Data

The Simulink model shown in Figure 4.7 was used to simulate the phase currents for the brushless DC motor discussed in this paper. The simulation results for a fault-

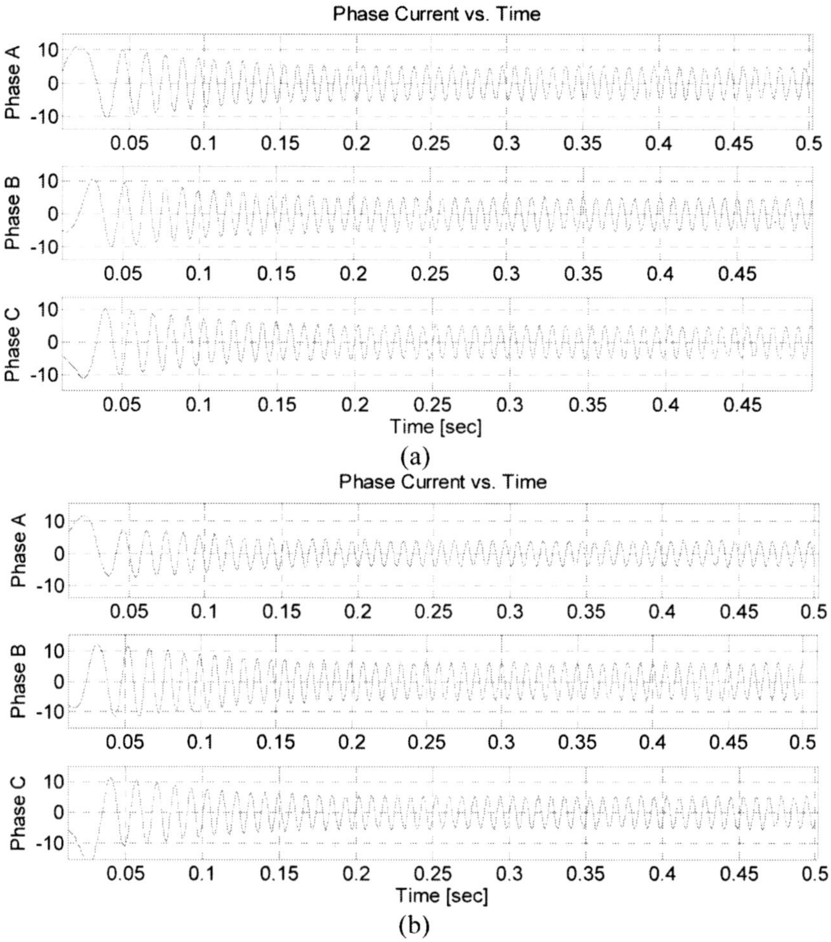

Fig. 4.8 Simulated phase currents for (a) no fault and (b) turn-to-turn winding insulation fault with fault dimension 0.800 inserted on phase winding A.

free and turn-to-turn winding fault placed on the phase winding A are provided in Figure 4.8. The simulated data was generated for a motor initially at rest with step input commanded to a constant motor speed. Notice for the case of the turn-to-turn winding insulation fault (Figure 4.8b) the amplitude of the phase currents at steady-state are not symmetrical.

4.3.4 Feature Selection and Extraction

Feature or condition indicator (CI) selection and extraction constitute the cornerstone for accurate and reliable fault diagnosis. The classical image recognition and signal processing paradigm of *data* $\rightarrow$ *information* $\rightarrow$ *knowledge* becomes most relevant and takes central stage in the fault diagnosis case, particularly since such operations must be performed on-line in a real-time environment.

Fault diagnosis depends mainly on extracting a set of features from sensor data that can distinguish between fault classes of interest, detect and isolate a particular fault at its early initiation stages. The remainder of this section evaluates a feature derived from simulated phase currents using the Hilbert transform to identify asymmetries as a result of turn-to-turn winding insulation faults.

Hilbert Transform – According to Johansson [19], the Hilbert transform defined in the time domain is a convolution between the Hilbert transformer $1/(\pi t)$ and a function $f(t)$. This results in phase shifting the original function $f(t)$ by $\pi/2$ radians. This is more rigorously defined by the following definition [20]:

Definition 1. The Hilbert transform of a function is defined for all t by,

$$\hat{f}(t) = \frac{1}{\pi} P \int_{-\infty}^{\infty} \frac{f(\tau)}{t-\tau} d\tau. \tag{4.11}$$

Provided the integral exists. The symbol P in front of the integrals denotes the Cauchy principal value. Note: refer to [21] for addition information regarding the Cauchy principle value theorem.

The Hilbert transform can be used to create an analytic signal $z(t)$ from a real signal $f(t)$ [19],

$$z(t) = f(t) + i\hat{f}(t). \tag{4.12}$$

Using this representation it is possible to describe the signal $z(t)$ as a rotating vector with an instantaneous phase $\varphi(t)$ and an instantaneous amplitude $A(t)$ in the time domain,

$$z(t) = A(t)\exp[i\varphi(t)], \tag{4.13}$$

where $A(t)$ and $\varphi(t)$ are related to $f(t)$ and $\hat{f}(t)$ by the following expressions,

$$\begin{cases} A(t) = \sqrt{f^2(t) + \hat{f}^2(t)} \\ \varphi(t) \ = \arctan[\hat{f}(t)/f(t)]. \end{cases} \tag{4.14}$$

Now consider the following real signal and its corresponding Hilbert transform (H.T.), provided in Equation (4.15), where the symbols A_0, ω_0 and φ_0 correspond to arbitrary constants for the amplitude, speed and phase shift respectively.

$$f = A_0 \cos(\omega_0 t + \varphi_0) \overset{H.T.}{\longleftrightarrow} \hat{f} = A_0 \sin(\omega_0 t + \varphi_0). \tag{4.15}$$

By applying Equation (4.14), the functions $f(t)$ and $\hat{f}(t)$ form an analytical signal, $z(t)$, where the instantaneous amplitude is shown to be phase invariant,

$$A(t) = A_0. \tag{4.16}$$

Primary Feature – Recall that Equations (4.15) and (4.16) demonstrated $|f(t) + \hat{f}(t)|$ is phase invariant provided that f is a sinusoidal signal of constant amplitude and speed. Extending this to a time-varying vector of three phase current signals $\mathbf{i}_{abcs}(t)$ provides a measure, or CI, of the amplitude of each phase during steady state.

Finally, by taking the standard deviation of the average amplitude of each phase current (a, b and c) over a finite time interval T, the following feature, denoted as $f_p(t, T)$ can be used to describe variations in winding symmetry,

$$f_p(t, T) = \underset{i \in a,b,c}{\text{std}} \left[\underset{t \in (0,T)}{\text{avg}} |i_{is}(t) + \hat{i}_{is}(t)| \right], \tag{4.17}$$

where the index i refers to the phase current of the ith phase winding. The simulated phase currents generated earlier in Figure 4.8 were used to evaluate the primary feature $f_p(t, T)$. The resulting features are provided in Figure 4.9 for both data sets.

4.3.5 Anomaly Detection and Prognosis

4.3.5.1 Particle Filter Framework

Particle filtering is an emerging and powerful methodology for sequential signal processing with a wide rage of applications in science and technology. Founded on the concept of sequential importance sampling and the use of Bayesian theory, particle filtering is particularly useful in dealing with difficult nonlinear and/or non-Gaussian problems. The underlying principle of the methodology is the approximation of relevant distributions with particles (samples from the space of the unknowns) and their associated weights. Compared to classical Monte-Carlo methods, sequential importance sampling enables particle filtering to reduce the number of samples required to approximate the distributions with necessary precision, and makes it a faster and more computationally efficient approach than Monte-Carlo simulation. This is of particular benefit in diagnosis and prognosis of complex dynamic systems, such as engines, gas turbines and gearboxes, because of the nonlinear nature and ambiguity of the rotating machinery world when operating under fault conditions. Moreover, particle filtering allows information from multiple measurement sources to be fused in a principled manner [1].

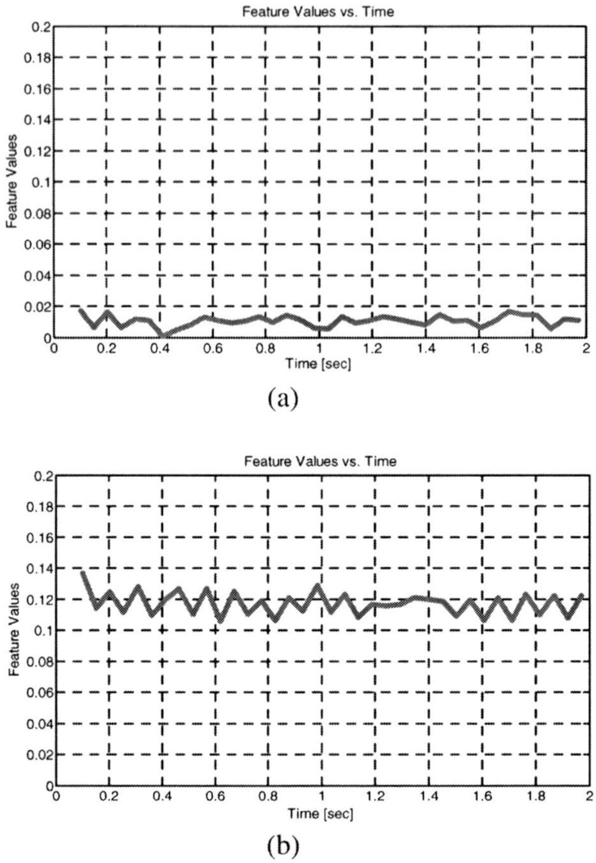

Fig. 4.9 Simulated feature values derived from phase currents for (a) no fault and (b) turn-to-turn winding insulation fault with fault dimension 0.800 inserted on phase winding A.

4.3.5.2 Particle Filtering in Real-Time Diagnosis Applications

Particle filtering has a direct application in the arena of fault detection and identification (FDI). Indeed, once the current state of the system is known, it is natural to implement FDI procedures by comparing the process behavior with patterns regarding normal or faulty operating conditions.

A fault diagnosis procedure involves the tasks of fault detection and identification (assessment of the severity of the fault). In this sense, the proposed particle-filter-based diagnosis framework aims to accomplish these tasks, under general assumptions of non-Gaussian noise structures and nonlinearities in process dynamic models, using a reduced particle population to represent the state pdf [22]. A compromise between model-based and data-driven techniques is accomplished by the use of a particle filter-based module built upon the nonlinear dynamic state model

$$\begin{cases} x_d(t+1) = f_b(x_d(t), n(t)), \\ x_c(t+1) = f_t(x_d(t), x_c(t), \omega(t)), \\ f_p(t) = h_t(x_d(t), x_c(t), v(t)), \end{cases} \tag{4.18}$$

where f_b, f_t and h_t are non-linear mappings, $x_d(t)$ is a collection of Boolean states associated with the presence of a particular operating condition in the system (normal operation, fault type #1, #2, etc.), $x_c(t)$ is a set of continuous-valued states that describe the evolution of the system given those operating conditions, $f_p(t)$ is a feature measurement, $\omega(t)$ and $v(t)$ are non-Gaussian distributions that characterize the process and feature noise signals respectively. For simplicity, $n(t)$ may be assumed to be zero-mean i.i.d. uniform white noise. At any given instant of time, this framework provides an estimate of the probability masses associated with each fault mode, as well as a pdf estimate for meaningful physical variables in the system. Once this information is available within the FDI module, it is conveniently processed to generate proper fault alarms and to inform about the statistical confidence of the detection routine. Furthermore, pdf estimates for the system continuous-valued states (computed at the moment of fault detection) may be used as initial conditions in failure prognostic routines, giving an excellent insight about the inherent uncertainty in the prediction problem. As a result, a swift transition between the two modules (FDI and prognosis) may be performed, and moreover, reliable prognosis can be achieved within a few cycles of operation after the fault is declared. This characteristic is, in fact, one of the main advantages of the proposed particle-filter-based diagnosis framework. Customer specifications are translated into acceptable margins for the *type I* and/or *II* errors in the detection routine. The algorithm itself will indicate when the *type II error* (false negatives) has decreased to the desired level.

Recently, a renewed interest has surfaced in the detection of an *anomaly* or *novelty* (incipient failure) associated with critical components/subsystems. The intent is to detect a fault as early as possible after its initiation allowing for a sufficient time to track its evolution and determine the RUL of a failing component. Figure 4.10 depicts the major modules of the proposed architecture for an anomaly detector. In this architecture, real-time measurements (from sensors conveniently located and designed to respond to a large class of fault modes) and information about the current operational mode are provided on-line. Data are pre-processed and denoised before computing the features that will assist to efficiently monitor the behavior of the plant. Statistical analysis applied to this set of features, which compare their evolution in time with respect to baseline data, is performed to simultaneously arrive at the probability of abnormal conditions, the probability of false alarms, and a simple on/off indicator. This indicator exhibits the exact time instant when the presence of a fault can be confirmed for a given confidence level (for example, 95%). If time and computational resources allow for further analysis, feature information can be used to complete the tasks of fault isolation, identification, and failure prognosis [23].

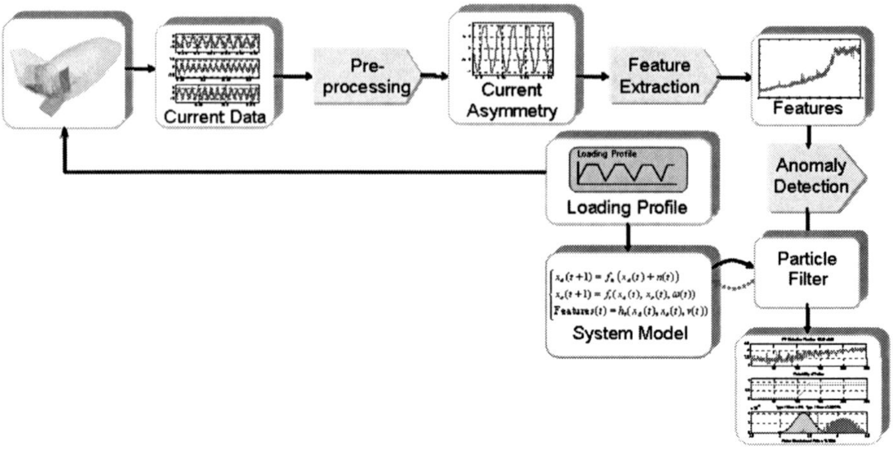

Fig. 4.10 Proposed architecture for the anomaly detector.

The modules of the architecture were evaluated using simulated phase currents generated earlier from Simulink. The results of the particle-filter based anomaly detector using the simulated data from Figure 4.8 are provided in Figure 4.11.

4.3.5.3 The Particle Filter Framework for Failure Prognosis

Prognosis on long-term prediction of the failure evolution is based on both an accurate estimation of the current state and a model describing the fault progression. If the incipient failure is detected and isolated at the early stages of the fault initiation, it is reasonable to assume that sensor data will be available for a certain time window allowing for corrective measures to be taken. At the end of the observation window, the prediction outcome is passed on to the user (operator, maintainer) and additional adjustments are not feasible since corrective action must be taken to avoid a catastrophic event.

Corrective terms are estimated in a learning paradigm to improve model parameter estimates and/or update load profiles thus resulting in better accuracy and precision of the algorithm for long-term prediction. A two-level procedure is introduced to remedy this problem. It intends to reduce the uncertainty associated with long-term predictions by using the current state pdf estimate, the process noise model, and a record of corrections made to previously computed predictions. In the first prognosis level, a p-step ahead prediction is generated on the basis of an *a priori* estimate, adjusting probabilities that are associated with the prediction according to the noise model structure. At the second prognosis level, these predictions are used to estimate the RUL pdf, also referred to as the time-to-failure (TTF) pdf.

Particle-filter based prognosis was demonstrated for the turn-to-turn winding insulation fault using artificial data generated from the simulation model discussed

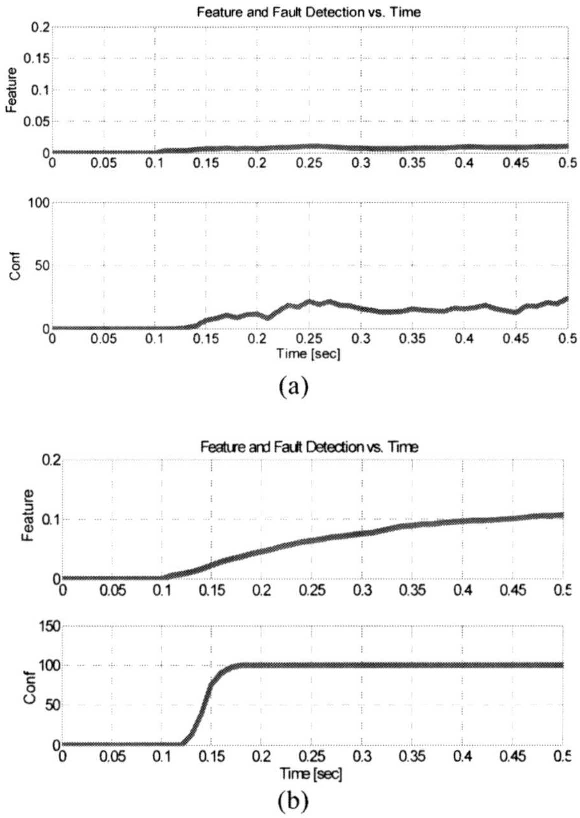

Fig. 4.11 Simulated feature values derived from phase currents for (a) no fault and (b) turn-to-turn winding insulation fault with fault dimension 0.800 inserted on phase winding A.

earlier. A result showing the particle filter updates, diagnostic assessment and long-term predictions of RUL are shown in Figure 4.12. A fault-growth model describing the rate of degradation for the fault driven by thermal stress was developed using Arrhenius law and a simplified thermal model. Details regarding the development of the fault-growth model are beyond the scope of this paper and are therefore omitted.

4.4 Experimental Evaluation

A scaled version of the EMA, shown in Figure 4.13, was experimentally evaluated in a "proof-of-concept" test-setup to evaluate the experimental methodology before applying it to the full-scale actuation system in Figure 4.2. The actuator test setup consists of two COTS motors where one motor simulates the actuator and the other

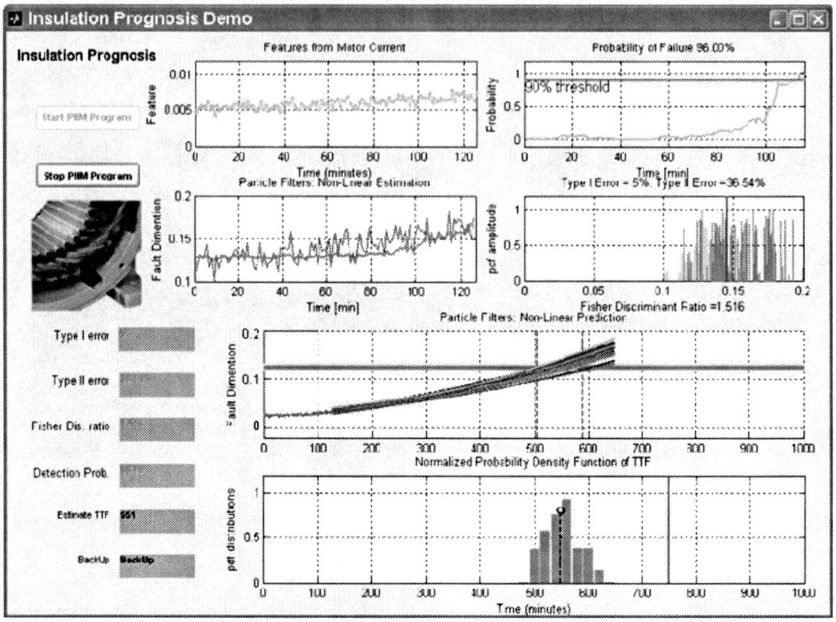

Fig. 4.12 Example of particle filter based prognosis using artificial data.

motor provides dynamic loading. The actuator and dynamic loader are tied directly back-to-back using metal bellow couplings with a high precision torque sensor. Both motors are controlled using a digital-space-vector pulse width modulation (PWM) sinusoidal drive. This setup allows the motor under investigation to operate under dynamic loading conditions for both opposing and aiding loads.

4.4.1 Seeded Fault Insertion

The fault insertion test setup simulates turn-to-turn winding insulation faults caused by thermal degradation of the insulation material. The motor chosen for the experiment has an accessible center winding tap to allow independent measurements of the phase-to-neutral voltage and phase current for each phase winding. Winding faults are inserted, or seeded, external to the motor by placing a resistor, also referred to as external fault resistance, in parallel to a selected phase winding terminal and the center tap as shown in Figure 4.14.

To simulate varying degrees of damage, or fault dimensions, a resistor network was used to control the particular value used for the external fault resistance for a controlled amount of time. To manage the selection of applied resistor values and turn-on time, the resistor network was connected to the motor winding using

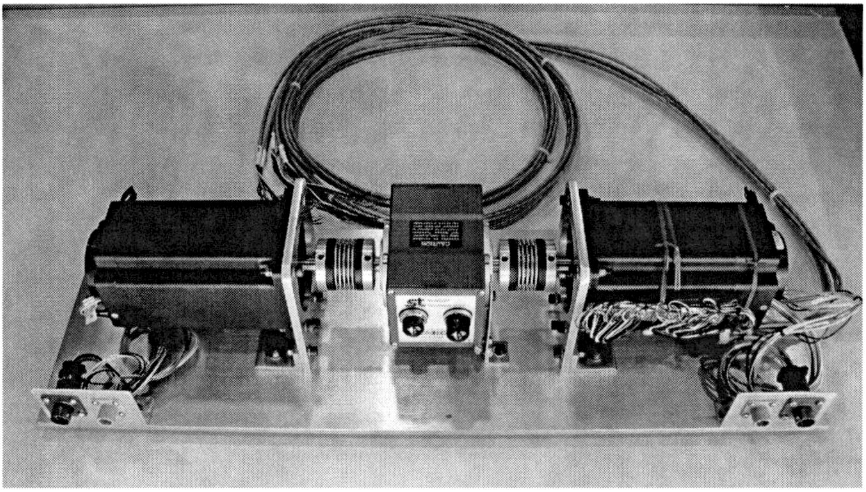

Fig. 4.13 Actuator and loader experimental setup.

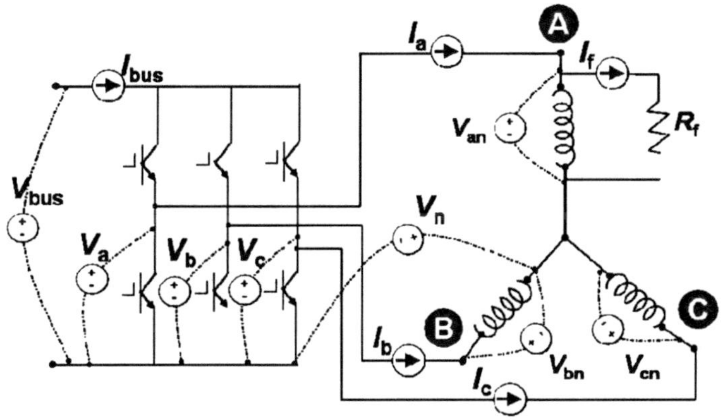

Fig. 4.14 Fault insertion setup.

high-speed power relays. The relays were controlled by a programmable microprocessor to trigger and simulate the duration and sequence of a particular fault. LabVIEW was used to initiate the fault insertion and synchronize the high speed data collection. This imposed minimal stress to the motor providing the capability for a repeatable and reliable seeded fault testing environment to evaluate different motor speeds and applied loads for a whole host of fault values. A graph of the experimental fault insertion profile used in this study is shown in Figure 4.15.

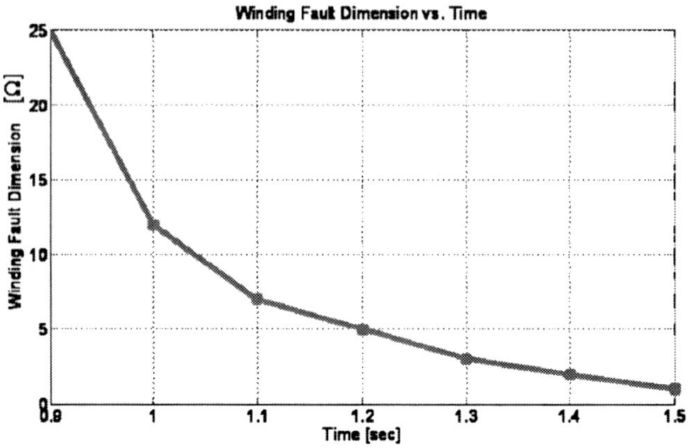

Fig. 4.15 Experimental fault insertion profile.

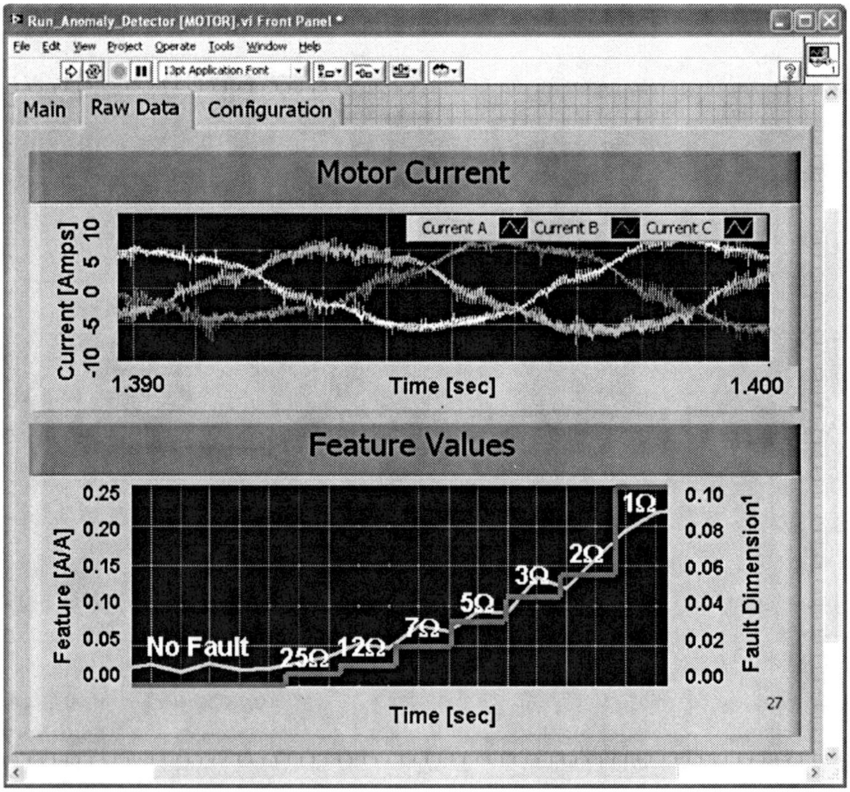

Fig. 4.16 Plots of experimental data (top) and computed features (bottom) acquired from the experimental test setup using LabVIEW.

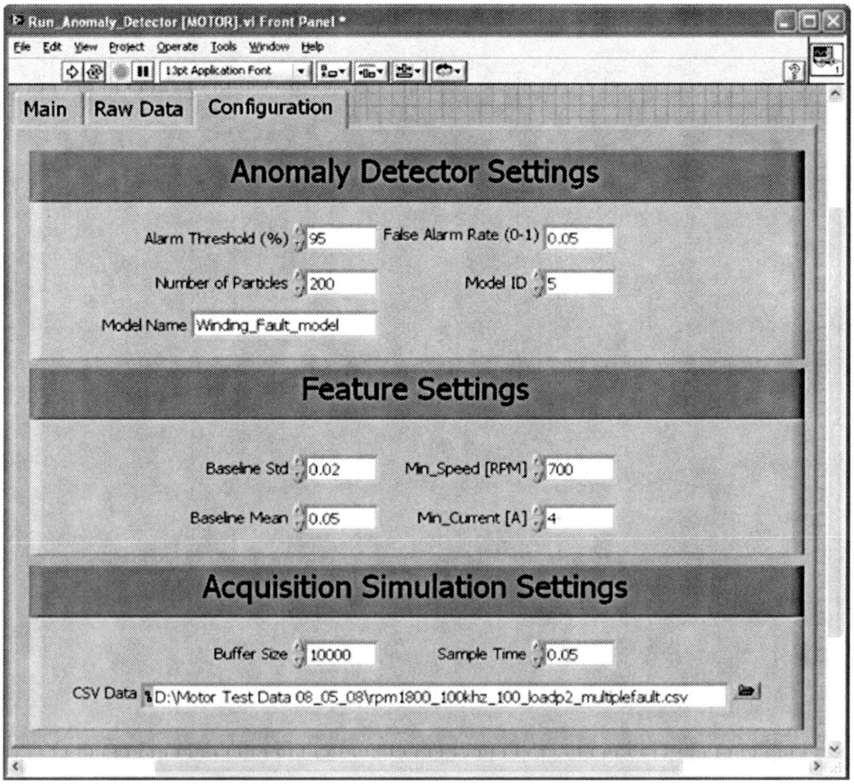

Fig. 4.17 LabVIEW interface used to define particle filter settings.

4.4.2 Experimental Results

Experimental results were used to validate the brushless DC motor model, the turn-to-turn winding insulation fault model, primary feature values and the anomaly detector algorithms investigated in the previous section.

4.4.2.1 Feature Validation

The primary feature identified in Equation (4.17) was calculated in real-time for experimental data obtained from the industrial motor test setup. Data was acquired using the experimental fault profile from Figure 4.15. A snapshot of raw data, including the calculated feature values, is provided in Figure 4.16.

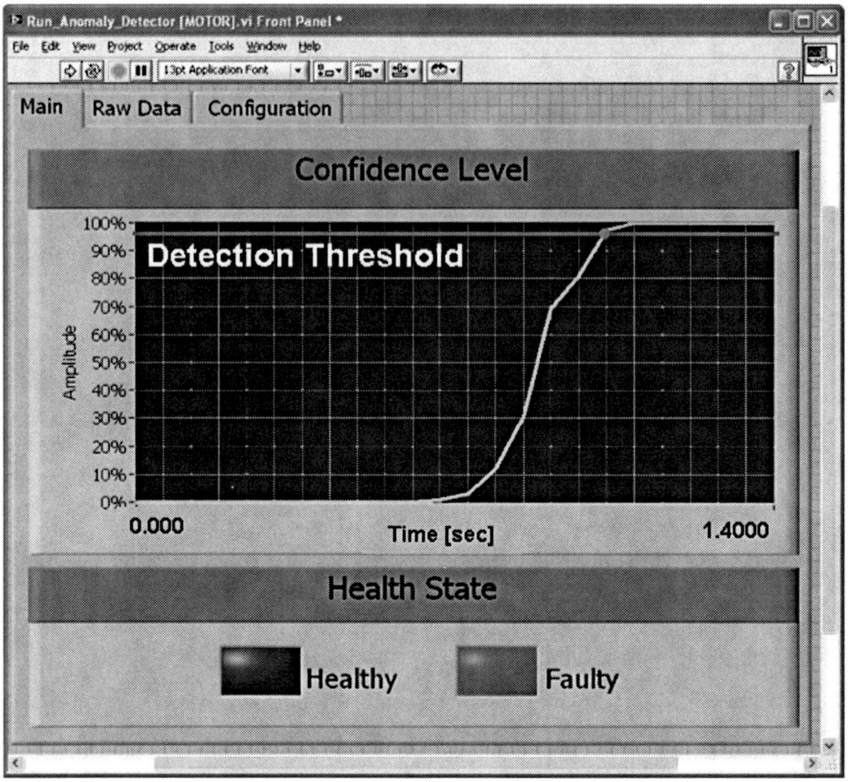

Fig. 4.18 Plot of anomaly detection confidence (top) and health state indicator (bottom) acquired from the experimental test setup using LabVIEW.

4.4.2.2 Anomaly Detection

The particle filter framework was integrated into LabVIEW to evaluate the real-time performance of the anomaly detector during data acquisition. Shown in Figure 4.17 is a GUI developed in LabVIEW to manage the anomaly detector settings which include the false alarm rate (type I error) and false alarm threshold (type II error) [24]. Additional configuration settings include the number of particles used by the particle filter, baseline statistics, fault model, and particle filter update/sample time.

The results from anomaly detector, using the data shown in Figure 4.16, are provided in Figure 4.18. The white (green) line represents the confidence of detection based on the false alarm rate and the red line represents the alarm, or detection, threshold as defined earlier in Figure 4.17. The dark (red) dot designates the crossing point during which the red health-state light illuminates to indicate an anomaly has occurred.

4.5 Conclusion

This paper presented a particle-filter based anomaly detector capable of detecting the occurrence of a major fault mode in its incipient stages for a safety critical aircraft actuation system. An overview of a generic PHM architecture were presented and applied to a particular EMA fault mode based on an internal FMEA study. The fault mode investigated in detail was a turn-to-turn winding insulation short for a brushless DC motor. The primary fault mode was modeled using physics-of-failure mechanisms indicating the primary failure effect, phase current asymmetry. A feature was derived using a Hilbert transform and statistical analysis to quantify the primary failure effect. Then, experimental data was acquired to validate the Simulink simulation model. Finally, real-time evaluation of the anomaly detector was performed using LabVIEW.

Although a specific system, an EMA, was selected as the test-platform with a specific fault mode, the overall PHM architecture can be applied to an entire range of systems and application domains. In fact, similar techniques which allow for early fault detection during acceptable performance in the presence of faults are being developed for a wide variety of system in both manned and unmanned air vehicles. Therefore, the concept of using system health information (diagnosis and prognosis) in conjunction with reconfigurable control is at the forefront of modern space and avionics applications requiring increasingly sophisticated fault-tolerant systems that are robust, reliable, and relatively inexpensive. Therefore the need arises for a fault-tolerant control architecture to further maximize performance and extend the operating life of COTS components in future avionics systems.

Acknowledgements

This work was performed under the United States Air Force (USAF) contract number FA8650-06-C-7628. The program is sponsored by the USAF Air Force Research Laboratory (AFRL) Air Vehicles Directorate (RB), Wright-Patterson Air Force Base, Ohio. Mr. J.B. Schroeder, AFRL/RBCC is the Air Force Technical POC and Robert Chen is the NGC Program Manager and Principal Investigator. Georgia Institute of Technology (GA Tech) lead the algorithm development and MOOG Aerospace provided actuator and motor data. Tests presented herein were conducted at NGC.

References

1. M. Orchard, A particle filtering-based framework for on-line fault diagnosis and failure prognosis, Ph.D. Thesis, Electrical and Computer Engineering, Georgia Institute of Technology, Atlanta, 2007.
2. M. Orchard and G. Vachtsevanos, A particle filtering approach for on-line fault diagnosis and failure prognosis, *Transactions of the Institute of Measurement and Control*, in print.

3. M. Orchard, B. Wu and G. Vachtsevanos, A particle filter framework for failure prognosis, in *Proceedings of WTC2005, World Tribology Congress III*, Washington DC, USA, 2005.
4. D. Brown, D. Edwards, G. Georgoulas, B. Zhang and G. Vachtsevanos, Real-time fault detection and accommodation for COTS resolver position sensors, in *Proceedings of 1st International Conference on Prognostics and Health Management (PHM)*, October 6–9, 2008.
5. A. Murray, H. Bruce and A. Hirao, Resolver position sensing system with integrated fault detection for automotive applications, *Proceedings of IEEE Sensors* **2**, 864–869, 2002.
6. B. Zhang, G. Georgoulas, M. Orchard, A. Saxena, D. Brown, G. Vachtsevanos and S. Liang, Rolling element bearing feature extraction and anomaly detection based on vibration monitoring, in *Proceedings of Mediterranean Conference on Control and Automation*, pp. 1792–1797, June 2008.
7. B. Wu, A. Saxena, R. Patrick-Aldaco and G. Vachtsevanos, Vibration monitoring for fault diagnosis of helicopter planetary gears, in *Proceedings of 16th IFAC World Congress*, July 2005.
8. R.R. Schoen, T.G. Habetler, F. Kamran and R.G. Bartheld, Motor bearing damage detection using stator current monitoring, *IEEE Trans. Ind. Appl.* **31**(6), 1274–1279, November/December 1995.
9. M. Baybutt, S. Nanduri, P.W. Kalgren, D.S. Bodden, N.S. Clements and S. Alipour, Seeded fault testing and in-situ analysis of critical electronic components in EMA power circuitry, in *Proceedings of IEEE Aerospace Conference*, March 2008.
10. N.H. Malik, A.A. Al-Arainy and M.I. Quereshi, *Electrical Insulation in Power Systems*, Marcel Dekker, Inc., New York, 1998.
11. S. Nandi and H.A. Toliyat, Condition monitoring and fault diagnosis of electrical machines – A review, in *Proceedings of the 34th Annual Meeting of the IEEE Industry Applications*, pp. 197–204, 1999.
12. P.J. Tavner and J. Penman, *Condition Monitoring of Electrical Machines*, Res. Studies Press, 1987.
13. C. Xianrong, V. Cocquempot and C. Christophe, A model of asynchronous machines for stator fault detection and isolation, *IEEE Trans. Ind. Electron.* **50**(3), 578–584, June 2003.
14. J. Penman, H.G. Sedding, B.A. Lloyd and W.T. Fink, Detection and location of interturn short circuits in the stator windings of operating motors, *IEEE Trans. Energy Convers.* **9**(4), 652–658, December 1994.
15. D. Hanselman, *Brushless Permanent Magnet Motor Design*, The Writers' Collective, 2003.
16. S.E. Lyshevski, *Electromechanical Systems, Electric Machines, and Applied Mechatronics*, CRC Press, Boca Raton, FL, 1999.
17. D.Y. Ohm, *Dynamic Model of PM Synchronous Motors*, Drivetech, Inc., Blacksburg, Virginia. www.drivetech.com.
18. L. Youngkook and T.G. Habetler, A stator turn fault tolerant strategy for induction motor drives in safety critical applications, *IEEE PESC '06*, June 18–22, 2006.
19. M. Johansson, The Hilbert transform, Master Thesis of Mathematics/Applied Mathematics, Växjö University, March 1999.
20. E.B. Saff and A.D. Snider, *Complex Analysis for Mathematics, Science and Engineering*, Prentice-Hall, Inc., New York, 1976.
21. P. Henrici, *Applied and Computational Complex Analysis*, Vol. 1, John Wiley & Sons, Inc., New York, 1988.
22. M. Orchard, G. Kacprzynski, K. Goebel, B. Saha and G. Vachtsevanos, Advances in uncertainty representation and management for particle filtering applied to prognostics, in *Proceedings of PHM08 Conference*, Denver CO, October 2008.
23. B. Zhang, C. Sconyers, C. Byington, R. Patrick, M. Orchard and G. Vachtsevanos, Anomaly detection: A robust approach to detection of unanticipated faults, in *Proceedings of PHM08 Conference*, Denver CO, October 2008.
24. R. Patrick, M. Orchard, B. Zhang, M. Koelemay, G. Kacprzynski, A. Ferri and G. Vachtsevanos, An integrated approach to helicopter planetary gear fault diagnosis and failure prognosis, in *Proceedings of 42nd Annual Systems Readiness Technology Conference, AUTOTESTCON 2007*, Baltimore, USA, September 2007.

PART II

UNMANNED AERIAL SYSTEMS

Chapter 5
Design of a Hardware and Software Architecture for Unmanned Vehicles: A Modular Approach

Richard Garcia, Laura Barnes and Kimon P. Valavanis

Abstract Unmanned Vehicles (UVs) have migrated from simple manipulators in mass production factories to air, land, and sea vehicles that are now common place on the battlefield and even in our homes. With the vast expansion of these vehicles' capabilities has come the equally vast increase in research and development for these platforms. To date, most research and development is performed with proprietary machines that severely limit researchers' abilities to design, implement, and test state-of-the-art technologies. There has also been a tendency for robotic platforms to be overly integrated which ultimately prevents the addition of new hardware, modification of existing hardware, or extraction of specific behaviors. These facts alone have had a staggering effect on the quality and timeliness of research and its ultimate transition to products. This work attempts to relieve these issues by presenting a methodology for developing a hardware and software architecture for unmanned systems that is usable across multiple vehicles and their varying payloads. This methodology is demonstrated through implementation spanning heterogeneous ground and air vehicles. Field experiments with the developed test-bed of vehicles demonstrate autonomous navigation, including take-off and landing, as well as multi-robot formation control.

Richard Garcia
Army Research Laboratory, Aberdeen Proving Grounds, MD 21005, USA;
e-mail: richard.d.garcia@arl.army.mil

Laura Barnes
Automation and Robotics Research Institute, University of Texas at Arlington, Fort Worth, TX 76118, USA; e-mail: lbarnes@arri.uta.edu

Kimon P. Valavanis
Department of Electrical and Computer Engineering, University of Denver, Denver, CO 80208, USA; e-mail: kvalavan@du.edu

K.P. Valavanis (ed.), Applications of Intelligent Control to Engineering Systems, 91–115.

5.1 Introduction

The field of robotics, for the past few decades, has allowed humans to immensely expand their capabilities. This is possible through the shift from teleoperated robotics, the remote operation of robots by humans, to fully autonomous robotics designed to remove or reduce the need for the human component.

With accelerated technology advances, research facilities cannot afford to allocate the time required to develop a variety of testbeds. By creating highly modular testbeds, research facilities can drastically reduce the time required to develop sufficient testing equipment. The main benefits of using a modular testbed design are:

- *System Extensibility*: Modular systems are inherently designed to be lightly coupled. As such, modular systems typically utilize generic interfaces for internal and external interaction. Examples of this would be USB hardware interfaces and middleware such as CORBA and shared memory. These types of generic interfaces, if proper design principles are used, allow for future extensions and upgrades of hardware and software to be integrated with ease and speed.
- *Reusability*: Utilizing a modular approach to develop the testbed allows for the reuse of both hardware and software. Reusable components and software also aid in collaborative work and technical support of the system.
- *Reduced Down Time*: Inevitably hardware will be damaged due to use, age, or accident. By utilizing modular system design damaged components can be easily separated and replaced. Parts from other testbeds can also be exchanged to keep critical systems in operation while awaiting parts. It should also be noted that down time can be further reduced by utilizing commercial components.
- *Reduced Training Time*: One of the main advantages of modular system design is the ability to utilize consistent hardware and software across platforms. By utilizing similar hardware and software architectures the need for extensive training for various testbeds is reduced. This allows researchers to easily span multiple projects.

Small unmanned vehicles are becoming popular due to their compact size, high maneuverability and high size-to-payload ratio. This is especially true with commercial, off-the-shelf (COTS) Radio Control (RC) vehicles due to their extensive development, mass production, low cost, availability, and ease of use. Autonomous RC cars and trucks have been used to research multi-vehicle control and coordination and mine detection [1, 12, 24] to name a few. RC planes have been used to study formation control, surveillance, target acquisition and tracking [2, 19]. Last, RC Vertical Take-off and Landing (VTOL) vehicles have led to new research in search and rescue [10, 18, 21, 22], traffic monitoring [4–6, 26], fire detection [3, 14, 17, 20], border patrol [9, 16], and failure control [8].

Although this type of vehicle has been used in a great deal of research, there has been very little done to develop a modular system design that can be utilized on a wide range of platforms. This fact is due to many issues including the vast differences in sensors, control software, and processing that must be performed to successfully and safely operate each type of vehicle. Further, the platforms themselves have limitations such as payload, run-time, and stability that make it difficult to create a modular system thus leading to specialized systems for each type of vehicle.

The remainder of this work is organized as follows: Sections 5.2 and 5.3 detail the testbed design methodology for both hardware and software. Sections 5.4 and 5.5 discuss the hardware and software implementation using these methodologies. Results are presented in Section 5.6. Section 5.7 summarizes and concludes this work.

5.2 Hardware Design

Advancements in hardware technology have opened the door to rapid and effective UV design. It has allowed UVs to expand to new areas of usage through decreased cost and hardware miniaturization.

When designing an UV testbed the future usability and adaptability of the system must be taken into account. A testbed system must be able to interface and utilize new and upcoming hardware, advanced software algorithms, and possibly operate on greatly varying platforms. Although this concept has, in the past, has led to designs that encompass every possible feature, this "all inclusive" design has major drawbacks and can severely reduce the robustness to various platform payloads, adaptability to new technology, and overall usability of the testbed.

As technology advances accelerate, UVs are becoming smaller, lighter, and cheaper. Adding non-essential hardware unnecessarily increases cost, power requirements, system complexity, size and weight of the testbed. The challenge is in creating a modular system that works on a variety of platforms, hardware, and software while remaining small, light, and cheap. To date, it is expected that your home computer will be able to interface with and utilize both current and future devices and software. This can be attributed to hardware and software standardization. This attribute is also desirable in a UV testbed and can easily be implemented with miniaturized processing and interface boards.

5.2.1 Minimum Design Requirements

There are a minimal set of hardware components that reside on almost all UVs. When designing an UV testbed, it is important to consider attributes such as cost, size, and weight while ensuring modularity. For example, a designer might feel that integrating GPS onto the processing board would reduce size and weight. This integration, although initially beneficial, stifles upgrades to the GPS and will be useless on testbeds operating exclusively in GPS denied areas.

The most obvious necessary components of an UV testbed are:

- *Processing System*: The processing system is the central point of device interfacing and is responsible for high level data processing and decision making. This hardware could simply gather and pass data to human operator or act as the brains of a fully autonomous vehicle.
- *Actuator Controller*: The actuator controller is responsible for directly or indirectly controlling the platform or component movements.
- *Sensors*: Sensors are responsible for measuring some attribute of the world such as position or heading. Examples of sensors that provide this type of information include GPS, IMU, range finders and encoders.
- *Communication*: Ultimately a UV will need to communicate with an external device. This may be as simple as sending progress data to a human or as complex as interfacing with a group of UVs. Typical devices include wireless serial modems, Ethernet and 802.11.

In order to create a modular hardware design, effort should be taken to include standard hardware interfaces on the testbed. This will allow for the greatest extensibility of the system. Adapters and hubs make it possible to utilize USB connections for external interfaces. To this date, most sensors communicate using some form of serial communication, but the availability of serial to USB adapters has made USB the most common interface.

One popular method for hardware selection is to select the smallest and lightest hardware that meets minimal requirements. This type of system is typically considered viable due to advancements in distributed computing. If the system needs more processing abilities you simply add another processing system. This has its own set of disadvantages including scalability and overhead demands not to mention that some algorithms cannot be distributed effectively.

5.2.2 Safety Measures

Safety is the most important aspect in UV design. Autonomous/semi-autonomous UVs are becoming more commonplace in both defense and civilian applications. This autonomy may be as simple as obstacle avoidance or avoiding extreme orientations. By default, these vehicles represent an inherent danger and must be designed with this mind.

A good design practice involves separating safety critical hardware from non-safety critical hardware. By doing this the designer can over engineer critical components and can ensure that failures in non-critical hardware do not impede critical components. Examples of this type of design include implementing secondary battery systems, audible warning systems, human override control and emergency shutdowns.

5.3 Software Design

Although hardware is an integral part of designing an autonomous vehicle testbed it is of very little use without corresponding software. Coherent and modular software architectures can decrease both training and integration time, simplify maintenance and vastly reduce testing requirements.

5.3.1 Operating System

To date there have been many advanced UVs developed that operate quite successfully without a traditional Operating System (OS) [25,27]. These embedded systems generally have the advantage of being minimal in both size and weight and operate with great computational efficiency. On the contrary, these UVs are typically difficult to upgrade and require in-depth knowledge of the system being maintained. Consequently, these restrictions are non-optimal for a testbed design where constant modifications and upgrades can be expected.

The traditional OS is simply a "middle man" providing an abstract interface between hardware and user designed applications. The OS effectively abstracts the hardware layer from the end user. This abstraction allows developers to ignore the low level hardware interfaces and simply code algorithms to interact directly with the OS. As most OSs are designed to operate on a vast array of processing hardware, code can simply be moved between processing hardware with little if any software modifications. This is further simplified if code is written in a run-time language but otherwise only requires that the software be recompiled for the specific architecture.

OSs come in various types offering an array of options and capabilities. Although many of these options are inconsequential, a select few can have significant effects on development. Typically the most crucial of these options is the use of a real-time kernel. Real-time operating systems utilize preemption and specialized scheduling algorithms to minimized missed deadlines (soft real-time) or deterministically guarantee deadlines are not missed (hard real-time). Although advantageous, real-time systems have increased OS overhead, require specialized programming techniques and support minimal types of hardware.

5.3.2 Software Modularity

Designing an UV with highly modular hardware is of little benefit if the corresponding software is not modular. In the past it has been very common to heavily integrate vehicle software to ensure minimal overhead and single process systems. This type of software integration has several drawbacks:

- Tight integration makes it very difficult to remove obsolete code. Many times obsolete code is simply left in place because removing it may have inadvertent affects on unrelated sections of the software.
- Upgrades become extremely difficult as they very often have to interact with multiple sections of code.
- Testing due to upgrades and maintenance increases significantly. Source code modification requires that any path to and from the altered code be subject to quality testing. Identifying paths can be very difficult and can quickly add up.
- Timing synchronization with external devices becomes very rigid or inefficient. Software may be forced to wait for communication with an external device consequently blocking other sections of software that could be running.

Software modularity can easily be implemented by utilizing middleware. Middleware is simply a layer of software that allows multiple processes running on one or more machines to interact. Although these processes need to communicate information, it may be incorrect to merge them. For example, suppose there are two processes. The first process, named *GPS_comm*, is responsible for communicating with a GPS sensor and parsing out position data. A second process, called *Planner*, is responsible for generating an obstacle free path to a goal. These two processes are obviously distinct from one another but *Planner* requires the vehicle's position to generate a feasible path.

Beyond simply allowing processes to communicate, middleware can significantly reduce timing issues. This is because individual processes can be dedicated to communicating with external devices. This ensures that processes that are ready to operate can do so. Continuing from the above example, if *GPS_comm* is blocked waiting for data from the sensor then *Planner* can continue to operate using data from other sources that have become available.

5.3.3 Prioritizing Software

Prioritizing software is a key feature to ensure system longevity and overall safety. Many processes will be considered absolutely critical and thus have higher priority than others. A clear distinction should be made between software that is critical to mission operation and software that is critical to general operation. For example, GPS is considered critical to mission operation of fixed wing autonomous UAVs operating in outdoor environments. Although this is currently true, it is relatively easy for this type of UAV to remain aloft and operating safely without GPS. Consequently, vehicle control and navigation software should receive a higher priority than the GPS software.

Priorities can obviously be used to justify software components that must receive extended testing and peer reviews. Beyond the development phase, priorities can be used during runtime to identify components that can be dynamically shutdown. This allows a monitoring system to automatically remove software that may currently be inhibiting higher priority processes.

Table 5.1 Testbed hardware components.

Hardware Components	Model
Processor	Pentium M 755 2.0 GHz
Motherboard	G5M100-N mini-ITX
Ram	2 Gigs 333MHz
Wireless	Intel Pro 2200 802.11B/G Mini-PCI
Frame Grabber	IVC-200G (4 channel)
Power Supply	120 Watt picoPSU-120
IMU Sensor	Microstrain 3DMG-X1 IMU
GPS Sensor	Novatel Superstar 2 GPS receiver (5 Hz, 5V model)
Range Sensor	URG-04LX Hokuyo laser range finder
Camera	Sony FCB-EX980S module camera
Control Board	Microbotics Servo/Safety Controller (SSC)

5.3.4 Protecting Your Hardware

Software can and should be used to protect your hardware. This can be accomplished through primitive behaviors or simply through failure identification and recovery. Some commonly used primitive behaviors include obstacle avoidance, orientation limiting and return home routines. These behaviors generally represent the lowest levels of control and are designed to protect the vehicle from user inputs, unknown situations, and malfunctioning higher level software routines.

Failure identification is also very useful when trying to protect equipment. Although it is very difficult to identify and classify failures, many modern sensors supply error and failure messages. A good rule of thumb is to have a plan for any point of failure that can be correctly identified and classified. This information can be used by software to compensate for failures. For example, a helicopter UAV may use a laser range finder to identify its vertical distance to a landing zone. If the laser has failed it may not be reasonable to have the vehicle attempt an autonomous landing. Consequently, it may be reasonable for the vehicle to refuse an autonomous landing and simply hover at its location.

5.4 Hardware Architecture

The above methodology is a collection of experience from design and redesign of UV testbeds for several platforms. This section will detail one of those implementations which has been integrated onto two distinct UAV vehicles and four identical UGV vehicles. The main hardware components of the designed modular testbed are outlined in Table 5.1.

Figure 5.1 depicts a group of UVs all utilizing the single testbed implementation. Figures 5.2 and 5.3 closely detail the integrated testbed on the individual UAV, a Maxi Joker 2 helicopter, and UGV, a Traxxas Emaxx truck. The above hardware

Fig. 5.1 Completely assembled testbed with UAV and UGVs.

Fig. 5.2 Assembled UAV testbed vehicle.

configuration was selected due to its high computational capabilities, various Input/Output (I/O) ports, small size, low heat emission, and cost. Figure 5.4 depicts the overall conceptual design for each of the testbed vehicles.

Fig. 5.3 Single testbed UGV.

5.4.1 Enclosure

Autonomous vehicles typically have multiple components that must be protected from their natural operating environment. The most typical method of protection is to encase these items within some type of enclosure. Enclosure design is generally specific to the choice of UV platform as well as the sensing/computing hardware that one is attempting to utilize. A system that is exposed to harsh environmental conditions must be designed and built to withstand those circumstances. For example, an enclosure designed for a turbine helicopter must be able to withstand the heat and vibration common to this type of platform.

The enclosure utilized for this work is a simple basswood box designed to encase and protect the processing system, GPS receiver, and safety switch. The utilization of wood has two major flaws. First, wood is an extremely poor conductor of heat. Thus, the internal components had to be sufficiently resilient to heat. Second, there is no natural Radio Frequency (RF) shielding effects in wood. Through experimentation it was discovered that frequencies from the motherboard interfere greatly with GPS reception. Several variations of motherboards were tested at varying frequencies and GPS degradation varied from 20 to 100% based on the type of hardware and relative location, up to 30 inches, from the GPS antenna and receiver. In order to avoid the effects of Electromagnetic Interference (EMI) the enclosure and GPS receiver were shielded. This was accomplished by encasing both the GPS receiver and enclosure in EMI foil creating a Faraday shield. This effectively shielded all GPS components including the receiver and antenna, from EMI produced by the motherboard.

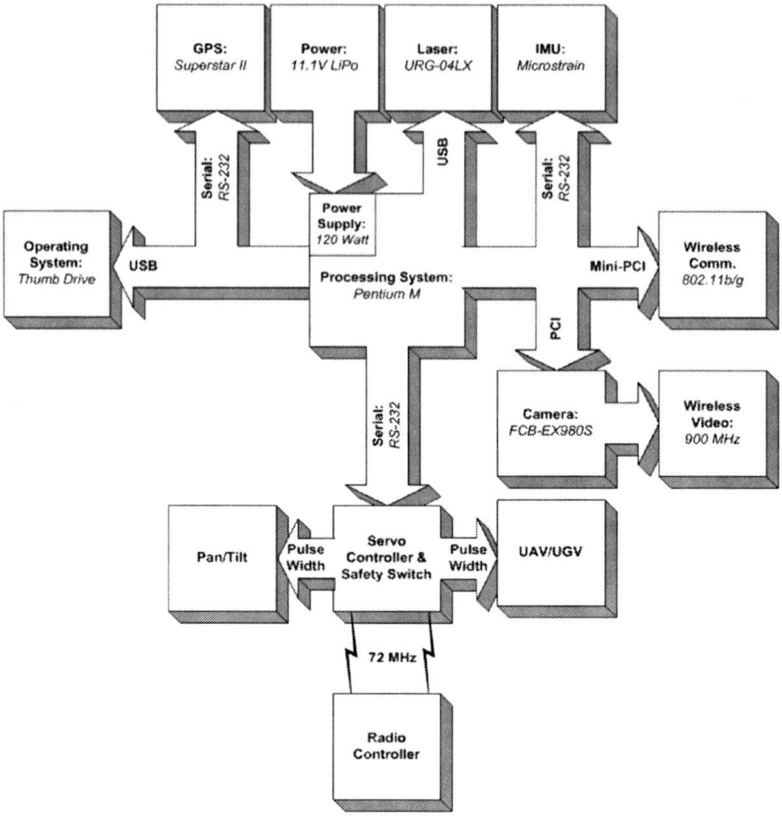

Fig. 5.4 Conceptual hardware diagram.

5.4.2 Servo/Switch Controller

Typical RC vehicles are controlled by Pulse Width Modulation (PWM) servos. As such, hardware must be present that can communicate efficiently and effectively with both the processing system and PWM servos. It should also be noted that typical servos require constant refresh to hold position. Thus servo control hardware must be capable of refreshing positions at an appropriate rate, typically 50 Hz or more.

Besides control, there are numerous safety concerns inherent to many research testbeds. It is inevitable that a software or hardware bug will cause the vehicle to endanger its self and possibly others. For this reason, any autonomous vehicle should be equipped with either a manual takeover or stop switch. A stop switch is defined as a device that removes the vehicle's ability to function. This type of device can simply cut power to the motor or immediately destroy the vehicle. Although a stop switch is effective, it is much more cost effective, and in many cases safer, to utilize a takeover switch. A manual takeover switch allows a human or possibly secondary system, to immediately take full control of the vehicle. Note that safety devices such as the takeover and stop switch must be functional in the event of a failure. Thus, these safety devices must be designed in a manner that allow them to function even if the rest of the system does not.

To adhere to these requirements the modular testbed is equipped with the Microbotics Servo/Switch Controller (SSC). This controller allows the on-board processing system to output servo commands via serial RS232 to all PWM devices used on the vehicle. The SSC is designed to allow a single switch on any common radio controller to take immediate control of the vehicle. The SSC also allows the on-board processing system to query the state of control, autonomous or human, and the current PWM signals to and from the SCC.

Due to the safety issues surrounding UAV research and the vast importance of manual takeover control, the SSC is designed to use a dedicated secondary battery. This design provides a human operator with complete control even in the event that the main power supply fails.

5.4.3 Orientation and Position Sensors

Vehicle state data, typically orientation, velocity, and/or position, are crucial to controlling any autonomous vehicle. Although there are numerous ways to obtain a system's orientation, velocity, and positional information, only a select few are widely accepted and heavily utilized.

The most popular solution for securing vehicle state data is the utilization of a complete sensor suite. This method utilizes multiple sensors that directly provide software with all desired data. Although optimal, this type of method is rarely feasible on small UVs due to payload limitations and current sensor technology.

For modular testbeds the most feasible design is to utilize a suboptimal sensor suite and attempt to calculate desired state data values from the provided information. This method can significantly reduce the size and complexity of the hardware components but can lead to large error accumulation and increased software complexity.

To satisfy the need for orientation, velocity, and positional data required to successfully control the testbed vehicles, a Microstrain 3DMG-X1 IMU and Superstar II GPS receiver (5 Hz model) were chosen. The Microstrain IMU allows the user access to orientation (Euler angles or Quaternions), accelerations, and angular rates at a rate of up to 100 Hz. The Superstar II GPS receiver provides latitude, longitude, and altitude (from sea level) position at 5 Hz. The Superstar II GPS is also Wide Area Augmentation System (WAAS) capable which allows the system to receive correction data without being locked to a ground station. Velocities are obtained through calculations using fused and filtered GPS and IMU data, available in [7].

5.4.4 Range Sensor

Although positional data provides information regarding the location of a vehicle, it cannot supply positional details about other obstacles such as trees or people. This becomes an issue when attempting to navigate a cluttered environment or autonomously take-off and land a vehicle. For UAVs, take-offs and landings require that the location of external objects such as the ground or landing platforms be known to a high degree of accuracy. GPS devices, similar to the one used here, have been shown to have altitude errors of three to five meters which is far too inaccurate for take-off and landing [11].

Range data, or distance from the vehicle to an object, can be acquired using stereo vision, optic flow, infrared sensors, sonar, radar, planar lasers, etc. Vision algorithms such as optic flow and stereo vision have shown great advancements in recent years but are still considered highly error-prone in outdoor dynamic environments [10, 15, 23]. Sonar and infrared type sensors are by far the lightest and cheapest of these devices but only provide distance in a single dimension and are commonly disrupted by uneven terrain, color and texture. Radar by far provides the widest range of distance measurements but due to its technology is typically too heavy for use on small UVs. Scanning/Planar lasers provide multiple distance readings over a two dimensional plane but are typically disrupted by both direct sunlight and heavy vibration.

To ensure that the testbeds are capable of obstacle avoidance and landing identification they are equipped with the Hokuyo URG-04LX scanning laser. This sensor provides the processing system with range data of up to four meters with single millimeter resolution at 10 Hz. This data can also be used for localization in the event that GPS fails.

5.4.5 Processing Board

The single most difficult component to be selected for the UV testbeds is the main processing board. This is mainly due to the abundant variations of form factors and the types and number of available peripherals. Processing boards come in many shapes and sizes and can be COTS or designed in-house. Popular form factors of processing boards include PC-104, Mini and Nano Integrated Technology Extended (ITX), and Advanced Technology Extended (ATX). Along with variations of form factor, the type and number of peripherals is also an important design choice. Popular peripherals include Universal Serial Bus (USB), Recommended Standard 232 (RS232), Recommended Standard 422 (RS422), Transistor-Transistor Logic (TTL), parallel, Firewire (IEEE1394), Ethernet, Super Video Graphics Array (SVGA), infrared, Peripheral Component Interconnect (PCI), and Mini-PCI.

When deciding on peripheral interfaces one should generally take into account the type of components used on the testbed vehicles and the overall most common interfaces. Since most sensors designed today utilize either serial or USB, it would

Table 5.2 G5M100-N interface support.

Port Type	# Available	Interface Type
USB	6	4 × 5 Pin Standard, 2× Board Pinout
Serial	3	2× RS232, 1× RS232/RS422
Ethernet	2	RJ45
VGA	1	VGA
PS2	2	1× Keyboard, 1× Mouse
PCI	1	PCI Slot
Mini-PCI	1	Mini-PCI
IDE	2	1× 40 Pin IDE, 1 × 44 Pin IDE
RAM	2	PC 2700

beneficial to ensure that the chosen processing board have several of these integrated into the system. Peripheral choice, although important, is typically eased by the abundance of converters and adapters available that allow the user to plug almost any device into any port.

The testbed is equipped with a G5M100-N Mini-ITX motherboard. This motherboard was chosen based on its type and number of peripheral interfaces (Table 5.2), support for a low power processor (Pentium M), overall size, and low cost.

5.4.6 Communication

In the implemented testbed, communication refers to any wireless transfer of information from the UV to any disconnected system. This includes data transferred via the radio controller, on-board processing system, and video transmitter. Although there are hundreds of variations of communication protocols, only a few are commercially available and even fewer are not restricted by the Federal Communications Commission (FCC) [13].

To date, publicly accepted and commercially available forms of wireless communication include aeronautical radio navigation, maritime mobile, TV and FM broadcasting, and satellite. Although all of these communication devices are available commercially, many frequencies typically require special authorization by the FCC and are restricted in certain locations. Furthermore, the FCC typically restricts the power outputs of unregulated frequencies.

For hobbyist type RC vehicles the typical frequency is 72 MHz, ranging from 72.010 MHz (channel 11) to 72.990 MHz (channel 90). This frequency is utilized by the radio controller to transmit PWM request to the vehicle. This data is received by the radio receiver and converted into PWM signals distributed to the appropriate servos. In the case of the developed testbeds, the 72 MHz frequency is also used to grant or remove control of the vehicle to and from the on-board computer. More recently there have been several models of radio controllers and receivers that utilize a 2.4 GHz frequency. The 2.4 GHz frequency is also utilized by 802.11 and

Bluetooth protocols as well as many cordless phones. Make note that the wide use of the 2.4 GHz frequency could potentially cause radio control failure and should be utilized with caution in industrial or residential areas.

Utilization of distinct wireless frequencies for video transmission is currently extremely popular on UVs. This is mainly due to the bandwidth required to wirelessly transmit high quality streaming video. To date, there are several non-regulated variations of wireless video transmitters being utilized by UVs. These devices typically operate on the 900 MHz, 2.4 GHz, and 5.0 GHz frequencies and are typically used to relieve bandwidth limitations imposed on the 802.11 protocol. The main differences between the varying frequencies are the average operational distances, which also varies based on power output, and the amount of bandwidth available.

The testbed is equipped with an Intel Pro 2200 Mini-PCI wireless card that allows the on-board processing system to communicate with any computer network operating on the 802.11 protocol, 2.4 GHz. The testbed is also equipped with a 72 MHz receiver that receives commands from the radio controller. The last piece of communication hardware present on the testbed is the wireless video transmitter. This device transmits a single video signal, via the 900 MHz 100 mW transmitter, to any receiver within broadcast range.

5.4.7 Camera

Commercial cameras are available that utilize multiple interfaces including Firewire (IEEE1394), USB, and composite and vary greatly in size, weight, accuracy, and cost. When developing an UV testbed one must take into account several limiting factors of cameras during the selection process.

First, due to the payload limitation of RC vehicles, especially aerial vehicles, the payload loss for a particular camera must be considered. This includes the weight of the camera and the weight of the extra electrical power required to operate the camera. The mounting location of the camera must also be considered. If the camera is mounted towards an extremity of the vehicle it may need to be compensated for with a counter weight which will further deplete the payload of the vehicle.

The second factor that must be considered is the camera's ability to function correctly in its operational environment. RC vehicles typically experience high frequency vibrations or large gravitational forces as well as operate in varying light intensity environments. As such, a designer must consider the vehicle's need to compensate for varying light intensities, focus dynamically, and zoom on objects of interest. Last, the interface utilized by the camera must be considered. The most common of these interfaces include Firewire, USB and Composite and have varying advantages and disadvantages.

The testbeds are equipped with Sony FCB-EX980S block cameras and an IVC-200G frame grabber. These block cameras support auto-stabilization, 26× optical zoom, low-light operation, and iris/zoom/focus control via an RS232 interface. The

PCI frame grabber allows for data acquisition at 30 frames per second on four channels simultaneously.

5.4.8 Data Storage

Data storage, in reference to the implemented testbed, describes the area or device where the OS, data acquisition software, control software, and collected data are stored. The most common data storage devices include magnetic drives, magnetic disks, and solid state memory. Although magnetic drives and disk are the cheapest, per gigabyte, they require mechanical devices to read and store data. This not only limits the speed at which data can be stored and retrieved, it requires that the device be virtually stable at all times to prevent damage to the device. This coupled with the size and weight of magnetic devices typically prevents their usability on small UVs.

Solid state memory, due to its size, weight, and operational characteristics, is the typical utilized device for data storage on UVs. One major short coming of solid state memory is its degradation due to usage. Individual sectors of solid state memory are typically only good for a few hundred thousand writes. Although this seems sufficient for almost any system one must consider that many OSs utilize permanent storage as virtual memory and may perform thousands of write operations during a single procedure.

Data storage for the testbed is handled utilizing both solid state memory, in the form of a USB thumb drive, and volatile Random Access Memory (RAM). The solid state memory is used to store the OS and any software necessary to make the system functional, including device drivers and communication protocols. The testbed is first booted from the USB drive where the operating system is copied to a RAM drive, sometimes referred to as a virtual drive. From this point all data storage is performed on the virtual drive. This allows the USB drive to be removed from the system after bootup preventing its loss or damage during vehicle operation. This also allows for a modular hardware design where alterations to the platform, sensors, or software only require the system to be rebooted utilizing a properly configured USB drive.

5.4.9 Hardware Chassis

The hardware chassis refers to the frame to which all of the hardware is mounted. Although this is somewhat inconsequential for the ground vehicles due to their passive suspension systems, it is of special importance for air vehicles. Helicopter based platforms naturally create high frequency vibrations caused by the motor, main rotor, and tail rotor. This vibration, if not isolated, will create noise, error, and possibly damage to hardware mounted to the platform. For these reasons it is advisable that some type of vibration isolation or reduction be built into the chassis. This would be similar to the suspension system built into modern ground vehicles to reduce road vibration.

Another important item that must considered when designing or modifying a small UV chassis is the material used in the design. As previously mentioned, most UVs have low payload capabilities thus a hardware chassis must balance the desire for stability and protection with payload restrictions. Also, sensors such as magnetometers, typically used to sense heading, are heavily influenced by devices that create and/or influence magnetic fields. This can include sensors, batteries, actuators, electrical current and ferrous materials. With this in mind the designer must not only ensure that sensors, like magnetometers, are mounted in areas that isolate them from interference but that the hardware and chassis used to mount these sensors do not create interference. Typically materials that do not adversely affect magnetometers are Styrofoam, wood, plastic, aluminum and brass.

The testbeds utilize custom designed chassis and braces made completely of aluminum. Aluminum allows the chassis to be lightweight and sturdy without adversely affecting the Microstrain IMU utilized on the vehicles. The UAV testbed chassis contains two distinct layers of vibration isolation mounts. Both layers utilize rubber isolation mounts and are designed to separate the chassis from the platform and the processing system and the sensors from the chassis.

5.5 Software Architecture

The software architecture described in this section covers the overall structure of the software developed for the testbed systems. It also discusses the details of the individual processes and how these processes interact to form a single entity.

5.5.1 Operating System

The OS is the backbone of any software architecture and provides a base for all supporting software. Although almost any modern OS would be sufficient, the testbed has been strictly operated by Linux distributions. Linux was chosen due to stability, ease of modification, and heavy use in both the academic and industrial arenas. The utilized distributions have included multiple versions of Slackware, versions 10.0 through 12.0, and Gentoo. The currently utilized distribution is Slackware 12.0. Slackware was chosen due to ease of installation and ability to install and boot from a USB thumb drive. Also note that the OS operates completely out of a RAM drive. Due to the physical limitations of RAM, this requires that the OS be minimal in size to allow for sufficient RAM for operating processes.

It should be noted that the boot up time is directly associated with the size of the OS. The configuration used on the testbeds takes between 180 and 200 seconds to boot. This boot up time can be significantly reduced if the OS is loaded using the USB 2.0 protocol or by using a solid state drive via a SATA or IDE interface.

5.5.2 Source Code Architecture

Source code describes all of the in-house developed software used to interface with and control the testbed systems. All of the source code for the testbed is developed in the C programming language.

From conception the software for the testbed was designed to be highly modular and to support an operating structure that could be dynamically modified. Modularity in the design allows for code reuse, quick integration, and ease of understanding. Dynamic modification simply means that the system is able to remove or add functionality during operation. To support these design requirements, software is developed as a set of processes that run concurrently and pass information through shared memory structures. The advantage of this software structure is threefold:

- *Reduced Integration Time*: Software simply needs to check and update a single location in memory rather than passing the information to multiple distinct sources. This reduces coding time, maintenance complexity and helps ensure data consistency throughout.
- *Simplifies Major Software/Hardware Changes*: Inevitably UVs will receive hardware upgrades and software modifications. A shared memory structure allows process modules to simply be replaced without affecting the entire system's software. This is especially beneficial when sensors are upgraded as they typically provide higher rates and greater accuracy but ultimately provide the same information. As such, only a single module must be altered greatly reducing software testing.
- *Online Functionality Adaptation*: By utilizing this method of development a monitoring process can be designed to start and stop processes as needed by the testbed. This is beneficial when a critical processes is operating at a lower than optimal rate. A monitoring process could shutdown lower priority processes freeing up both memory and CPU for the higher level process.

Although the vehicles described in this work do not utilize the ability to dynamically alter the software structure, support for it is maintained throughout software development. Figure 5.5 details the process architecture for the UV testbeds.

It should also be mentioned that all of the developed software is written and tested on a soft real-time OS. This means that the OS attempts to minimize delay but does not guarantee run time. Although there is no guarantee of run time there is a reasonable expectation that process will be handled efficiently and should operate within some boundaries. Using this assumption many of the processes were designed to attempt to operate within a given window of time. If they fail to operate within this window they report a warning detailing how excessive their delay is.

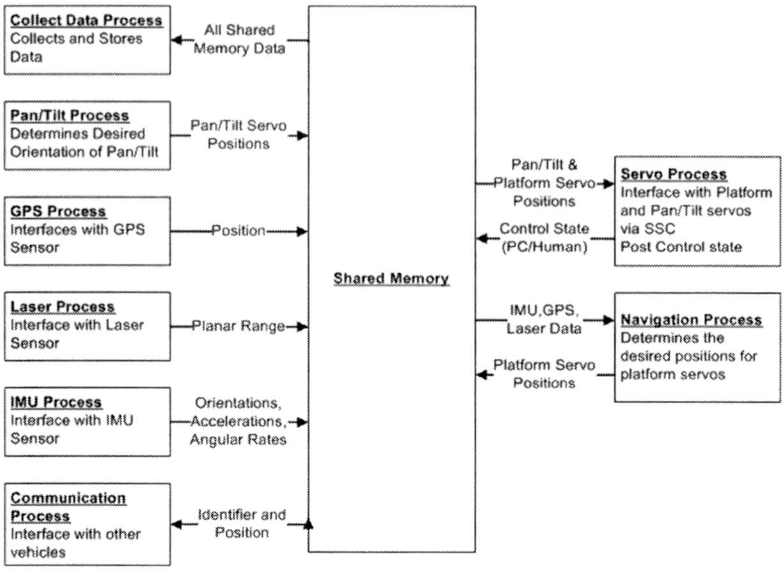

Fig. 5.5 Process architecture.

Due to the minimalistic design of the testbeds OSs, source code is kept and compiled on a ground station laptop. During field testing the source code can be modified, compiled, and uploaded directly to the testbed. Note that the GCC compiler, version 4.1.0, is utilized to compile all of the software before it is uploaded to the testbed.

5.5.2.1 GPS Process

The GPS process is responsible for reading and parsing data from the GPS receiver as well as placing the data in shared memory. This process will also inform the user in the event that a significant change has occurred in the accuracy of positional data. Significant changes in accuracy include no lock (zero), lock without WAAS correction (one), and lock with WAAS correction (two).

5.5.2.2 IMU Process

The IMU process is responsible for accessing the Microstrain IMU. This process gathers Euler angles, angular rates, and accelerations at approximately 80 Hz and posts their values, along with an associated time stamp, to shared memory.

5.5.2.3 Laser Process

The laser process is responsible for gathering range data from the laser. This information is used for both take-off and landing on the UAV helicopter as well as obstacle avoidance on the UGV. This process also monitors the state of the laser and attempts to identify failures. When failures are identified the process attempts to compensate for the particular failure, typically using a soft shutdown and restart of the laser. If failures cannot be compensated the process sets a failure flag. This flag is ultimately used by other processes to compensate for the loss of laser range data. For example, if the laser fails during a UAV flight the vehicle will refuse autonomous landing and must be human piloted to the ground.

5.5.2.4 Servo Process

The servo process is responsible for generating packages which will request that the SSC move individual servos to particular locations. It is also responsible for retrieving and parsing state data from the SSC. This data allows processes to determine if the human safety pilot is in control or if the on-board processing system is in control.

Unlike previous processes, the servo process creates three separate shared memory locations and three corresponding semaphores. The first two shared memory locations are dedicated to position requests for the platform's servos and the pan/tilt's servos. This is done to allow pan/tilt controlling software and vehicle controlling software to be separate modules. As such, pan/tilt control software will not affect platform control software. The third shared memory location is dedicated to SSC state data. This shared memory location stores a single variable which identifies who is providing vehicle control, the human operator or the on-board computer.

5.5.2.5 Pan/Tilt Process

The pan/tilt process was developed as a functional template. The process is fully operational but is only coded to hold the pan/tilt in a neutral position. This was done to support quick integration of vision code that may need to control the pan/tilt. This program, in its empty template state, will only connect to the shared memory location created by the servo process and output pan and tilt neutral positions.

5.5.2.6 Collect Process

The collect process is responsible for data collection during operation of the testbed. This process collects the vehicle's position, orientation, accelerations, approximated velocities, control outputs, and SSC state from shared memory and commits them to a file, data.txt, on the RAM drive. Data is collected and stored at approximately 4 Hz.

5.5.2.7 Communication Process

In order to exchange data in multi-vehicle experiments, a communication process is used. Mobile Mesh, dynamic ad-hoc networking software, is utilized for routing and a custom server/client process is used to send and receive data. The communication model is a simple UDP client/server protocol.

The communication process encompasses two threads, the server and client, which operate simultaneously in the background. The server thread listens, parses, and posts all incoming data to shared memory. The client thread gathers both positional data and a vehicle identifier and propagates this data to all vehicle members.

In addition to listening, parsing, and posting data, the server thread also keeps track of communication data rates with individual members. This data is used to determine whether a particular member is dead or alive signalling the swarm to either wait or move on due to falling behind members.

5.5.2.8 UAV Navigation Process

The navigation process is the focal point for all of the software developed for the testbeds and is the one process that typically varies from platform to platform. This process calculates and maintains state data and interfaces with the vehicle controllers. With respect to the UAV testbed, the navigation processes is also responsible for stage selection (waypoint navigation, take-off, and landing), see Figure 5.6.

The take-off procedure is broken into three phases. These phases are responsible for spinning up the motor, prepping the collective, and lifting off the vehicle. Note that prior to lift off this procedure will lock in the current heading, latitude, and longitude as the desired heading, latitude, and longitude. This method ensures that the vehicle will attempt a completely vertical take-off and will minimize the possibility of extreme maneuvers at low altitudes. Note that this procedure will not start until a valid GPS lock has been attained.

The first phase of the take-off procedure spins up the main rotor while holding the collective pitch negative. This is done by incrementally increasing the throttle PW to its predetermined flight value. Phase two prepares the collective for flight by slowly increasing the collective command from its minimum value to its approximate hovering value. The last phase of the take-off procedure is responsible for lift-off. This is accomplished by simply relinquishing control of the vehicle to the flight controllers. Once the vehicle has reached an altitude of two meters normal waypoint navigation is engaged.

The navigation procedure is simply responsible for navigating the testbed to desired locations. Once a desired location has been reached within a small threshold and maintained for a short period of time the next location is proceeded to. Once the

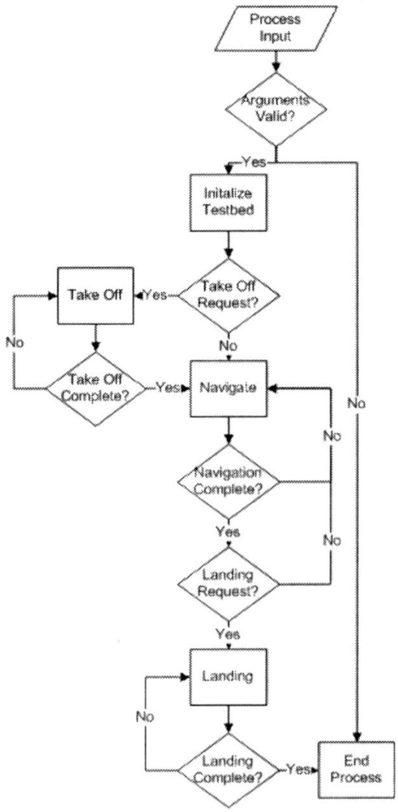

Fig. 5.6 Flow chart for the UAV navigation process.

final position has been reached the process will either initiate the landing procedure or hover at the last location awaiting user input.

The landing procedure is also broken down into three phases. These phases are responsible for touching down, removing lift, and powering down the vehicle. The first phase of the landing procedure is responsible for lowering the vehicle to within approximately one inch of the ground. This is done by decreasing the vehicle's hovering altitude to the desired relative height. The second phase of the landing procedure incrementally decreases the collective pitch to its lower limit removing all vertical force from the vehicle. The final phase of the landing procedure simply powers down the vehicle's motor.

The specifics of the controllers and the state data calculations are outside the scope of this work, they are described in great detail in [7]. These descriptions include velocity calculations, servo cyclic and collective pitch mixing, GPS antenna translation and flight path and orientation control.

5.5.2.9 UGV Navigation Process

Navigation and obstacle avoidance for the UGV testbed platforms are achieved utilizing potential fields to create a vector in the x and y directions. In multi-vehicle experiments, formation control is achieve by imposing additional restrictions on the vector field to hold the UGVs in a desired formation [1]. For avoidance of UGV members, GPS coordinates are sent and received via the communication process and utilized in order to calculate collision free paths. The generated vector from the navigation process is converted into two servo values corresponding to the throttle and steering inputs. With respect to the UGV testbed, the navigation process runs until the final goal is reached. When the goal is reached, the vectors will be zero denoting neutral throttle and steering values.

5.6 Field Experiments

It should be noted that the UV testbeds are designed for field testing technology but they are not designed to replace fundamental testing such as unit testing and simulation. Field testing requires a large amount of logistics including moving the equipment to testing locations and prepping vehicles for operation. Field testing is also inherently dangerous. Any testing performed in the field, no matter what safety precautions are taken, inherently places the vehicle, onlookers, and surrounding environment in a potentially hazardous situation.

For these reasons simulation testing was performed on both the UAV and UGV vehicles. Air vehicle simulations were performed by integrating a Simulink control system, designed to mimic the actual vehicle controllers, with the commercially available X-Plane simulator. This was done to take advantage of the heavily developed and highly refined world and vehicle models supplied with X-Plane. Simulink is also used to model the swarm of UGVs as well as the formation controller.

5.6.1 Autonomous Flight, Take-off and Landing of a VTOL

Ultimately, the goal of this research is to develop a fully operational outdoor, or indoor if desired, testing vehicle. Experimentation in the field is the key to showing practical results. The Maxi Joker 2 UAV has performed over 500 hours of autonomous flights including take-off, landing, and waypoint navigation. These experiments included several common flight patterns including the square-s, straight line, and vertical steps. Note that vertical steps are transitions in the longitudinal and vertical axes. Figure 5.7 details typical results from these flight paths.

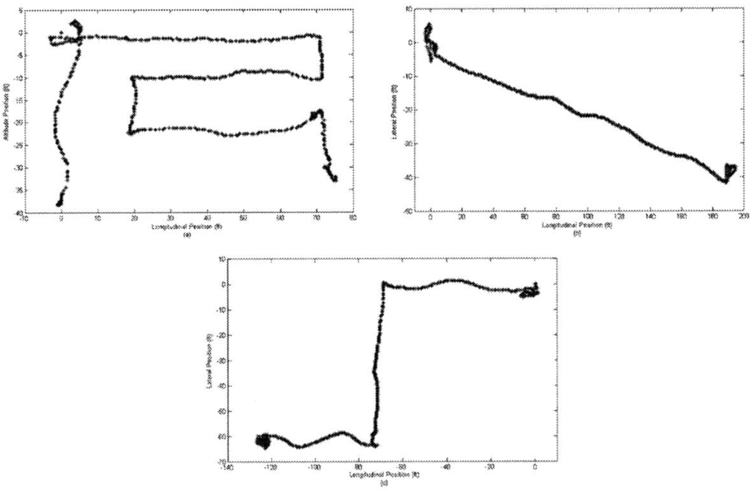

Fig. 5.7 UAV flight paths (a) vertical steps, (b) straight line, (c) square-S.

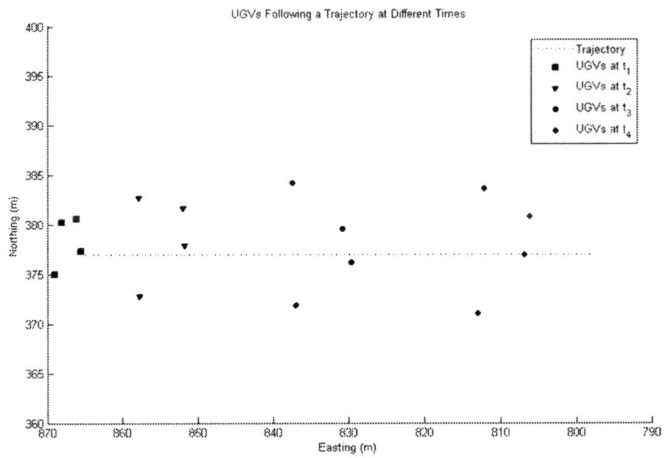

Fig. 5.8 Ellipse UGV formation at t1 , t2 , t3, and t4/

5.6.2 UGV Formation Control

Multiple experiments were performed to demonstrate UGV coordination. In one specific experiment, four UGVs autonomously navigate a trajectory while holding an ellipse formation [1]. The UGVs receive GPS coordinates which are used to calculate trajectories. These coordinates are a function of time which all UGVs receive periodically. Each UGV broadcasts its position to the other UGVs for obstacle avoidance purposes. At each time step, the UGVs compute their vectors based on their current position, trajectory, and nearby obstacles. Based on the output of the vector fields, a desired speed and heading are computed which are used to control the vehicles servos. The four UGVs follow the trajectory in formation while avoiding each other. Figure 5.8 shows the UGVs formation over time.

5.7 Conclusions

This work attempts alleviate the temporal and fiscal burden associated with designing, building, and testing UV testbeds. Although UVs can be readily purchased from a variety of manufactures, they typically contain proprietary hardware and software, have minimal modularity, and are very difficult to upgrade and maintain. As such, this work describes the methodology and justification for building in-house research UVs collected over four years of development and refinements of our own research. Future work includes applying these design standards to a variety of platforms ranging from micro-vehicles to manned vehicles.

References

1. L. Barnes, M. Fields, et al., Unmanned ground vehicle swarm formation control using potential fields, in *Proceedings of Mediterranean Conference on Control & Automation*, Athens, Greece, 2007.
2. R.W. Beard, T.W. McLain, et al., Decentralized cooperative aerial surveillance using fixed-wing miniature UAVs, *Proceedings of the IEEE* **94**(7), 1306–1324, 2006.
3. D.W. Casbeer, D.B. Kingston, et al., Cooperative forest fire surveillance using a team of small unmanned air vehicles, *International Journal of Systems Science* **37**(6), 351–360, 2006.
4. B. Coifman, M. McCord, et al., Surface transportation surveillance from unmanned aerial vehicles, in *Proceedings of the 83rd Annual Meeting of the Transportation Research Board*, pp. 11–20, 2004.
5. B. Coifman, M. McCord, et al., Roadway traffic monitoring from an unmanned aerial vehicle, *IEE Proc. Intell. Transp. Syst.* **153**(1), 11, 2006.
6. F. Dovis, P. Mulassano, et al., HeliNet: A traffic monitoring cost-effective solution integrated with the UMTS system, in *Vehicular Technology Conference Proceedings*, VTC 2000-Spring Tokyo, IEEE 51st, 2000.
7. R.D. Garcia, Designing an autonomous helicopter testbed: From conception through implementation, Ph.D. Thesis, Computer Science Engineering, University of South Florida, Tampa, 2008.
8. R.D. Garcia, K.P. Valavanis, et al., Fuzzy logic based autonomous unmanned helicopter navigation with a tail rotor failure, in *Proceedings of 15th Mediterranean Conference on Control and Automation*, Athens, Greece, 2007.
9. A.R. Girard, A.S. Howell, et al., Border patrol and surveillance missions using multiple unmanned air vehicles, in *Proceedings of 43rd IEEE Conference on Decision and Control*, CDC, 2004.
10. S. Hrabar, G.S. Sukhatme, et al., Combined optic-flow and stereo-based navigation of urban canyons for a UAV, in *Proceedings of the 2005 IEEE International Conference on Intelligent Robots and Systems*, 2005.
11. S. Ireland, T. Sheppard, et al., Measurement of altitude at hot air balloon competitions, Retrieved November 23, 2007, from ftp://www.fai.org/ballooning/meetings/pc/2007/altitude%20measurement.doc, February 10, 2007.

12. M. Kontitsis, K.P. Valavanis et al., A simple low cost vision system for small unmanned VTOL vehicles, in *Proceedings of IEEE/RSJ International Conference on Intelligent Robots and Systems 2005 (IROS2005)*, pp. 3480–3486, 2005.
13. Management, O. o. S., United States frequency allocations: The radio spectrum, U.S. Department of Commerce, U.S.A. Frequency Allocation Table, allochrt.pdf, 2003.
14. J.R. Martínez-de Dios, L. Merino, et al., Fire detection using autonomous aerial vehicles with infrared and visual cameras, in *Proceedings of the 16th IFAC World Congress*, 2005.
15. L. Mejias, S. Saripalli, et al., Visual servoing of an autonomous helicopter in urban areas using feature tracking, *Journal of Field Robotics* **23**(3), 185–199, 2006.
16. J. Morris, DHS using Northrop Grumman UAV in Arizona border patrol flights, Retrieved July, 18, 2008, from www.aviationweek.com/aw/generic/story_generic.jsp?channel=hsd&id=news/HSD_WH1_11104.xm, 2004.
17. A. Ollero, J. Ferruz, et al., Motion compensation and object detection for autonomous helicopter visual navigation in the COMETS system, in *Proceedings of 2004 IEEE International Conference on Robotics and Automation, ICRA'04*, Vol. 1, pp. 19–24, 2004.
18. M. Quigley, M.A. Goodrich, et al., Semi-autonomous human-UAV interfaces for fixed-wing mini-UAVs, in *Proceedings of the 2004 IEEE/RSJ International Conference on Intelligent Robots and Systems*, 2004.
19. M. Quigley, M.A. Goodrich, et al., Target acquisition, localization, and surveillance using a fixed-wing mini-UAV and gimbaled camera, in *Proceedings of the 2005 IEEE International Conference on Robotics and Automation*, pp. 2600–2605, 2005.
20. G. Rufino and A. Moccia, Integrated VIS-NIR hyperspectral/thermal-IR electro-optical payload system for a mini-UAV, in *InfotechAerospace*, pp. 1–9, 2005.
21. A. Ryan and J.K. Hedrick, A mode-switching path planner for UAV-assisted search and rescue, in *Proceedings of 44th IEEE Conference on Decision and Control, 2005 European Control Conference, CDC-ECC'05*, pp. 1471–1476, 2005.
22. A. Ryan, M. Zennaro, et al., An overview of emerging results in cooperative UAV control, in *Proceedings of 43th IEEE Conference on Decision and Control, 2004 European Control Conference, CDC-ECC'04*, 2004.
23. S. Saripalli, J.F. Montgomery, et al. (2003), Visually guided landing of an unmanned aerial vehicle, *IEEE Transactions on Robotics and Automation* **19**(3), 371–380.
24. K. Schutte, H. Sahli et al., ARC: A camcopter based min field detection system, in *Proceedings of the Fifth International Airborne Remote Sensing Conference*, San Francisco, USA, 2001.
25. N. Slegers and M. Costello, Model predictive control of a parafoil and payload system, *Journal of Guidance, Control, and Dynamics* **28**(4), 816–821, 2005.
26. S. Srinivasan, H. Latchman, et al., Airborne traffic surveillance systems: Video surveillance of highway traffic, in *Proceedings of the ACM 2nd International Workshop on Video Surveillance & Sensor Networks*, pp. 131–135, 2004.
27. A. Wang, The development of unmanned vehicle control system based on embedded microprocessor, in *Aeronautics & Astronautics*, National Cheng Kung University, Tainan, Taiwan, M.S., p. 78, 2006.

Chapter 6
Designing a Real-time Vision System for Small Unmanned Rotorcraft: A Minimal and Cost-Effective Approach

M. Kontitsis and K.P. Valavanis

Abstract This chapter outlines the design process of a real-time, cost effective, vision system under the constraints imposed by a small unmanned rotorcraft vehicle and the environment in which the latter operates. The system is capable of detecting and tracking arbitrarily chosen objects using features such as color and size for detection and an adaptive template matching mechanism for tracking. The evaluation process included search and detect missions as well as traffic monitoring in real settings where adequate performance was demonstrated. Finally some design limitations and future enhancements are discussed.

6.1 Introduction

6.1.1 Motivation

In recent years unmanned aircraft systems (UAS) have been successfully used in a wide variety of applications. Their value as surveillance platforms has been proven repeatedly in both military and civilian domains. As substitutes to human inhabited aircraft, they fulfill missions that are dull, dirty and dangerous [1]. Representative examples of successful use of UAS are in areas including battlefield assessment, reconnaissance, port security, wildlife protection, wildfire detection, search and rescue missions, border security and patrol, resource exploration and oil spill detection, to

M. Kontitsis
Department of Computer Science & Engineering, University of South Florida, Tampa, FL 33620-5399, USA; e-mail: mkontits@mail.usf.edu

K.P. Valavanis
Department of Electrical and Computer Engineering, University of Denver, Denver, CO 80208, USA; e-mail: kimon.valavanis@du.edu

K.P. Valavanis (ed.), Applications of Intelligent Control to Engineering Systems, 117–167.

name just a few. The main common component among all those diverse applications is that they are practically variations of remote sensing and surveillance missions.

The reliance of almost every Unmanned Aerial Vehicle (UAV) application on the ability to sense, detect, see and avoid from a distance has motivated this work, attempting to further investigate this issue. In particular, among the various types of UAS, small scale unmanned rotorcraft or Vertically Take-off and Landing (VTOL) vehicles have been chosen to serve as the sensor carrier platforms because of their operational flexibility. Having the advantage of being able to operate from almost anywhere, since they require little to none preexisting infrastructure, outweighs their deficit of speed and endurance when compared to their fixed wing counterparts. Their ability to hover and fly in relatively confined spaces makes them almost ideal for deployment at low altitude and in urban settings in which the majority of fixed wing platforms would be challenged to operate. Therefore, and although reported research findings are general enough, the focus of this work is on designing and implementing an object detection system for a small unmanned custom made VTOL vehicle.

6.1.2 Introducing the Problem

To accomplish the aforementioned tasks/goals autonomously, any UAS must be equipped with the appropriate sensors to collect data and have enough on-board processing power for data interpretation and decision making. It must also employ a collection of algorithms capable of dealing with a variety of tasks.

A commonly used sensor suite includes inertial measurement units such as accelerometers, Global Positioning System (GPS) receivers, laser or barometric altimeters and cameras operating at various areas of the electromagnetic spectrum. In several occasions, synthetic aperture radars have also been utilized to provide a more detailed description of the ground below from a relatively high altitude. Although inertial sensors (IMU) and GPS measurements alone provide a relatively good estimation of the UAS's location and pose, these sensors are subject to failures and faulty readings on top of not always being available. The GPS signal is susceptible to interference by naturally occurring obstructions such as tree tops or clouds, as well as by artificial ones such as buildings, not to mention intentional malicious interference. As a result, readings provided by the GPS unit may be inaccurate or even non-existent in cases when not enough satellites are visible by the receiver at that particular location. A typical inertial measurement unit can provide acceleration and velocity vectors in three axes as well as Euler angles and quaternions at a rate of 5 to 100 Hz depending on the model. The inherent problem of this kind of sensors is the presence of measurement drift. If left uncorrected, the IMU estimate of the location tends to drift away within a few seconds. Several correction mechanisms exist that mitigate the problem usually by modeling the error, by using GPS measurements for correction or a combination of both. Laser or barometric altimeters have been used to provide additional information regarding the position of the UAS, namely the altitude, and they are a good complement to a sensor suite but nevertheless not enough to ensure autonomous navigation in most cases.

Cameras have been used as part of the UAS's sensor suite primarily as data collection equipment rather than navigational aids. Their function usually is to passively sense the environment for the presence of a specifically defined object of interest, record and transmit visual data back to a ground station for evaluation. Alongside with their role as data collectors, cameras can be very useful in estimating the relative or even absolute position of the carrying vehicle [39]. This method is known as visual odometer or visual simultaneous localization and mapping (VSLAM). It usually amounts to an attempt at reconstructing the three dimensional environment around the vehicle. The result is a map which, if accurately constructed, allows for autonomous navigation of the vehicle.

Another essential ability of an autonomous aerial vehicle is that of recognizing and tracking objects of interest, thus, keeping them within the field of view of the camera while recording their trajectory. This enhances the utility of the unmanned vehicle and facilitates the work of the ground control personnel. It allows the UAV to be used as a surveillance tool that expands the covered area without requiring constant attention. However, identifying arbitrary objects from an overflying moving platform in an uncontrolled environment can be extremely challenging given the variability of the parameters that influence the process. In an outdoors environment varying lighting conditions, unstructured clutter, motion, random occlusions and visual obstructions must be dealt with by the detection and tracking algorithms.

A very important design directive for an autonomous UAV is the requirement of real-time operation. All the tasks, especially the ones related to navigation, must be completed as fast as possible. In the worst case, the computation time of the decision making components must not exceed the 33 ms barrier that is considered to denote real-time performance.

An additional constraint is imposed on the algorithm by the carrying platform itself. Small aerial vehicles set a bound on the electrical power that can be carried along, which in turn limits the available processing power. With limited processing power at hand, the complexity of the algorithms becomes an important factor. Between algorithms that accomplish the same task, the one with low complexity is always desirable. In this case, it is crucial that the selected algorithm be able to run in real-time on less than state of the art equipment.

Another line of distinction exists between systems that are designed so that the processing takes place on-the-ground station and the ones that use an on-board computer. Obviously the former are not affected by any payload limitations therefore allowing for more powerful computers to be used.

In this chapter we address the problem of object identification and tracking in a largely unknown dynamic environment under the additional constraint of real-time operation and limited computational power.

6.1.3 Method of Solution

In order to successfully address the aforementioned problem a series of simple, relatively low complexity techniques have been employed. Specifically the object detecting algorithm is based on the application of appropriate thresholds on the image once the latter is converted to the Hue, Saturation, Intensity (HSI) colorspace. The final decision is made by using an accumulator that reflects the number of recent frames in which an object has been detected. When this exceeds a predefined limit the system produces an alarm and notifies of the object's presence. To address the tracking problem a simple template matching algorithm based on a similarity measure such as the sum of absolute differences or the normalized correlation coefficient, was implemented. The template is being continuously updated to maintain its relevancy throughout the time period that the object it describes is being tracked. The updated template at any time k is a linear combination of the best matching image patch and the template at time $k - 1$. Finally the tracking system has been designed so that it can concurrently accept input from both a human operator and an automated source such as another program.

6.1.4 Contributions

The contribution to the area of vision systems for unmanned aerial systems is the design and implementation of a cost effective system capable of performing object identification and tracking in real-time that:

- requires minimal information about the dynamic environment in which it operates;
- uses a single uncalibrated, not stabilized camera;
- tracks multiple objects without requiring a-priori knowledge of or using any assumptions about their trajectories; and
- Does not require an IMU.

The result is a system that can be assembled by commercially available hardware and can be configured to perform surveillance without calibration of the camera or detailed knowledge of the operating environment. It becomes evident that the use of an uncalibrated, not stabilized camera makes the problem very challenging and to some extend limits the accuracy of obtained results. However, this is one major point addressed in this work: even with an uncalibrated, unstabilized camera, results are sufficient to complete assigned missions.

6.1.5 Outline of Chapter

This chapter consists of six sections. Section 6.1 introduces the work and briefly describes the problem, the method of solution and the contributions. Section 6.2 provides a review of related work and some remarks on them. Section 6.3 is devoted to the detailed description of the proposed solution and the implemented system. The performance evaluation is presented in Section 6.4 along with a description of the actual scenarios where the system was deployed and the specific tasks that it carried out. Concluding remarks follow in Section 6.5 and future research topics that can enhance the current implementation are given in Section 6.6.

6.2 Literature Review

6.2.1 Related Work

Vision systems, techniques and algorithms suitable for UAVs range in complexity from simple color segmentation to statistical pattern recognition. This literature review considers a publication as being related to this work if the implemented system is specifically designed for use by UAVs. Furthermore, a work is considered directly comparable if the resulting vision system is physically placed on an unmanned VTOL and has been shown to function under real operating conditions in an outdoors environment.

Published related work and proposed machine vision architectures indicate the use of both "on-board" [2–4, 6, 19, 21, 24, 34] and "on-the-ground" processing setups [5, 7–15], with the latter being preferred most of the times. For on-board vision systems, due to the limited processing power provided by the on-board computer, derived algorithms have the additional constraint to run in real-time, requiring reduction of the computational load sometimes accomplished by processing selected image windows instead of the entire image. Table 6.1 presents a summary of machine vision techniques used by University research groups, the main processing unit (on-board, on-the-ground) and the unmanned VTOL vehicle platform they have been implemented on. Table 6.2 summarizes functionality and capabilities of existing fully operational vision systems, including techniques employed by each one of them.

The problem of object identification and tracking has been studied extensively in computer vision. Although several methods exist that can identify objects in a controlled setting, special reference will be made to those of them that have been adapted for use in unmanned aerial systems since it is believed that they relate more closely to the problem at hand. One such example is illustrated in the work of Ludington et al. [27] that presents a method for target tracking using a technique based on particle filters. Each target is described as a four dimensional particle containing the image coordinates and the dimensions of the rectangle that surrounds it. The as-

Table 6.1 Existing vision systems for VTOL vehicles.

	Machine Vision techniques used	Processing unit	Vehicle
Berkeley University [5]	No details provided	No details provided	BEAR
Georgia Tech [17, 18]	Edge detectors, morphing, statistical pattern matching	On- board	Rmax by Yamaha
Standford [11]	YUV color segmentation, signum of Laplacian of Gaussian (sLoG)	On-the-ground	Hummingbird Aerospace Robotic Laboratory at Standford
MIT [20]	Template matching	On-the-ground	Black Star by TSK
Rose Hulman IT (RHIT) [21]	Template comparison	On-board	Bergen Twin
IT Berlin [14]	No details provided	On-the-ground	MARVIN by SSM Technik
University of Texas [12]	Edge linking matching	On-the-ground	XCell .60
Swiss Federal Institute of Technology (ETH) [22]	No details provided	On-board integrated in camera	Huner Technik
Carnegie Mellon University [23]	Template matching and RGB color	On-the-ground	Rmax by Yamaha
USC [2, 4, 6]	Omnidirectional, optic flow	On-board	Bergen Twin
Southern Polytechnic State Univesity [13]	Stereo vision, Sobel edge detector	On-the-ground	Vario Robinson R22
Linkoping University, Sweden (WITAS) [24]	No details provided	On-board	Rmax by Yamaha

sumption for the system to operate is that the target moves smoothly between frames and that the frame rate remains sufficiently high. The motion is modeled as Gaussian random walk and the measurement model relies on color and motion cues. Finally, a neural network is responsible for constructing a performance estimate according to which adaptations are made to the particle filter.

Another example is the system described in [28, 29] where features such as rectangles are first extracted using the Douglas–Peucker algorithm [30] and then tracked using a Kalman filter.

Although not explicitly designed for an unmanned vehicle the system presented in [31] is addressing the problem of tracking moving objects in aerial video. It employs two different trackers, one for short and another for long term tracking. The short term tracking is accomplished by first registering successive frames using an affine transformation to correct for background motion and then extracting and matching Kanade–Lucas–Tomasi [32] features between successive image pairs.

Table 6.2 Summary of system characteristics and functionality.

		Institution					
		Berkeley University	Georgia Tech	Univ. of S. California	COMETS[a] [25]	WITAS[b] [24]	CNRS[c] [26]
Experimental setup	Dynamic observer	X	X	X	X	X	X
	Dynamic environment				X	X	
	Static/man-made environment	X		X			X
	Known landmarks	X	X	X			
	Natural landmarks				X		
	Calibrated cameras				X		
Capabilities	3D reconstruction/ depth mapping		X	X			
	Object identification	X	X	X	X		
	Object tracking		X	X	X		
Methods used	Optic flow			X		X	X
	Motion estimation	X			X	X	X
	IMU data						X
	Template matching	X		X	X	X	

[a] COMETS is a multi-national effort supported by the European Commission
[b] Wallenberg laboratory for research on Information Technology and Autonomous Systems (WITAS)
[c] Centre National de la RecherchĆ Scientifique (CNRS) in France

The long term tracker relies on model matching to follow a specific pattern through the sequence. The Lucas–Kanade tracker is also utilized by Kanade et al. [33] in conjunction with a motion prediction scheme that relies on Kalman filtering, an image pyramid and two dimensional affine motion models to deal with large motion in image sequences taken from a micro-unmanned aerial vehicle.

Motion tracking using low cost off-the-shelf components is also investigated in [38] where a fixed wing UAV is relaying data back to a ground station where the processing takes place. The authors use a proprietary vision system which according to their accounts "lost target track on a regular basis". Tracking salient points is demonstrated in [40] using SIFT features and the RANSAC algorithm where the authors are reporting correct projection in 55.4 to 82.5% of the frames while spending 0.63 to 1.93 s per frame.

Another area where tracking a ground target is important is that of autonomous landing. As shown in [35] the helipad is usually marked by an "H" which is extracted from the images using fixed threshold segmentation and tracked during the landing maneuver by means of second and third order moments of inertia.

6.2.2 Remarks

For completeness it must be stated again that there is a vast portion of the machine vision literature relating to the problems of object identification, tracking, motion estimation that cannot be presented here since this work was not intended to be a literature survey. A more detailed presentation of such techniques can be found in [36].

Lastly, it should also be noted that the problem to be addressed in this chapter can be classified as a dynamic vision problem with a moving camera and moving objects, arguably among the most general and difficult ones [37].

6.3 Restating and Addressing the Problem

In this section the challenges associated with detecting and tracking objects from an unmanned helicopter are described along with the proposed solution. Before doing that, it is necessary to review the definition of the overall problem that is discussed in this chapter. In brief, it can be stated as follows:

Design a vision system for a small autonomous helicopter that will be able to:

- identify arbitrary objects using a minimal description model and a-priori knowledge;
- track objects of interest;
- operate in real-time; and
- operate in a largely unknown, dynamically changing, outdoors environment.

The system will operate under the following constraints:

- Limited processing power and payload.
- Low cost, off-the-shelf components.

The main design directives remain that of real-time execution and low price, high availability components. It is in a sense an investigation for the minimum required hardware and algorithmic complexity to accomplish the desired tasks.

6.3.1 Challenges of Implementing a Vision System for a VTOL

Helicopters are attractive as unmanned vehicles due to their ability to take off from almost anywhere without the requirement of a runway. Furthermore, their ability to hover makes them ideal for surveillance since they can keep the onboard sensors pointed towards the same area without having to execute elaborate maneuvers. The price to pay for that flexibility is their lower speed, limited endurance and inherent instability when compared to fixed wing aircraft. Small unmanned helicopters are

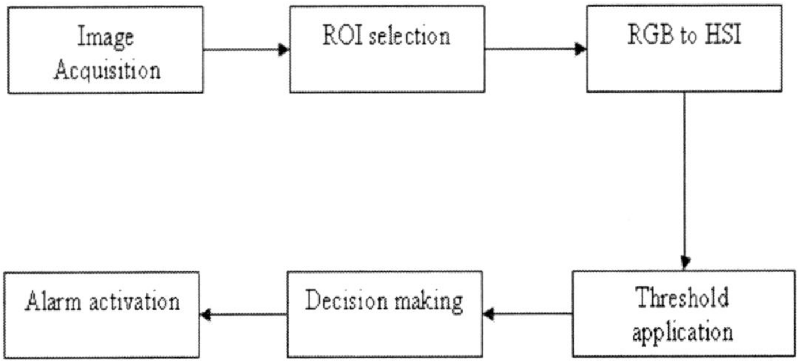

Fig. 6.1 Block diagram overview of the object detection system.

even more unstable and susceptible to even modest changes in environmental conditions. This instability affects the images that any onboard camera would acquire and requires either the use of stabilization hardware which adds weight or stabilization software, which adds in turn complexity and demands more processing power.

The unmanned aerial vehicles that were used for the purposes of this work are designed to operate outdoors. Such an environment is notoriously difficult for computer vision primarily due to variations in lighting. Furthermore, there is limited availability of a-priori knowledge of the environment and certainly no three dimensional map which leaves little room for helpful assumptions to be made.

The low cost requirement and the low payload allow only for a single camera to be carried on-board. A second camera could have been utilized to allow for a stereoscopic system and provide additional data for verification purposes leading to a more robust detection. Its absence can be viewed as an additional design constraint for the vision algorithms.

6.3.2 Module Overview

Taking into consideration the aforementioned design directives and constraints a system was implemented that alerts of the presence of objects of prescribed color. There were two individual implementations of the design, with the second having a better more energy efficient processor and an analog camera with more powerful optics. The computer platforms selected were two different variants of the x86 architecture that fit the profile of low power consumption and off-the-shelf availability. Based around the 1.2 GHz EPIA and later the Pentium M processor, the system processes the data captured by either a firewire IEEE1394 digital camera or an analog color CCD camera connected to a framegrabber for analog to digital conversion. Both camera types are widely available, have a low cost and require little power to operate. The operating system of choice was Linux. A block-diagram overview of the system is given in Figure 6.1. Briefly stated, the system acquires images from either a firewire camera or a framegrabber (see Figure 6.2), selects the regions of

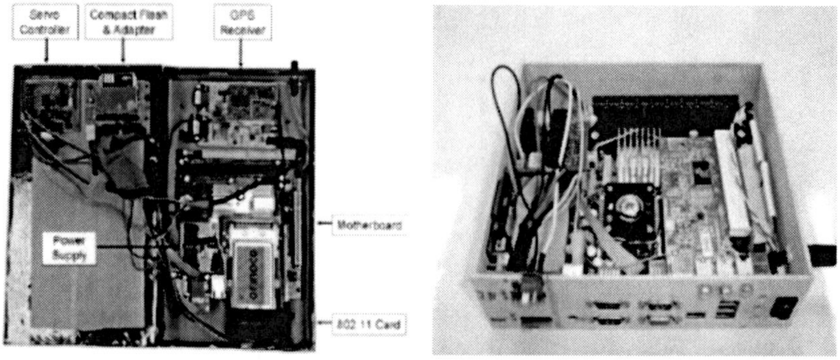

Fig. 6.2 First (left) and second (right) iterations of the on board processing system (assembled and tested by Richard Garcia).

interest on that image, converts them to an appropriate colorspace, selects the areas that meet certain criteria and finally makes a decision on whether to activate an alarm indicating the presence of an object of interest. The overall object detection system consists of the following parts:

- Image acquisition.
- Selection of region of interest (ROI).
- Color conversion.
- Application of thresholds.
- Decision making.

The following sections explain how each of the modules operates.

The image acquisition module reads the image from either a firewire port or a framegrabber. It provides access to the appropriate hardware via the open source libraries libdv, librwaw1394, libavc1394, libdc1394 for the firewire port or video for linux (v4l) for the case of the analog camera attached to the framegrabber. It also allows for the setup of various hardware parameters of the cameras such as white balance, resolution, frame rate etc. For our purposes the frame rate was set at the maximum allowable by the hardware (30 fps).

The region of interest (ROI) selection module is employed as a measure to reduce the overall computational load by reducing the number of pixels that are considered as possibly belonging to the object that we are trying to identify. This selection is based on previous decisions of the system. If an area has caused the decision making module to activate the alarm at frame t then that particular area is selected for the subsequent Delta_t frames while the rest of the image is not. Every Delta_t frames the image is selected in its entirety to allow for the introduction of new areas. It was experimentally determined that a good compromise between computational load reduction and the ability to incorporate new areas for classification occurred when the value of Delta_t was set to 15.

The design requirements called for a minimal description model and limited a-priori knowledge about the object of interest. To satisfy that we relied on the color of the object as the sole feature on which to base the classification. This decision was based on the assumption that if an object is sufficiently different than the environment then the color alone should be adequate to identify it. In their raw form the images are acquired in either YUV or RGB format. Although sufficient for displaying images these colorspaces do not allow for a simple direct definition of color as it is perceived by humans. Furthermore, in the case of the RGB model the information about intensity and color is entangled in the triad of values for the Red, Green and Blue components. The colorspace that was deemed appropriate for the task of providing a simple description of color closer to that of human perception was the Hue, Saturation and Intensity (HSI).

Theoretically, the Hue and Saturation values should not be affected by variations in lighting. This is a very important property especially when operating in an uncontrolled environment subject to varying illumination. The conversion from the RGB to HSI color model is straightforward and described by the following equations [37]:

$$\text{Intensity} = \frac{1}{3}(R + G + B), \tag{6.1}$$

$$\text{Saturation} = 1 - \min(R, G, B), \tag{6.2}$$

$$\text{Hue} = \left\{ \begin{matrix} \theta, B \leq G \\ 360 - \theta, B > G \end{matrix} \right\}, \quad \theta = \cos^{-1} \left\{ \frac{\frac{1}{2}[(R - G) + (R - B)]}{\sqrt{(R - G)^2 + (R - B)(G - B)]}} \right\}, \tag{6.3}$$

where R, G, B are the values for Red, Green and Blue component.

After the conversion is completed the Hue Saturation and Intensity components are forwarded to the threshold application module which selects the pixel(s) that may belong to an object of interest. The object for which the vision system is searching is defined by a series of upper and lower bounds on Hue, Saturation and Intensity. The image is scanned and pixels that fall within those bounds are selected as belonging to the object in question. From the segmented image we extract connected components likely belonging to objects and bounding rectangles are drawn that enclose them. From the list of the extracted areas the ones that fall outside the bounds for size are discarded. Again the main implementation concern is simplicity.

The decision making module is responsible for the final decision regarding the presence of a target. It raises or lowers an alarm signal indicating that something of interest may be present in the image. In detail, if the number of the selected pixels exceeds a threshold, then the decision mechanism classifies the frame as containing an object and increases a counter by a constant value. In the case that nothing is detected the same counter is decreased by a quantity relative to the exponential of its current value. When the value of the counter is greater than a certain threshold then the alarm is raised. This can be viewed as a "leaky bucket" that fills gradually every time a frame is found to contain an object of interest and drains rapidly when an object is not present. The operation of the decision making module may be described in pseudo-code as given in Table 6.3.

The constant C1 is related to the rate at which the counter is increased with each "detection", while C2 controls the descent of the counter's value when an object is

Table 6.3 Pseudo-code for raising the alarm.

IF object==detected THEN counter=counter + C1
Else counter=counter - exp (counter / C2) IF counter>activation threshold THEN alarm=ON Else alarm=OFF.

not present. By selecting those two constants it is possible to tune the behavior of the decision making mechanism in terms of its tendency to raise the alarm. For the same threshold value a larger value for C1 will result in an easier activation of the alarm since the counter will be increased by a larger amount. Similarly, a smaller value of C2 will lead to a steeper descent of the counter when an object is not present resulting in a faster deactivation. Typical values for C1 and C2, found after some experimentation, are 2 and 40, respectively.

To avoid extremely high counter values that would make proper deactivation of the alarm almost impossible, an upper bound, typically 100, is introduced and the counter is not allowed to exceed that bound even if the object of interest is continuously present in the image. There is also a lower bound, usually 0 to 20, which the counter never goes below in order to avoid very low values that would prevent the alarm to be activated properly. In this way a scale from 0 to 100 is created for the values of counter with higher ones corresponding to a higher number of recent frames containing an object of interest.

6.3.3 Complexity

As it is apparent during the design of the system, simplicity has been the primary constraint. This has resulted in an algorithm having to apply a threshold on the pixels of the image, making it of order $O(n^2)$, where n is the dimension of a square image. The main computational burden is posed by the conversion of the image into the HSI color space. More specifically, the calculation of the Hue component of the image includes a call to the inverse cosine and the square root function.

With the incorporation of the region selection mechanism this number is drastically decreased which allows the on-board vision system, despite having less computational power than a ground based computer, to achieve a processing rate of 30 to 80 fps. The region selection algorithm consists of a series of iterations, each of which expands a bounding box around a given pixel. The computational time that it needs depends on the number and the size of the targets present as well as the size of the image. The worst case is again $O(n^2)$, where n the size of the $n \times n$ image. The common case though is to have a small number of regions, usually one or two. Since every region is not allowed to exceed a certain size, if it is to be considered a valid object, it can be said that the computational time is bounded by a constant. The Decision making module is a simple equality test in the on-board system and

Table 6.4 Computational complexity of the modules used by the object identification system.

System Configuration	Modules		
	Conversion to HSI	Region Selection	Decision Making
On-Board system	$O(n^2)$	Constant	Constant

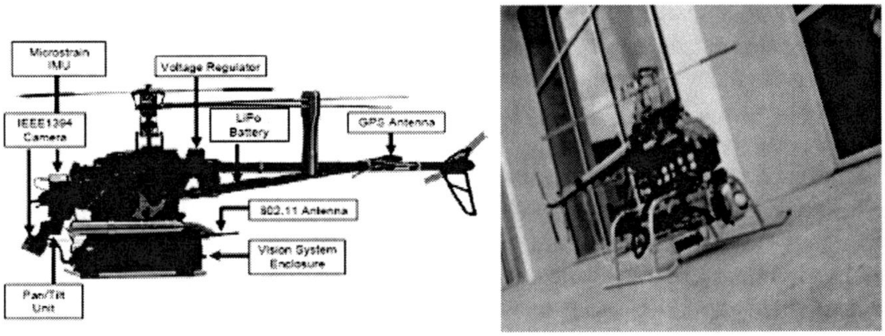

Fig. 6.3 The Raptor 90 SE (left) and the Maxxi Joker 2 (right).

a little more complex leaky-bucket mechanism for the system with the processing unit on the ground. In both cases the delay is negligible and independent of the size of the image. The evaluation of both systems in terms of complexity is summarized in Table 6.4. As it will be shown in the next section its simplicity allows real-time execution while exhibiting fairly robust detection.

6.3.4 Tracking System

Besides being able to detect objects given a minimum description model, the system is also required to be able to track them through time. It is designed in a way that allows the user to designate the object to be tracked during runtime. The main challenges include trajectory prediction as well as occlusion handling and motion segmentation. The demand for low computational cost, real-time execution, and a minimal object description model still applies.

Although the helicopter is perceived to be stationary when hovering and attempting to track ground bound objects, this is rarely the case. Given the relatively small size of both the Raptor 90 and the Maxxi Joker 2 (see Figure 6.3), even slight variations in the wind's direction or speed can result in an unstable flight profoundly influencing the quality of the acquired images. This translates to relatively high disparities between corresponding objects in subsequent image frames. Furthermore, it makes tracking objects close to the boundaries of the image almost impossible because they may appear and disappear erratically due to the relative motion of the camera with respect to them.

Occlusions present a significant challenge when attempting to track a specific object in a dynamically changing environment. Objects, be them robots, cars or anything for that matter, are expected to move almost randomly and therefore occlude each other. The background environment although static can contribute to this problem whenever it includes obstacles comparable in size with the objects of interest. Tree lines for example are such typical obstructions. Also, since the camera is mounted on a moving platform it is possible for any object to become occluded even by remaining stationary. As one might expect, occlusions are greatly reduced in frequency when the optical axis of the camera is perpendicular to the terrain. However hovering directly above the area of interest may not always be feasible or desirable due to safety concerns.

Motion or background segmentation is another challenge due to the nature of the environment that the unmanned vehicle operates in. Typical background extraction techniques such as frame differencing or Kalman filtering do not cope well with rapidly varying scenes. In particular, frame differencing degenerates to a crude edge detector when applied to a sequence of images acquired by a moving camera. On the other hand, motion estimation algorithms like the ones used in the MPEG standard that were also considered, although relatively accurate, found to be highly demanding in terms of processing power. However with dedicated hardware that accelerates MPEG encoding this could be a viable choice for motion estimation.

6.3.4.1 Tracking System Module Overview

Having made a review of the typical problems related to object tracking within the context of small unmanned VTOLs, we now describe the operation of our system along with the modules that constitute it. At first the object to be tracked has to be specified. This information can either come from the object detection system that is automatically searching for a pre-specified object or from a user who activates the tracking system by manually selecting an object in the image. The target selection module then creates a description of the selected object and forwards it to the matching module. The latter will attempt to find a corresponding template in the subsequent image. Once such a match is found the original template that describes the object is updated with information from the most currently acquired frame. Finally the Pan-Tilt controller signals the Pan-Tilt mechanism to move accordingly so that the tracked object is at the center of the image. Briefly stated the tracking system is comprised of:

- the target selection module;
- the matching module;
- The template update module; and
- The Pan-Tilt controller.

A block diagram showing the interconnections between the aforementioned modules can be seen in Figure 6.4. The operation of each of the modules is described in the following paragraphs as well as depicted in Figure 6.5.

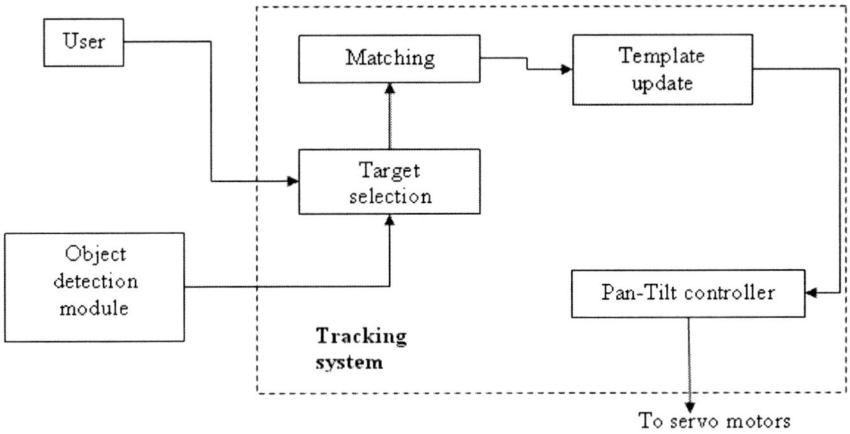

Fig. 6.4 A block diagram of the tracking system.

6.3.4.2 Target Selection Module

This component is responsible for receiving the user's or the object detection system's input and creating a description for the object to be tracked. In both cases the input to this module is a series of coordinates (x, y) representing the object's position in the image's column and row coordinate system as seen in Fig. 6.6. The design choice has been made to employ a Nt x Nt area taken around the image point (x, y) as a template for matching into subsequent image frames.

6.3.4.3 Matching

This module attempts to find a match for the template within the current frame. To reduce the search space for this match the search is limited within a ($N_{\text{search}} \times N_{\text{search}}$) area around the latest previously known position. Assuming continuity of motion for the object under tracking, it is reasonable to expect that it will appear in the next frame relatively close to its current position.

The similarity measure for the match is a normalized correlation coefficient between the template and all the $N_t \times N_t$ square sub-images within the search space. The correlation coefficient $r(d_x, d_y)$ between the template and the region that is displaced by d_x, d_y from the origin of the search space is given by [37]:

$$r(d_x, d_y) = \tag{6.4}$$

$$\frac{\sum[\text{template}(x, y) - \overline{\text{template}}][\text{image}_k(x + d_x, y + d_y) - \overline{\text{image}_k}]}{\sqrt{\sum[\text{template}(x, y) - \overline{\text{template}}]^2 \sum[\text{image}_k(x + d_x, y + d_y) - \overline{\text{image}_k}]^2}}.$$

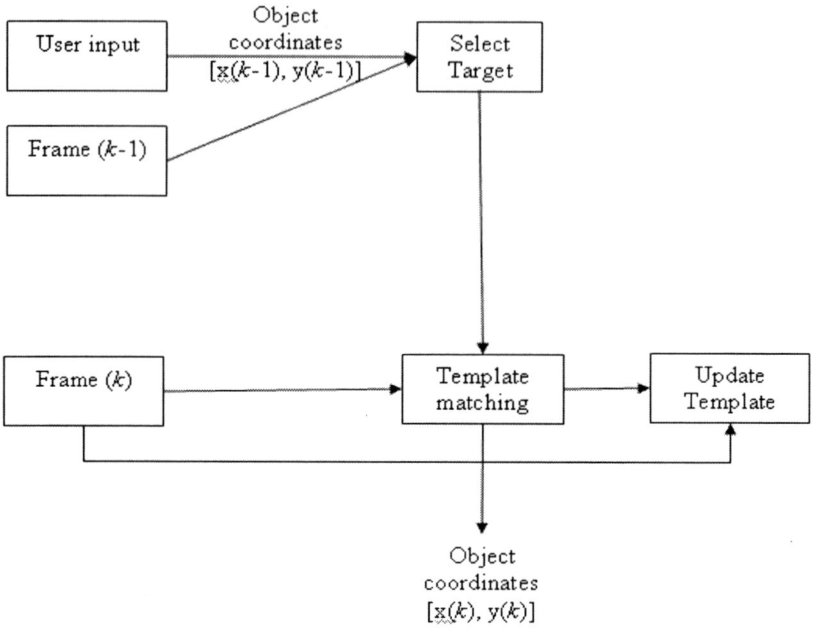

Fig. 6.5 Operation of the tracking systemat a given time k.

The output of this module is the row and column position for which the aforementioned similarity measure is maximized:

$$\arg\max_{dx,dy} (r(d_x, d_y)). \tag{6.5}$$

Alternatively, the sum of absolute differences:

$$\text{SAD}(d_x, d_y) = \sum |\text{template}(x, y) - \text{image}_k(x + d_x, y + d_y)| \tag{6.6}$$

can be used with similar results. In that case the row and column position that minimize the similarity measure become the output of this module:

$$\arg\min_{dx,dy} (\text{SAD}(d_x, d_y)). \tag{6.7}$$

Obviously a trade-off exists when selecting the size of the search space. A large value for N_{search} will allow for a larger disparity between the positions of the object in subsequent frames. The penalty for a larger search space is obviously the extra computational cost which increases with the square of N_{search}. On the other hand, decreasing the search space may save some computing time but it entails the possibility of not finding a proper match just because the object moved more than $N_{\text{search}}/2$ pixels in any direction. A good compromise was achieved by making a

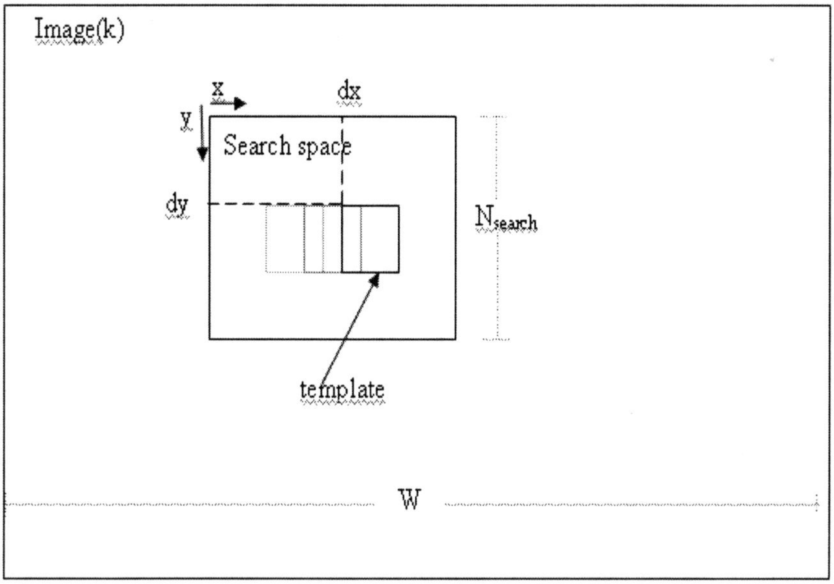

Fig. 6.6 The template matching process.

selection based on the maximum expected apparent motion Max_disparity. For our applications we selected: $N_{\text{search}} = 2 * \text{Max\_disparity} = W/10$, where W is the width of the captured image. Our implied assumption is that the apparent motion of the observed object will not be exhibiting inter-frame displacements of more than $W/20$ pixels. Figure 6.6 shows the relation between the sliding template, the image and the search space.

6.3.4.4 Template Update Module

The typical weakness of a tracking system based on template matching is the fact that with time the template may become irrelevant. Objects are moving and their pose with respect to the camera is almost constantly changing. As a result, the projected two dimensional image of any given object differs significantly within the time span of a few seconds making any attempt for a match with the original template almost impossible.

To mitigate this effect the template is updated at every cycle of the algorithm's execution. Every new captured image, within which a match was found, contributes to the template by introducing information to it thereby forming a new one. The new template is a linear combination of the existing one and the neighborhood of the best match. In most cases the linear combination of the current template and the best matching patch is sufficient to maintain the relevancy of the template without incurring a significant processing power penalty. If Template(k) is the template at

Fig. 6.7 Tracking results for $a = 0.005$.

time k and Match(k) the $N_t \times N_t$ neighborhood around the coordinates of the best match then:

$$\text{Template}(k+1) = a\,\text{Match}(k) + (1-a)\,\text{Template}(k), \quad \text{where } a \in [0, 1]. \quad (6.8)$$

The design decision to be made when calculating this new template is about the amount of new information that will be incorporated versus the amount that will be retained from the "old" template. Apparently there is a trade-off. If one chooses to retain more of the older template the result will be a slower changing template unable to accommodate pose variations that happen within the time span of a few frames.

This, however, will make the template impervious to short duration random illumination variations as well as to the occasional mismatch. On the other hand if the choice is made to aggressively update the template with new information then

Fig. 6.8 Tracking results for $a = 0.155$.

it has a better chance of remaining relevant and being able to cope even with objects whose pose and appearance vary rapidly. The caveat in this case is that the template becomes susceptible to noise and to the fact that even a single mismatch by the matching module can easily throw-off the tracking system by forcing it to follow the new mismatched object. The balance between the new and old information is controlled by the constant a. Lower values of a place more weight on the old template rather than the newly acquired image and higher values of course have the opposite effect. After some experimentation the value $a = 0.025$ was chosen as the one that yielded the best compromise between robustness and adaptation.

6.3.4.5 Evaluating the Template Update Module

To further investigate the way the values of a influence the output of the tracking system, a series of experiments were conducted. Each time, the system was instructed

Fig. 6.9 Tracking results for $a = 0.025$.

to track the same object through the sequence while the value of a remained constant. This was repeated for values of a ranging from 0.005 to 0.995 with a step of 0.01. The sequence selected for that purpose is one that contains a car executing a u-turn maneuver. The choice was made because that particular video excerpt contains an object that changes its pose relatively fast allowing the opportunity to validate the effectiveness of the way the template is updated to accommodate the changing appearance of the target. However, the template must also retain some of the past information to ensure the identity of the target. During the experiment, it was verified that for low values such as 0.005 the template does not adapt fast enough to maintain the track. Figure 6.7 shows exactly that. Conversely, values above 0.155 forced out of the template enough past information so that the tracking system abandoned the initial target as shown in Figure 6.8.

Finally, the value of a for which the system exhibited the desired behavior was 0.025. As shown in Figure 6.9 the vehicle is consistently tracked through the sequence. Having established this trajectory as the ground truth, the root mean square

Table 6.5 Control rules for the Pan-Tilt.

If $\text{error}_x < -N$ then pan left by 5 degrees
If $\text{error}_x > N$ then pan right by 5 degrees
If $\text{error}_y < -N$ then tilt up by 5 degrees
If $\text{error}_y > N$ then tilt down by 5 degrees

(RMS) error was calculated for the ones produced by each of the 100 different values of a. The results are shown in Figure 6.10. Isolating some characteristic values of a and plotting the error for them yields Figure 6.11. One can notice the sharp increases that correspond to the time that the tracking failed.

6.3.4.6 Pan-Tilt Controller

It is usually desirable, if not required, that the tracked object remains in the camera's field of view (FOV). This task is accomplished by the Pan-Tilt controller, which as the name implies, sends control signals to the servos that adjust the pan and tilt angles of the camera. The goal is to keep the tracked object approximately at the center of the captured images. For that reason an error vector is calculated between the center of the image and the image point where the object is located. The vertical and horizontal components of this error vector are then used to adjust the pan and tilt angles so that the error is minimized. The control rules for the Pan-Tilt are shown in Table 6.5. To avoid oscillation and constant corrections the object is kept within a $N \times N$ window centered around the center of the image.

6.4 System Implementation and Performance Evaluation

6.4.1 Applications

After the development of the system we had the opportunity to evaluate its performance in an array of applications. The cases that were chosen for that purpose were:

- Detection of semi-concealed objects such as mines.
- Detection of a group of ground robots.
- Traffic monitoring.

The design was also evaluated on some sequences of images that were slightly out of the scope for which it was originally intended such as infra-red video and car racing footage. As will be shown in the following paragraphs our design coped with all the challenges presented to it while requiring only minor application specific adjustments and modifications.

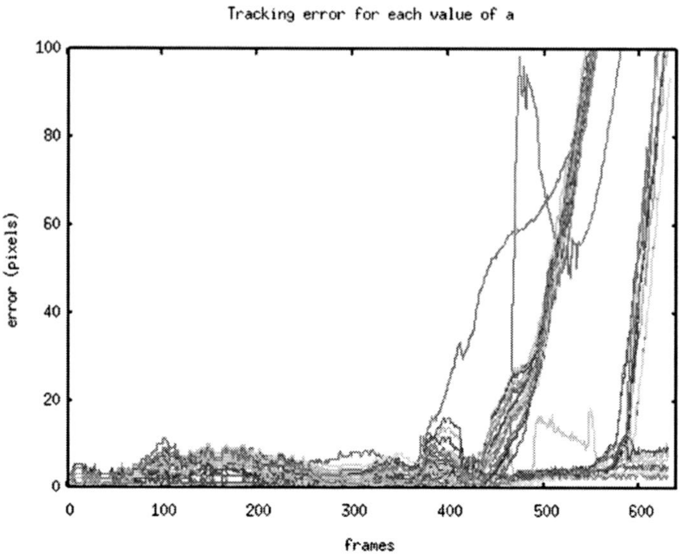

Fig. 6.10 Plot showing the RMS error for each value of a.

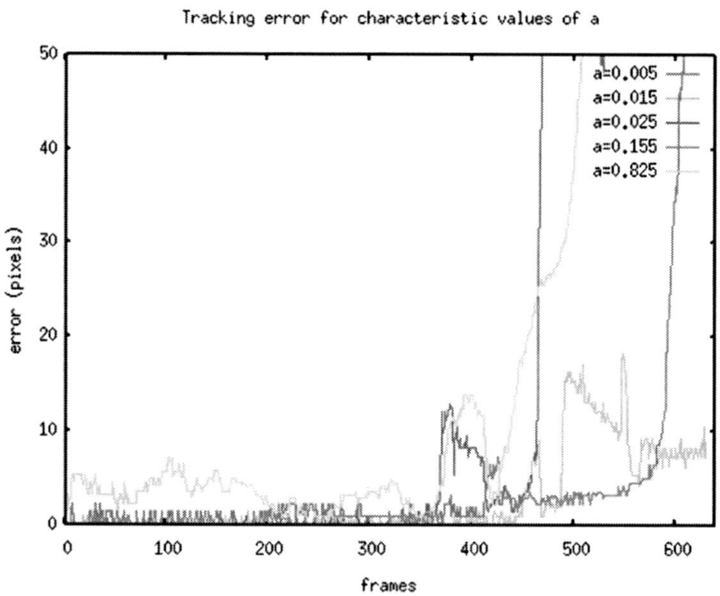

Fig. 6.11 Plot showing the RMS error for some characteristic values of a.

Fig. 6.12 Input image and resulting response by the system. Pixels belonging to the object of interest are highlighted in the second image.

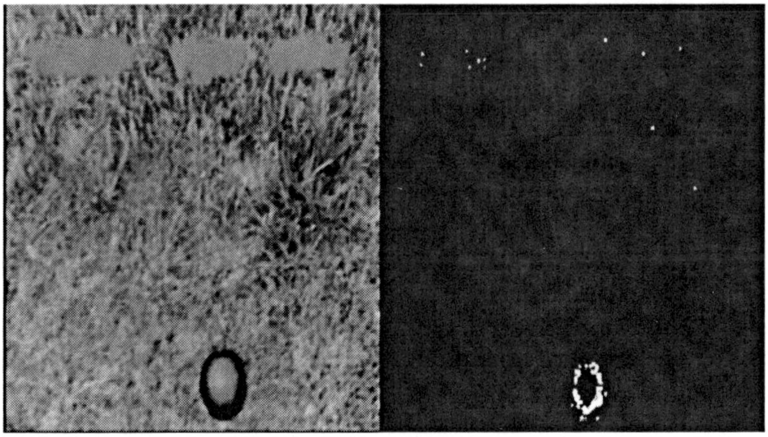

Fig. 6.13 Another input-output pair of images. Notice that some pixels on the top of the image have been misclassified as belonging to the object of interest.

6.4.2 Application 1 – Mine detection

The first application, for which we evaluated the suitability of our design, is based on a hypothetical scenario in which an unmanned VTOL is scouting an area trying to identify semi-buried objects on the ground. The vehicle flew in an unknown outdoors environment with several artificial landmarks that needed be identified. For the purpose of our experiment, various hemispherical and almost spherical black objects were randomly placed, some half-buried in the ground, imitating mine types. The choice of black was made to test the limits of the threshold application module. The latter being based on HSI has a distinct disadvantage when evaluating pixels

Fig. 6.14 Another set of input image and resulting response by the system. Pixels belonging to the object of interest are highlighted in the second image.

with low color content. Furthermore, black is, theoretically, a singularity for the Hue component. Figures 6.12, 6.13 and 6.14 show input images, objects of interest, captured by the camera while the vehicle was flying, as well as the resulting system output. In Figures 6.12, 6.13 and 6.14, pixels belonging to the object are painted white, while all others are black. Shadows and other disturbances had little effect on the overall system performance, since only few frames were misclassified as containing objects of interest. Therefore, the final output of the decision making module was not affected. The angle at which the camera was directed towards the ground had no effect on the system as illustrated in Figure 6.13 which shows a view of the object from directly above. Figuer 6.15 demonstrates a case where the system erroneously classified some image pixels as belonging to the object of interest. The object of interest was intentionally placed in the shadow of a tree in an attempt to test the system's sensitivity to changes in lighting conditions. This can be remedied by the introduction of a module in the algorithm that will compensate for the various lighting conditions. In Figure 6.16 another test under different lighting conditions is demonstrated. The minimum bounding rectangle containing the object is superimposed over the actual input image.

The overall performance was deemed satisfactory with a correct identification rate that ranged between 85 and 90%. False positives were present especially when illumination approached either extreme. In very dark images differentiating a black object from its environment is difficult since pixels not belonging to the object will register as such since their color information is almost lost. In cases were brightness was at or close to maximum, the washout of the colored image caused the algorithm to either miss the object or wrongly identify some other area of the image as being the object of interest. The behavior of the system regarding alarm activation was

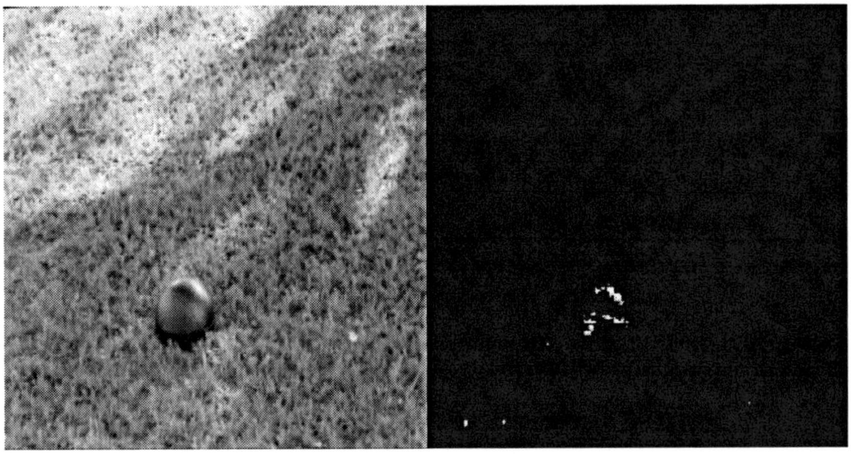

Fig. 6.15 Images that show how the system performs in different lighting conditions. Some pixels are misclassified due to the presence of shadow.

tuned in a way that maximizes activation sensitivity at the cost of false alarms. During the test it was observed that the helicopter would sometimes fly in an erratic pattern due to the prevailing weather conditions. As a result there were cases where the object would come into the camera's field of view and then exit again within a few tens of milliseconds. To compensate for this variability the alarm deactivation was intentionally delayed. This was performed by adjusting the constants C1 and C2 of the decision making module. It allowed the system to achieve a high identification rate while eliminating the effect of oscillating alarm signals.

It is also important to notice that the operation was always in real-time (30 frames per second).

To further evaluate the system's performance a series of flights were conducted. During this experiment the Unmanned Aerial System was tasked to identify and alert for the presence of an object of interest. The latter being a dark spherical object arbitrarily placed in the field on which ground vehicles roamed in a way unbeknownst to the VTOL.

In addition, the object was moved around randomly between flights to ensure that the detection rate was not the result of a particular placement with respect to sunlight and background. A total of five flights were attempted in one of which the VTOL was flown over an area that did not contain the object in question (see Table 6.6). The system succeeded in identifying the target in all other four attempts. Sample instances of its operation are shown in Figures 6.17(a) through (d) where the correctly identified object of interest is enclosed in a minimum bounding rectangle.

Figures 6.17(a) and (b) show images that were taken at a different time of the day than Figures 6.17(c) and (d) as evidenced by the length of the shadow the object casts. These samples demonstrate the system's ability to successfully operate even when a significant component of lighting conditions, such as the direction of the main source of illumination, varies. Another property illustrated by the quartet of

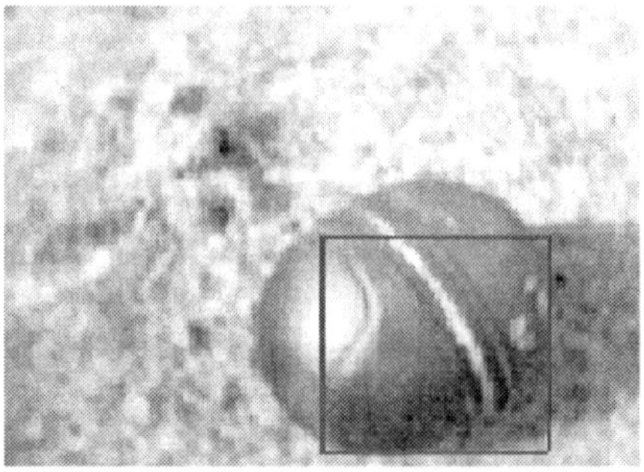

Fig. 6.16 Test image under different lighting conditions. A minimum bounding rectangle superimposed over the area that the algorithm detected the object.

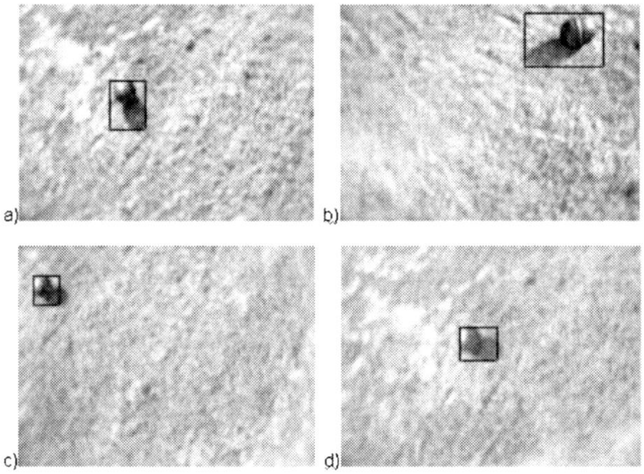

Fig. 6.17 Correct identification of the object of interest under different conditions.

images in Figure 6.17, is that of scale invariance to some degree. Although the aerial vehicle did not maintain a steady altitude through the course of each flight, the object was identified successfully both in instances that it was relatively close to the camera such as in Figure 6.17(b) and in those in which it was further away such as in Figures 6.17(a), (c) and (d).

As expected the vast majority of the processed frames did not contain the object of interest and were correctly rejected. Such a frame is shown in Figure 6.18 where simply the background is seen blurred by a stream of exhaust fumes emanating form the VTOL's internal combustion engine. Unfortunately, there were some cases in which the object was missed as shown in Figure 6.19. False alarms were not entirely avoided as seen in Figures 6.20 through 6.22.

In Figure 6.20 the false alarm shown is produced by the exhaust fumes of the helicopter. When sufficiently concentrated they registered on the camera with a bluish hue that was close to that of the object of interest. Some random background

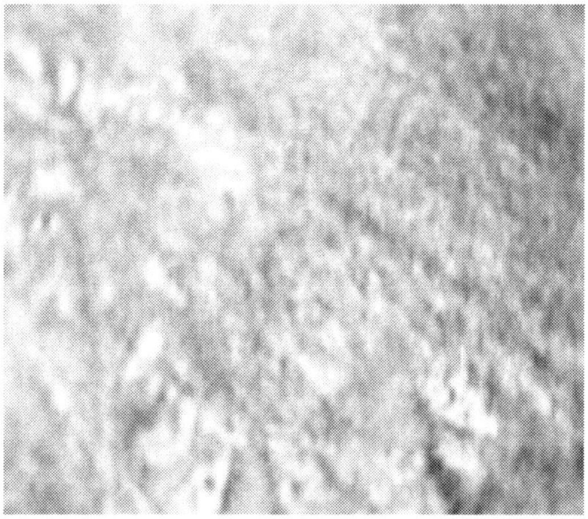

Fig. 6.18 A correctly rejected image showing just the background.

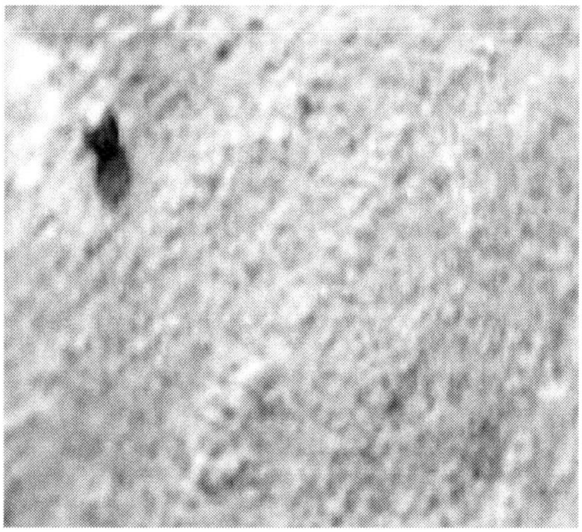

Fig. 6.19 The object is not detected.

formations when clustered together also spurred false alarms such as the one shown in Figure 6.21.

Objects that had components sharing color characteristics with the object of interest also resulted in false alarms to be raised. A typical example was one of the ground robots which had a blue body and black wheels. Under certain lighting conditions part of the body was darkened enough to be close to the hue value of the object of interest.

The typical hit, miss, false alarm and correct rejection rates are presented for each attempt in Tables 6.6 through 6.10. Lastly, the results from all five flights were aggregated to produce the data displayed in Table 6.11. The overall performance of

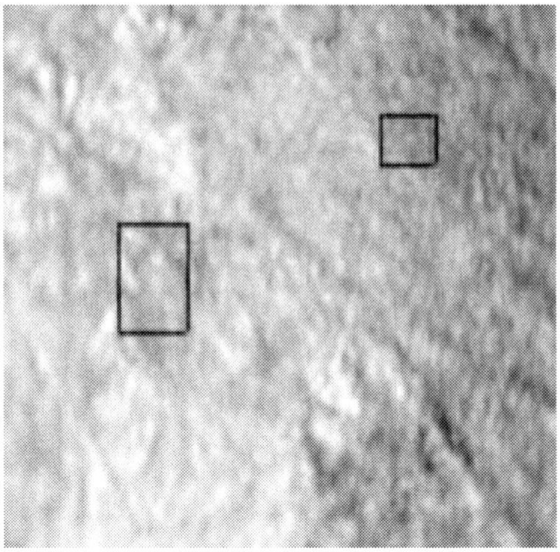

Fig. 6.20 False alarms caused by the presence of the helicopter's exhaust fumes.

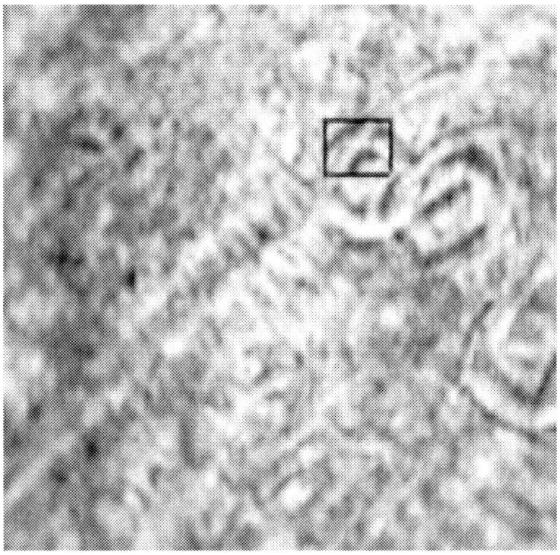

Fig. 6.21 False alarms caused by random background formations.

the system can be characterized as adequate since it exhibits an average detection rate of 89.60% while producing an acceptably low false alarm rate of 6.40%.

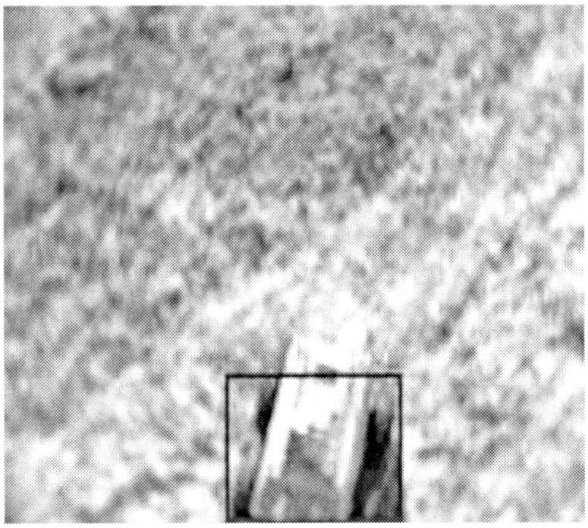

Fig. 6.22 False alarm caused by an object with similar hue to the object of interest.

6.4.3 Application 2 – Tracking a Team of Robots on the Ground

A trend with rising popularity among operators of unmanned vehicles is to use them in teams. Usually there is team of ground robots assigned to perform some task while an aerial vehicle inspects the environment well beyond the range of the sensing capabilities of its ground bound teammates. When operating in such a scenario the UAV should be able to verify the presence of each of the ground robots. To explore the ability of the system to identify multiple targets of different color we arranged a team of four (4) robots and placed them in a typical outdoors environment. The VTOL flew along with the team of the robots and attempted to hover over them.

The images in Figures 6.23 to 6.26 show the results of the identification process where each ground vehicle is enclosed in a bounding rectangle that matches its color. What may not be easily conveyed by the still images is the fact that the helicopter's altitude was fluctuating significantly, constantly changing the point of view of the camera. Despite these abrupt pose changes, the algorithm was able to detect all four of the ground robots.

6.4.4 Application 3 – Traffic Monitoring

6.4.4.1 Introduction

A very interesting application for an unmanned VTOL is that of traffic monitoring. The VTOL is a surprisingly suitable platform since it can hover over a particular traffic node for varying periods of time depending on configuration. A medium sized autonomous helicopter such as the Bergen Observer can hover for a period

Table 6.6 Performance evaluation for flight 1.

	Object present	Object absent
Alarm ON	88.24%	16.23%
Alarm OFF	11.76%	83.77%

Table 6.7 Performance evaluation for flight 2.

	Object present	Object absent
Alarm ON	89.66%	1.79%
Alarm OFF	10.34%	98.21%

Table 6.8 Performance evaluation for flight 3.

	Object present	Object absent
Alarm ON	N/A	2.22%
Alarm OFF	N/A	97.78%

Table 6.9 Performance evaluation for flight 4.

	Object present	Object absent
Alarm ON	100.00%	11.34%
Alarm OFF	0.00%	86.66%

Table 6.10 Performance evaluation for flight 5.

	Object present	Object absent
Alarm ON	77.78%	0.55%
Alarm OFF	22.22%	99.45%

Table 6.11 Performance evaluation for all flights.

	Object present	Object absent
Alarm ON	88.60%	6.40%
Alarm OFF	11.40%	93.60%

of 45 minutes and up to more than an hour and a half if equipped with extra fuel tanks. However the electric powered Maxxi Joker 2 can only provide a 12 to 15 minute hover, although this is expected to increase as new, higher density battery technology becomes available. Furthermore, a traffic monitoring system based on autonomous or semi-autonomous VTOLs can be deployed very rapidly in areas that provide little or no infrastructure. Once deployed, it can provide real-time data to operators on the ground or to traffic simulation models which can then provide more accurate predictions for traffic parameters based on more current observations. The

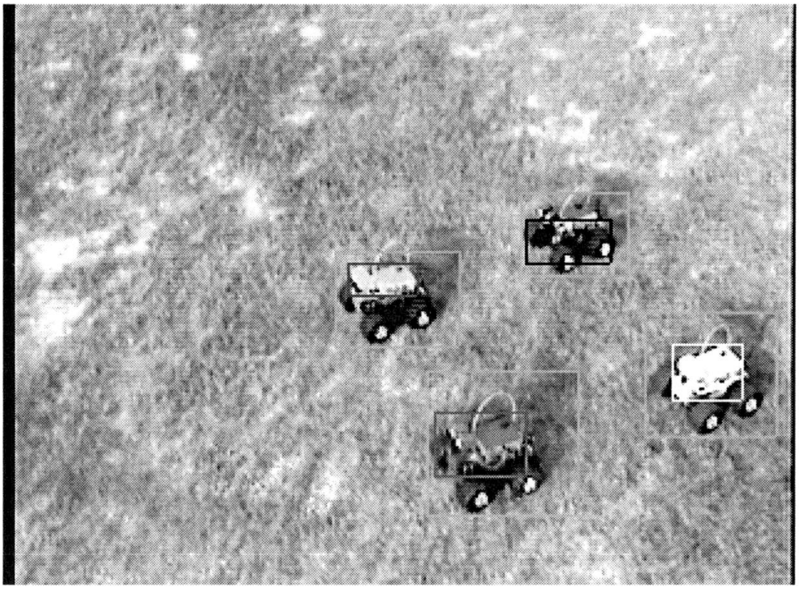

Fig. 6.23 Identification of a team of four UGVs at a close distance.

operator of the system can also direct the system to track a certain vehicle by manually selecting it with a pointing device (e.g. a mouse). Additionally, such a system can serve as an ad-hoc replacement for already existing traffic monitoring infrastructure in case the latter has failed or has been accidentally destroyed. In emergencies such as a hurricane evacuation or a large scale automotive accident an autonomous VTOL deployed from a first responder's vehicle can provide helpful information.

Briefly stated, the autonomous VTOL provides the capability of a traffic monitoring system without the requirement of extensive infrastructure. It has two modes of operation, one automatically extracting traffic data and another tracking manually selected vehicles.

6.4.4.2 Description of the System

As stated above one of the objectives of the system is to extract meaningful real-time data from the video stream captured by the onboard camera on the VTOL. Such data include the total number of vehicles that pass through a given part of the road network, the number of vehicles that follow a certain path and the overall traffic flow. This task can be separated into three distinct steps. Initially the image areas that correspond to vehicles have to be differentiated from the environment. Secondly, the extracted vehicles must be consistently tracked as they traverse the portion of the road network that is being examined. Lastly, the result of the tracking procedure is converted to meaningful traffic measures. A block diagram showing the

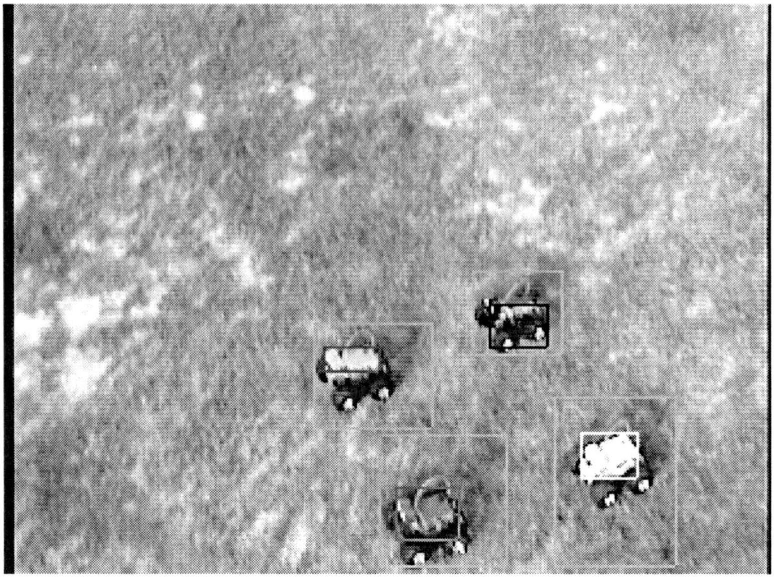

Fig. 6.24 Identification of a team of four UGVs at a medium distance.

Fig. 6.25 Identification of a team of four UGVs when the image is subject to blur.

succession of these steps is shown in Figure 6.27. The following paragraphs provide a more detailed description of the whole process.

The first step towards extracting the desired traffic data involves identifying the vehicles. In keeping with the goals of this work, the vehicle extraction process must

Fig. 6.26 Identification of a team where only three out of four UGVs are visible.

be simple, computationally cheap and requiring minimal a-priori knowledge. The selected method takes advantage of the fact that the areas of the image corresponding to a paved road usually have extremely low saturation values. A simple sufficiently low threshold is then applied to suppress the part of the background that is the road. Following that, a pair of erode/dilate morphological operators are used to further distinguish the blobs that correspond to vehicles. A size filter is employed to avoid some residual areas of the background being categorized as vehicles. In particular, for any formation of pixels to be accepted as a vehicle it has to fit within a bounding rectangle no smaller than $W/10 \times H/10$ and no larger than $W/3 \times H/3$, where W and H are the width and height of the image respectively. Its effectiveness is based on the assumption supported by the observation that areas smaller than $W/10 \times H/10$ usually correspond to noise whereas areas with dimensions larger than $W/3 \times H/3$ are usually representative of various background formations. The application of such a filter enforces a measure of scale on the problem albeit a reasonable one. Figures 6.28 to 6.31 depict the image processing steps leading from the input image to the extracted vehicles.

In succession, a set track$(k-1)$ is constructed containing the center of gravity of all the regions that pass the size filter at a given time $k-1$. This set containing pairs of (i, j) image coordinates is presented as input to the tracking module described in Section 6.3. The output of the tracking module forms a set track that holds the respective matching coordinates at the current frame k. It is worth mentioning that time in this case has been discretized to coincide with the acquisition rate of the camera. So frame k and time k are interchangeable.

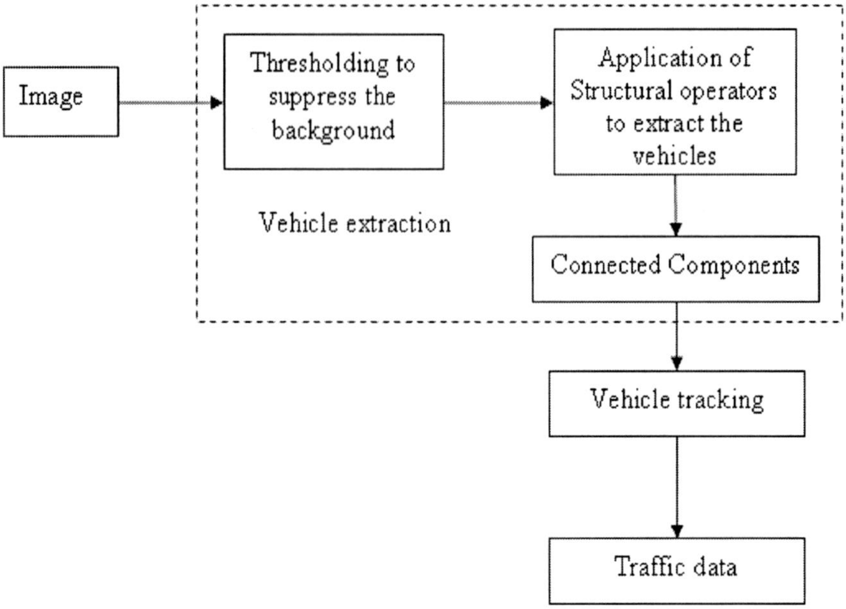

Fig. 6.27 A block diagram representation of the data extraction mode of the traffic monitoring system.

Fig. 6.28 The RGB input image in the left is converted to HSI. The saturation component is shown in the right.

Having completed the vehicle extraction and tracking the remaining task is to calculate the traffic parameters of choice. The number of vehicles currently present on the road is simply equal to number of objects represent in the tracking set. In other words the cardinality of track(k) gives the number of current vehicles. The amount of vehicles over any given period of time can be found by integrating the function of vehicles over time.

Fig. 6.29 The application of a threshold on the saturation component (left) eliminates most of the pixels belonging to the road (right).

Plotting the function of vehicles over time, results in the graph shown in Figure 6.32. The light gray (red) line represents the number of vehicles that the system estimates are currently in view, the black (blue) line is the actual number of vehicles present and the dark gray (green) line shows the execution time in milliseconds. On average the system estimated the current traffic with an accuracy of 81%. For comparison, traffic monitoring systems that are utilizing fixed cameras [42] have reported higher rates of 95 to 97%. They benefit from the more consistent background and motion extraction since the sensors remain static and all apparent motion in the image can only be attributed to vehicles. Note that the execution time remains well below 33 ms which signifies real-time operation. The occasional spikes correspond to write cycles during which the operating system logged the output of the vision program to the disk. They are not due to executing the core vision code and are not present during normal operation. The computer used in this case was a Pentium 4 at 3 GHz.

The system can also function in another mode of operation that relies on user input for target designation. It allows a certain number of manually selected vehicles to be tracked. The operator has only to point and click on the vehicles to be tracked. The image coordinates are gathered and presented as input to the tracking system described in Section 6.3. The block diagram in Figure 6.33 describes this process.

Results of this mode of operation can be seen in Figure 6.34. Figure 6.35 shows some targets being abandoned by the system as they exit the field of view in order to make the resources of memory and processing power available for reallocation to tracking newly acquired targets. Figure 6.36 shows that the tracking can be maintained despite the unpredictable motion of the VTOL and the parallax inducing motion of the vehicles. Further resilience to parallax is demonstrated in Figures 6.37 and 6.38 where vehicles are shown to be tracked while executing turns and u-turns that change their pose with respect to the camera.

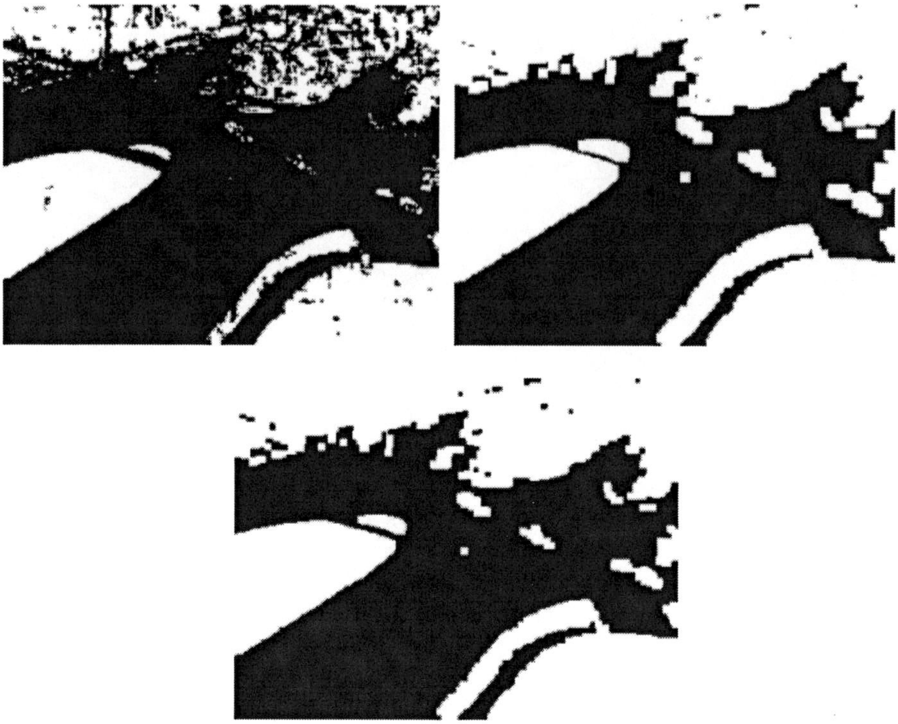

Fig. 6.30 Applying morphological operators to the binary image (upper left). Dilation (upper right) then erosion (lower center).

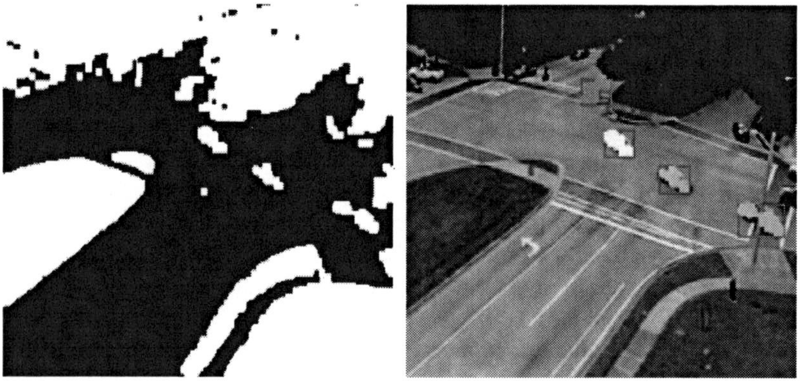

Fig. 6.31 Extraction of regions using connected components and a size filter. Notice the rejection of very large and very small blobs.

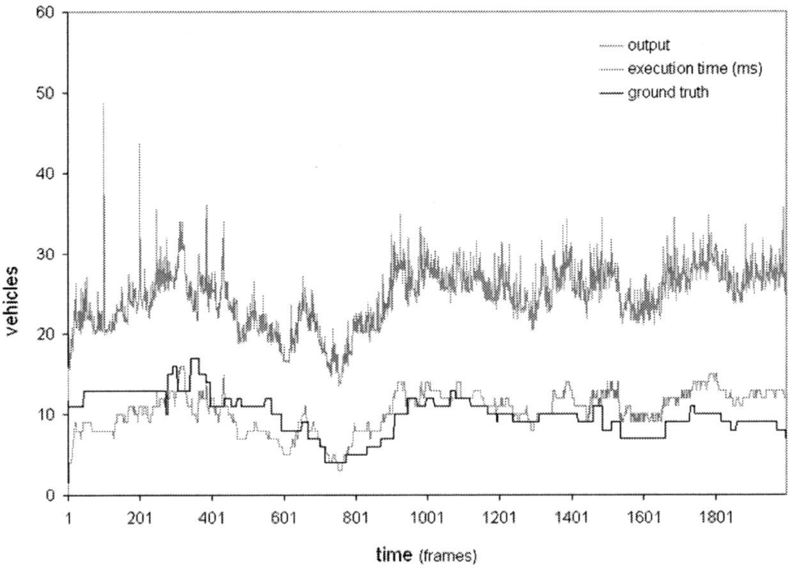

Fig. 6.32 The output of the traffic load estimator compared to the ground truth.

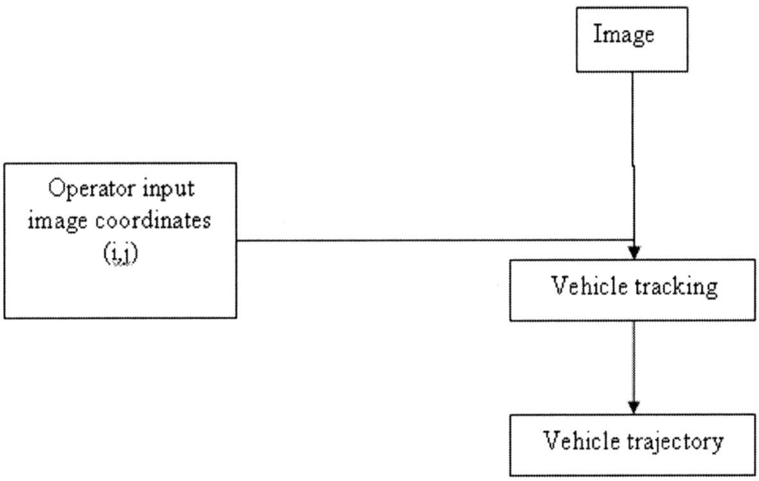

Fig. 6.33 A block diagram representation of the manual target designation mode of the traffic monitoring system.

Fig. 6.34 Tracking of multiple manually selected vehicles.

6.4.5 Additional Testing

To further examine its capabilities and limitations, the proposed design was also tested on an image sequence recorded by an infra-red aerial camera that was well outside the operational envelope as delimited by the scale of objects it can track and the disturbance for which it can compensate.

The infra-red video, found on the world wide web, depicts some buildings and static fixtures as well as some moving vehicles. Tracking the latter proved especially challenging because the apparent size of the moving objects is near the limits of the scale for which the system was designed. In other words the moving targets are too small to provide the template with enough information. Furthermore, there are some instances were the movement of the camera becomes so sudden and abrupt that the image itself is blurred for several frames (see Figure 6.39) resulting in an unrecoverable loss of track as seen in Figure 6.40. Figure 6.41 shows the tracking of several static, background objects while Figure 6.42 depicts the result of the vehicle tracking.

Fig. 6.35 Tracking is terminated when targets, such as the two vehicles at the right side of the frame, exit the field of view.

Fig. 6.36 Tracking is maintained despite the unpredictable motion of the VTOL and the parallax inducing motion of the vehicles.

Fig. 6.37 Tracking of vehicles executing maneuvers such as turns and u-turns.

6.5 Conclusion

6.5.1 Introduction

This chapter presented a system for object identification and tracking to be used by unmanned aerial systems. The object detection part is based on image thresholding and exploits the inherent theoretical invariability to lighting conditions of HSI's Hue component. The final decision to raise the alarm signifying the presence of an object of interest is made after the latter has been detected in sufficiently many recent frames.

Also in this work, a tracking system has been presented. It has been based on template matching using the sum of absolute differences as a similarity measure. A template update occurs in every loop of the algorithm by incorporating new information so that the former remains relevant allowing the track to continue despite changes in the object's appearance. Furthermore it allows for manual entry of targets by the human operator as well as receiving input from other image processing modules. The following paragraphs further discuss the results and contributions of this work as well as provide some final remarks.

Fig. 6.38 Tracking of a vehicle making a u-turn.

6.5.2 Discussion of Results

The proposed design was evaluated in several different scenarios and application contexts including detection of semi-concealed objects, robot teams and traffic monitoring. It performed adequately in all of them exhibiting a high 89.60% detection rate, for the object identification missions, while suffering an acceptable 6.40% false alarm rate.

Although color has proved a fairly good descriptor for objects, it may not be sufficient and has to be supplemented by features such as shape and texture. Most of the false alarms were due to objects of similar hue to that of the object of interest while most missed detections can be attributed to specular reflections and shadows. In an outdoors environment reflections are beyond the control of the designer of the system and can occur at any time given the fact that the observer (the camera carrying flying platform) moves with respect to the main lighting source (the sun). When

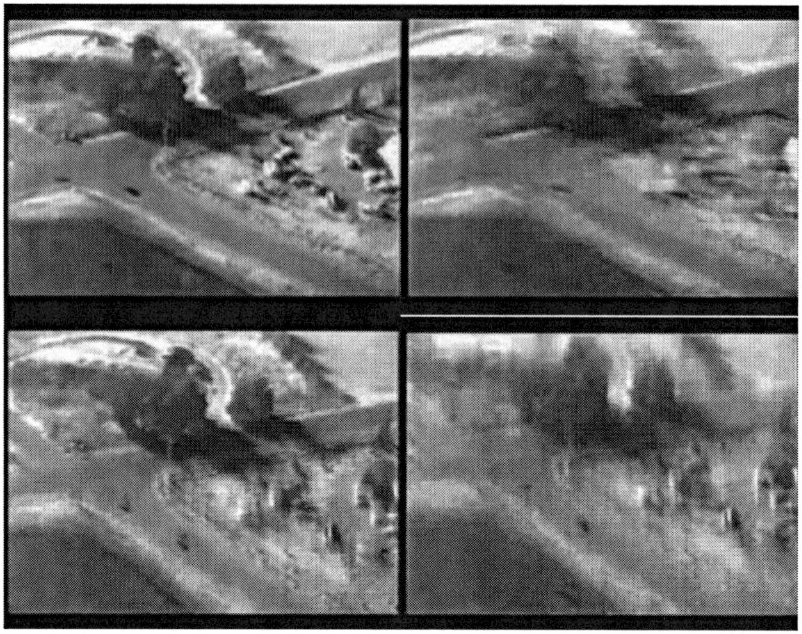

Fig. 6.39 Images blurred by the abrupt motion of the carrying platform. Upper left image is unaffected by blur for comparison.

Fig. 6.40 Severely abrupt motion results in image blur and loss of track.

Fig. 6.41 Tracking of static objects in an IR sequence.

Fig. 6.42 Tracking vehicles in an IR sequence.

enough of the object's surface appears to be covered by these specular reflections it becomes impossible to detect because the color information is almost lost for that area. Lastly, another parameter affecting the performance is the scale within which the system operates. As the VTOL's altitude varies during the flight the apparent size of the observed objects can change significantly. Since the design relies on an assumption about the apparent size of objects, misses and false alarms may occur if the object appears smaller than the lower or larger than the upper limit for size. When fixed optics are used, the scale is closely associated with the physical size of the object and the altitude at which the VTOL flies. Variable optics can provide some independence from those but again only within the mechanical limits of the zoom mechanism.

The tracking system developed in this work was able to correctly track on average 81% of the visible vehicles when assigned to the task of traffic monitoring. It was shown capable of coping with significant disturbance, resulting from the nature of the carrying platform and the unstabilized camera, as well as with occasional brief occlusion and change of pose of the tracked object. However, there are specific situations that are beyond the ability of the system to compensate for. A severe disturbance that results in an apparent motion of more than $W/20$ between successive frames, where W is the width of the frame, will also result in a lost track. For an image acquisition rate of 30 fps, this apparent motion amounts to $1.5W$ pixels/sec which is arguably a high enough limit. Reversely, assuming that the camera is perfectly stable, only objects able of covering the largest visible dimension in 2/3 of a second will avoid being tracked. Another case where the track can be lost is when an occlusion occurs that covers more than 50% of the template corresponding to the object being tracked. In such a circumstance, within a few repetitions of the template update process the template will lose enough information to make finding the proper match problematic.

6.5.3 Discussion of Contributions

This work contributes to the area of vision systems for unmanned aerial systems by proposing a monocular, uncalibrated, not gyro-stabilized design capable of identifying and tracking multiple objects in real-time, without strict assumptions about their trajectories, by relying on minimal information about the environment while forgoing the need for motion compensation usually requiring an inertial measurement unit (IMU). By selecting a monocular, non-stabilized system the design avoids the extra cost and weight of a binocular configuration along with that of a gyro-stabilized turret. The fact that the camera does not require calibration for the system to operate, further simplifies the assembly and setup process.

The ability to track multiple objects at a low computational cost allows the design to run in real-time on inexpensive, less powerful, power efficient computing platforms. This is especially critical for relatively small unmanned with limited payload capabilities for which carrying a large energy reserve to support a powerful computer is impossible. Although the prevailing trends in computer design and battery chemistry promise to alleviate this problem by providing more efficient, powerful hardware and higher energy density batteries respectively, the desire for the minimalist's approach in the design of a vision system will most likely continue to be relevant. This is especially true for the class of unmanned aerial systems that are even smaller than the Maxxi Joker 2 platform that was used in this work.

By keeping the requirement of knowledge about the environment to a minimum, the design is not explicitly required to perform elaborate background extraction thus further reducing the computational burden. Only the apparent size and color of the object of interest are required for the identification process. No a-priori known landmarks are used to facilitate detection and tracking and minimal assumptions, such as the one used in the traffic monitoring scenario about the tarmac being gray, are made. This allows for a fairly versatile system capable of performing in different environments with only a few, if any, modifications.

6.5.4 Final Conclusion

In summary, this work produced a design suitable for object identification and tracking tasks in real-time. It demonstrated that it is possible to endow relatively small, inexpensive VTOLs with capabilities usually reserved for more expensive, higher end unmanned systems.

6.6 Future Research

6.6.1 Introduction

In this section the discussion revolves around the possible future research related to the work presented so far. Several methods exist that require investigation as to their ability to provide better solutions to the central problem presented in this chapter. Some of them focus on improving the individual modules of the existing system while others require major revisions of the approach described here. In the following paragraphs a selection of the potentially beneficial modifications is presented. They have been chosen for their potential to improve the performance of the system while adhering to the real-time operation mandate.

6.6.2 Adaptive Thresholds and Constants

The current system relies on a series of manually selected thresholds and constants in order to perform the tasks of recognition and tracking. One logical approach to enhance its operation is to have these values being selected automatically by the algorithm itself. Implementing an efficient algorithm capable of reliably varying the thresholds in the HSI space that are associated with the object of interest will allow for operation under extremely varied lighting conditions. Several methods already exist that are based on local or global statistics of the intensity values in the image. Furthermore, the information from the GPS sensor that a typical UAS carries can be used to estimate the position of the sun and therefore the direction of the illumination. This will permit the disambiguation between the object of interest and its shadow as well as the shadows cast by other objects present in the scene.

6.6.3 Active Vision

The availability of relatively inexpensive, lightweight cameras with electronically controlled optics offers the opportunity to further explore the problem of target identification and tracking from a flying platform which already falls firmly in the area of active vision.

Varying the focal length in a known manner can facilitate the calculation of the essential or fundamental matrix for each camera. This will allow for better reconstruction of the three dimensional space leading to improved reliability for the trajectory predictions made by the system. Furthermore, varying optics can be employed by an attention focusing mechanism to extract more information in cases where disambiguation between objects is required. An intelligent mechanism of that sort could zoom in to areas that present some difficulty such as between objects that are too close or even when one partially occludes the other. Alternatively the camera can be instructed to zoom out in order to acquire a wider area and provide information about some global reference points or landmarks. Generally it can provide the object identification algorithm with information at different levels of scale. This zooming capability can also be of significant help to the human operator who can employ it to further investigate an object of interest without interfering with the flight of the VTOL.

6.6.4 Consistent Target Selection

As previously mentioned the operator of the unmanned system can manually select objects for tracking. To accomplish that the images captured by the camera on-board the flying platform must be transmitted to a ground station. As with any transmission there is some delay involved. Care should be taken ensuring that the delays caused by the communications channel are such that at any given time the same frame is available to both the human operator on the ground and to the algorithm running on the VTOL. Preferably, the time to transmit a frame to the ground station and have it displayed for the operator to see should not be more than the time elapsed between two consecutive frame acquisition cycles of the camera. Typically, this amounts to 1/30 seconds. Exceeding that can cause the two image sequences to be out of synchronization. As a result, the image coordinates chosen by the operator may not correspond to the intended object in the frame that the algorithm on the VTOL is currently examining. A mechanism should be put in place that detects this event and selects the appropriate object for tracking.

Time stamping the captured frames provides a straightforward solution to the problem. Every frame is given a monotonically varying number that determines its relative position in the stream and can be used to identify the frame in which the operator designated the intended object for tracking.

However, time stamping the captured frames is not always possible. Sometimes the video is transmitted to the ground through a channel that bypasses the computer on board the VTOL. In this case, it is safer to assume that there is always a discrepancy between the image viewed by the operator and the one presented to the algorithm. The problem now becomes one of finding an area in a buffer of recently captured images that matches the one selected by the operator.

6.6.5 Multiple UAVs

Employing a team of unmanned aerial systems instead of just one can improve overall effectiveness in several different ways. Firstly, using several UAS can extent the operational time of the system simply by using one flying platform at a time to cover the same area. Even this straightforward approach presents some planning problems. It requires a decision making mechanism to direct the lift-off and land cycles of the helicopters. Secondly, a team of VTOLs can cover a far greater area than a single unit could. Depending on their arrangement it is possible to have overlapping areas of coverage for purposes of redundancy. Having a specific area covered by multiple cameras can also help resolve some occlusion problems for the simple reason that an object may be occluded when viewed from a specific point in 3D space but visible from another. In general, the deployment of the aerial team can be seen as seeking a 3D configuration that optimizes any set of parameters such as area coverage, occlusion resolution and endurance under possible constrains of no fly areas imposed by the presence of immovable obstacles, forbidden by regulations or denied by an adversary.

For any deployment of the UAS team that allows overlapping areas establishing correct correspondences is important and remains a significant open problem. Regardless of the existence of overlap, the system must be able to maintain a coherent registry of the objects being tracked. When an object leaves the field of view of one sensor and enters that of another it should be identified and tracked by the system as being the same object. This is known as the hand-off problem and it is central in multi sensor distributed systems. To accomplish that some prediction and projection of every object's path should be available either centrally at a ground based coordinating node or distributed to the computer on-board each of the autonomous VTOLs. Every flying platform should be able to expect an area of possible appearance for each of the objects currently tracked by another team member or at least for those that are more probable, based on their current and predicted trajectory, to migrate between fields of view (areas of coverage). For the predicted trajectories to be relevant they have to be on a common coordinate system. Since the position of the observers varies with time, excellent localization is required for the flying platforms. In summary, the problem can be seen as one of tracking multiple targets given a multiple image, time varying, uncertain geometry.

6.6.6 Improved Hardware/Different Platforms

One trend that the computer science community is contributing to and has come to rely on is that of improving hardware. Central processing units with ever increasing processing power and are becoming available for lower cost. The power consumption penalty usually associated with increased performance has lately been mitigated by the manufacturers' efforts to improve efficiency. The result is a series of powerful, low consumption processors that can be used in mobile platforms such as unmanned vehicles. This can allow the proposed vision system to operate in real-time on smaller, more efficient processing platforms thus endowing even smaller VTOLs with object identification and tracking capabilities. Conditional on the availability of low power multi core systems, a possible future implementation can also explore the parallelization of each component which will further increase the number of objects that can been tracked simultaneously. During the evaluation process, the system described in this chapter exhibited some hints of versatility by being able to operate in different environments and in a scenario unrelated to the initial purpose for which it was designed. This warrants further investigation with the ultimate goal being the development of a generalized algorithm capable of operating equally well on a wide variety of vehicles such as ground, aerial, marine and submarine.

Acknowledgment

This research has been partially supported by NSF Grant, IIP-0856311 (DU Grant number 36563).

References

1. Unmanned Systems Roadmap: 2007–2032, Office of the Secretary of Defense, December 2007.
2. Vision Hardware, University of Southern Calafornia, online 2004, http://www-robotics.usc.edu/~avatar/vision_hw.htm (Accessed 27 January 2005).
3. L. Mejias, S. Saripalli, G. Sukhatme and P. Cervera, Detection and tracking of external features in an urban environment using an autonomous helicopter, Paper presented at IEEE International Conference on Robotics and Automation, 2005.
4. S. Hrabar and G.S. Sukhatme, A comparison of two camera configurations for optic-flow based navigation of a UAV through urban canyons, in *Proceedings of IEEE/RSJ International Conference on Intelligent Robots and Systems*, pp. 2673–2680, September 2004.
5. V. Sastry, Vision based detection of autonomous vehicles for pursuit evasion games, Paper presented at 15th Triennial World Congress, Barcelona, Spain, 2002.
6. R. Sattigeri and A.J. Calise, An adaptive approach to vision based information control, in *Proceedings of Guidance, Navigation and Control Conference*, No. AIAA-2003-5727, Austin, TX, August 2003.
7. S. Hrabar and G.S. Sukhatme, Omnidirectional vision for an autonomous helicopter, in *Proceedings of IEEE International Conference on Robotics and Automation*, pp. 3602–3609, 2003.
8. S. Saripalli, J.F. Montgomery and G.S. Sukhatme, Vision-based autonomous landing of an unmanned aerial vehicle, in *Proceedings of IEEE International Conference on Robotics and Automation*, pp. 2799–2804, May 2002.
9. B.R. Woodley, H.L. Jones, E.A. LeMaster, E.W. Frew and S.M. Rock, Carrier phase GPS and computer vision for control of an autonomous helicopter, Paper presented at ION GPS-96, Kansas City, Missouri, September 1996.
10. J. Debitetto, Modeling and simulation for small autonomous helicopter development, Paper presented at AIAA-1997-3511 Modeling and Simulation Technologies Conference, New Orleans, LA, August 1997.
11. B. Woodley, H. Jones, E. Frew, E. LeMaster and Stephen Rock, A aontestant in the 1997 International Aerial Robotics Competition, online, http://sun-valley.stanford.edu/papers/WoodleyJFLR:97.pdf (Accessed 20 December 2008).
12. N. Holifield, J. Lallinger and G. Underwood, International Aerial Robotics Competition 2004, University of Texas at Austin IEEE Robot Team Aerial Robotics Project June 1, online, 2004, http://iarc1.ece.utexas.edu/ lynca/final_documentation/utiarc2004.pdf (Accessed 10 June 2006).
13. W. Burleson, W. Salhany and J. Hudak, Southern Polytechnic State University Autonomous Remote Reconnaissance System, online, 2006, http://a-robotics.spsu.edu/SPSU_paper2005.pdf (Accessed 10 June 2006).
14. M. Musial, U.W. Brandenburg and G. Hommel, MARVIN – Technische Universität Berlin's flying robot for the IARC Millennial Event, in *Proceedings of Symposium of the Association for Unmanned Vehicle Systems*, Orlando, FL, USA, 2000.
15. RMAX Type I/ Type II G, Yamaha, online, 2004, http://www.yamahamotor.co.jp/global/business/sky/lineup/rmax/index.html (Accessed 27 January 2005).
16. E. Johnson and S. Mishra, Flight simulation for the development of an experimental UAV, Paper presented at AIAA Modeling and Simulation Technologies Conference and Exhibit, 5–8 August, 2002.
17. A. Proctor, B. Gwin, S. Kannan, A. Koller, H. Christophersen and E. Johnson, Ongoing development of an autonomous aerial reconnaissance system at Georgia Tech, online, 2004, http://controls.ae.gatech.edu/gtar/iarcpapers/git2004.pdf (Accessed 1 March 2009).
18. E.N. Johnson, A.J. Calise, A.R. Tannenbaum, S. Soatto, N. Hovakimyan and A.J. Yezzi, Active-vision control systems for complex adversarial 3-D environments, A tutorial, in *Proceedings of the American Control Conference*, 2005.

19. Enhanced Vision System (EVS) overview, CMC Electronics, online, 2001, www.cmcelectronics.ca/En/Prodserv/Commav/commav_evs_overview_en.html (Accessed 22 January 2005).
20. E. Johnson, P. DeBitetto, C. Trott and M. Bosse, The 1996 MIT/Boston University/Draper Laboratory autonomous helicopter system, in *Proceedings of the 15th Digital Avionics Systems Conference*, 1996.
21. J. Groven, E. Holk, C. Humbert, J. Krall and D. Schue, Rose–Hulman Institute of Technology autonomous helicopter for the 2004 International Aerial Robotics Competition.
22. J. Chapuis, C. Eck, H.P. Geering, R. Mudra, B. Schneuwly and R. Sommerhalder, The Swiss Entry into the 1996 International Aerial Robotics Competition, in *AUVSI'96 Proceedings*, Orlando, FL, pp. 947–953, July 1996.
23. O. Amidi, An autonomous vision-guided helicopter, M.A. Thesis, Carnegie Mellon University, 1996.
24. K. Nordberg, P. Doherty, G. Farneback, P.-E. Forssen, G. Granlund, A. Moe and J. Wiklund, Vision for a UAV helicopter, in *2002 IEEE/RSJ International Conference on Intelligent Robots and Systems, IROS 2002, Proceedings Workshop WS6 Aerial Robotics*, Lausanne, pp. 29-34, 2002.
25. A. Ollero, J. Ferruz, F. Caballero, S. Hurtado and L. Merino, Motion compensation and object detection for autonomous helicopter visual navigation in the COMETS system, in *Proceedings of IEEE International Conference on Robotics and Automation*, New Orleans, LA, USA, April 2004.
26. F. Ruffier and N. Franceschini, Visually guided micro-aerial vehicle: Automatic take off, terrain following, landing and wind reaction, in *Proceedings of IEEE International Conference on Robotics and Automation, ICRA'04*, April 26–May 1, Vol. 3, pp. 2339–2346, 2004.
27. B. Ludington, J. Reinmann and G. Vachtsevanos, Target tracking and adversarial reasoning for unmanned aerial vehicles, in *Proceedings of Aerospace Conference*, 3–10 March, IEEE, pp. 1–17, 2007.
28. L. Mejias, S. Saripalli, P. Campoy and G. Sukhatme, Visual servoing of an autonomous helicopter in urban areas using feature tracking, *Journal of Field Robotics* **23**(3/4), 185–199, 2006.
29. L. Mejias, J.F. Correa, I. Mondragon and P. Campoy, COLIBRI: A vision-guided UAV for surveillance and visual inspection, in *Proceedings of IEEE International Conference on Robotics and Automation*, Rome, Italy, pp. 2760–2761, 2007.
30. D.H. Douglas and T.K. Peucker, Algorithms for the reduction of number of points required to represent a digitized line or its caricature, *The Canadian Cartographer* **10**(2), 112–122, 1973.
31. W. Bell, P. Felzenszwalb and D. Huttenlocher, Detection and long term tracking of moving objects in aerial video, 1999.
32. J. Shi and C. Tomasi, Good features to track, in *Proceedings of IEEE Conference on Computer Vision and Pattern Recognition*, pp. 593–600, 1994.
33. T. Kanade, O. Amidi and Q. Ke, Real-time and 3D vision for autonomous small and micro air vehicles, in *Proceedings of 43rd IEEE Conference on Decision and Control, CDC 2004*, December, 2004.
34. J.D. Anderson, D.-J. Lee, B. Edwards, J.K. Archibald and C.R. Greco, Real-time feature tracking on an embedded vision sensor for small vision-guided unmanned vehicles, in *Proceedings of International Symposium on Computational Intelligence in Robotics and Automation*, CIRA, pp. 55–60, 2007.
35. S. Saripalli, J.F. Montgomery and G.S. Sukhatme, Visually guided landing of an unmanned aerial vehicle, *IEEE Transactions on Robotics and Automation* **19**(3), 371–380, June 2003.
36. Keith Price Annotated Computer Vision Bibliography, University of Southern Calafornia, online, 2008, http://iris.usc.edu/Vision-Notes/bibliography/contents.html (Accessed 23 April 2008).
37. R. Jain, R. Kasturi and B.G. Schunck, *Machine Vision*, McGraw-Hill, 1995.
38. V.N. Dobrokhodov, I.I. Kaminer, K.D. Jones and R. Ghabcheloo, Vision-based tracking and motion estimation for moving targets using small UAVs, in *Proceedings of the 2006 American Control Conference*, Minneapolis, Minnesota, USA, June 14–16, pp. 1428–1433, 2006.

39. M. George and S. Sukkarieh, Camera aided inertial navigation in poor GPS environments, in *Proceedings of Aerospace Conference, IEEE*, pp. 1–12, 2007.
40. I.F. Mondragon, P. Campoy, J.F. Correa and L. Mejias, Visual model feature tracking for UAV control, in *Proceedings of IEEE International Symposium on Intelligent Signal Processing*, pp. 1–6, 2007.
41. Open Computer Vision Library, online, 2008, http://sourceforge.net/projects/opencvlibrary, (Accessed 15 December 2008).
42. Wan, C.L.; Dickinson, K.W., "Computer vision and neural networks for traffic monitoring, in *Road Traffic Monitoring*, IEE Conf. Pub. 355, p. 143, 1992.

Chapter 7
Coordination of Helicopter UAVs for Aerial Forest-Fire Surveillance

K. Alexis, G. Nikolakopoulos, A. Tzes and L. Dritsas

Abstract In this article, a system using Unmanned Autonomous Quadrotor Helicopters (UqHs) for forest fire surveillance is presented. Quadrotor helicopters (equipped with inertial navigation systems, GPS, RF-transceivers and cameras) fly over the expanding perimeter of the fire and are used as the fire-front sensing systems. These helicopter-units patrol in a continuous manner and agree on varying rendez-vous with each other. These rendez-vous are set as waypoints in order to divide the surveyed fire-front length at equidistant portions. After the rendez-vous-meetings between these helicopters, and the anticipated spread of the fire, new temporal-spatial rendez-vous points are predicted and transmitted to each UqH. These rendez-vous waypoints rely heavily on the fire-front spread algorithm (wind condition, moisture, fire fuel and the firefighters' reaction. For more accurate observation, these UqHs fly at low-altitude and are prune to sudden wind-gusts. These wind-gusts can be detrimental not only to the flight performance of these units but also to their overall stability. In each helicopter, a back step propagation unit along with a stabilizing controller relying on the Constrained Finite Time Optimization Control (CFTOC) scheme is used. The CFTOC is responsible for attenuating the effects of the random wind-gust disturbances on tracking of the a priori reference trajectory. Simulation results are used to investigate the efficiency of the suggested concept.

7.1 Introduction

The scientific area of Unmanned Aerial Vehicles (UAV) fleets has received significant attention over the last few years. Recent technological advantages in UAV's autonomy in conjunction with the theoretical improvements in the field of position-

K. Alexis, G. Nikolakopoulos, A. Tzes · L. Dritsas
Department of Electrical & Computer Engineering, University of Patras, Patras, Achaia 26500, Greece; e-mail: tzes@ece.upatras.gr

K.P. Valavanis (ed.), Applications of Intelligent Control to Engineering Systems, 169–193.

ing and tracking control have boosted the applications of the UAVs in areas where their utilization in the past was not feasible. Examples of such applications are those that require aerial surveillance in civilian and military applications such as crowd monitoring, remote monitoring in oil fields and pipeline inspections, and forest fire monitoring.

The low costs of these aerial vehicles and their increased computational capabilities is creating intelligent robotic systems that can combine in a single autonomous aerial system features as perception learning, real time control, situation assessment, reasoning, decision-making and planning capabilities for evolving and operating in complex environments [1].

Although in the past years the trend was the utilization of a single 'expensive' UAV, its increased cost along with the need to operate in a hazardous environment have driven the scientific developments in this area, in automating aerial surveillance by a fleet of homogeneous low cost UAVs.

These fleets of UAVs are characterized by increased autonomy and group cooperative behaviors for accomplishing a certain task. These cooperative behaviors imply that the group-members share a common objective and act accordingly to the group's mutual interest.

From an application point of view, the surveillance of forest fires is a likeable scenario for demonstrating the operation of cooperative fleet of UAVs. The forest fires are considered from a system's point of view as highly complicated, non-structured environments where the utilization of multiple sources of spatiotemporal information is essential. The fire propagation and the presence of smoke require flexible re-planning and re-allocation of tasks for the fleet of cooperative UAVs.

In the suggested approach an intelligent cooperative control algorithm is presented relying on multiple rendez-vous points between these quadrotor helicopters. These UqHs fly over the expanding perimeter of the fire and act as the fire-front sensing systems through their continuous patrolling task based on a varying set of rendez-vous with each other. These rendez-vous are defined as way points in order to divide the surveyed fire-front length at equidistant portions.

After the UqH-rendez-vous meetings, and taking into account the anticipated spread of the fire, the spatiotemporal flight perimeter is predicted followed by the computation of the updated rendez-vous points. These points rely heavily on the fire-front spread algorithm (wind conditions, moisture, fire fuel/forest flammability, and the firefighters' reaction).

The proposed approach in the fire surveillance application with multiple UAVs decreases drastically the time needed for surveillance of the fire perimeter while transmitting more reliable data due to their low-altitude flights. Furthermore, from an operational point of view, the utilization of multiple low cost UAVs decreases dramatically the expenses needed for the same flight scenario, compared to that of using a typical helicopter.

These UqHs, when flying in low-altitudes are prone to sudden wind-gusts. During the propagation of a forest fire, the wind-gusts's density and direction can be detrimental in the performance and overall stability (crash) of the UqH. Handling of these sudden and unpredicted wind disturbances on the UqH's flight requires

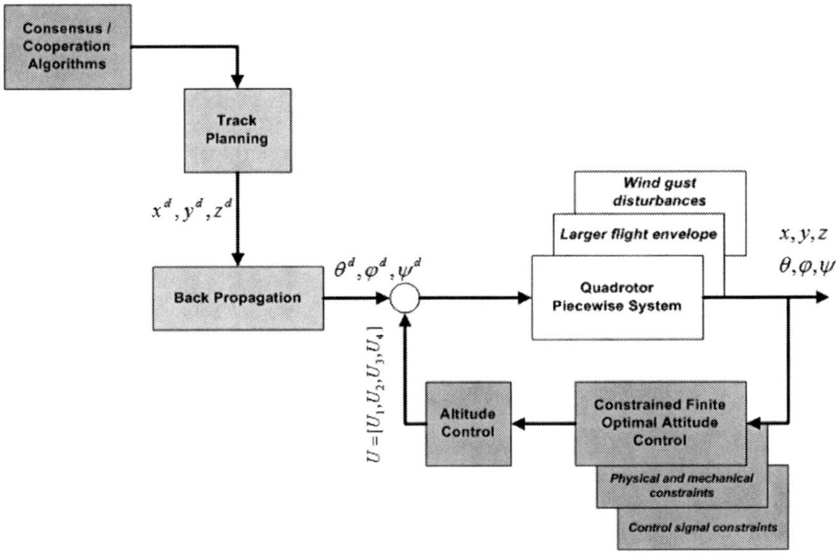

Fig. 7.1 Control architecture for UqHs for forest fire monitoring.

that the wind gusts are modeled appropriately as random disturbances affecting the nominal system description. Subsequently, a feed-forward controller along with a stabilizing controller relying on the Constrained Finite Time Optimization Control (CFTOC) [2–6] scheme is used. The feedback control action is responsible for attenuating the effects of the wind-gust disturbance on the tracking of the a priori reference trajectory. Compared to the existing UAV-control designs, including PID [7], Sliding Mode [8], Nonlinear Dynamic Inversion and Backstepping [9], the novelty in the proposed control approach is that the controller during the design phase can explicitly take into account: (a) the physical and mechanical constraints of the system, and (b) the disturbances from the environment. These factors, that degrade the system's performance, are generated from the physical and mechanical characteristics of the system and are embedded in the system model during the modeling phase for the control problem synthesis.

The integrated proposed control approach for the forest fire monitoring application based on a fleet of UAVs can be presented in the block diagram in Figure 7.1.

The remainder of this chapter is organized as follows. In Section 7.2, the different types of low-cost UAVs that can be used for tracking the fire perimeter are presented focusing on the currently developed UqH for the cooperative forest fire surveillance. In Section 7.3, the Forest Fire Propagation model, along with elements from Cooperation Strategies and UqHs Consensus will be presented. In Section 7.4 the dynamics of the UqH followed by the Constrained Finite Time Optimal Control and the rendez-vous scheme will be formulated. Simulation studies using the integrated scheme appear in Section 7.6 followed by conclusive remarks.

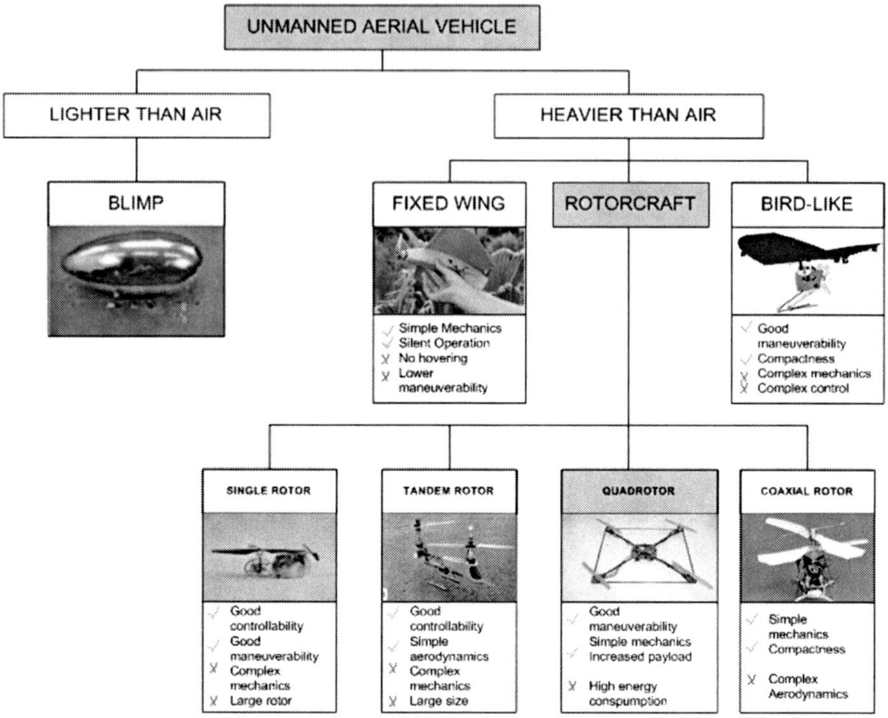

Fig. 7.2 Classification of low-cost UAVs.

7.2 Low-Cost Unmanned Aerial Vehicles

Low-cost UAVs are typically thought as mobile systems that carry a variety of sensors and the necessary communication equipment for maintain an RF-link with a ground-station. The shape and the characteristics of these vehicles are quite different in comparison to the conventional fixed wing aircrafts. More specifically the different constraints on overall size, the flying speeds and the aerodynamics at low speed are responsible for the shape of these vehicles. In the relevant literature, there have been several types and configurations of aerial vehicles that can generally be divided into two categories: (1) Lighter Than Air, and (2) Heavier Than Air, as shown in Figure 7.2 where a more detailed classification of UAVs is presented depending on the flying principle.

Fixed wing UAVs have the advantage of simple mechanics, silent operation and generally lower power consumption. On the other hand a fixed wing aircraft does not have hovering capabilities and has lower maneuverability, issues that are critical in surveillance operations.

Rotorcraft UAVs usually have complex mechanics, consume more power and are usually difficult to control, while the ability to hover and the higher maneuverability capabilities give a potential advantage for surveillance, inspection and almost every kind of operation in constrained, dynamically variable environments.

The classical single rotor, from an operational and deployment point of view, has the advantages of high controllability and maneuverability capabilities. On the other hand these configurations have significant drawbacks, including its complex mechanics, the need for a large rotor and long tail boom.

The tandem-rotor configuration is a viable option for several missions due to its high controllability, simple aerodynamics and high payload. Despite these advantages, the complex mechanics, large size and less maneuverability attributes in these tandem-rotorcrafts has narrowed their application width.

The quadrotor helicopter [10] is a special member of the rotorcraft family. Excluding its high energy consumption, the quadrotor outperforms almost every other UAV on the issues of maneuverability, survivability, simplicity of mechanics and increased payload. These advantages are of paramount importance in forest-fire monitoring applications, where hovering over certain "hardened" spots while carrying a large sensor-payload [11] is a typical maneuver.

7.3 Cooperation Strategies and Consensus for UAVs in Forest Fire Surveillance Operations

7.3.1 Cooperative Strategy Theory Elements

An important element of an autonomous-team of UAVs is its ability to make decisions and assign each member with a set of tasks, in a way that minimizes an overall objective cost and thus converging to an optimal cooperation [12–15]. For these tasks significant work has been reported formulating these problems using centralized approaches and Mixed-Integer Linear Programming [16]. However in order to use these approaches, the cost between consecutive tasks must be known a priori. This condition may not be known a priori in several operations leading to devise strategies for UAVs in a decentralized environment. The typical decentralized approach [17] can be split into the following steps:

- Define analytically the main *cooperation objective* and other local sub-objectives.
- Identify the essential information (*coordination variable*) that each vehicle needs to know in order to coordinate with the rest of team.
- Derive a cooperation strategy in order to minimize the team's objective function assuming that each member of the team has global (centralized) knowledge.
- Design a consensus – decentralized cooperation strategy, which accounts for *constraints* or limited communication abilities and knowledge at each UAV.

In the proposed algorithm for the UAVs' rendez-vous, the decentralized strategy is implemented as a single consensus algorithm [18] implemented in each UAV

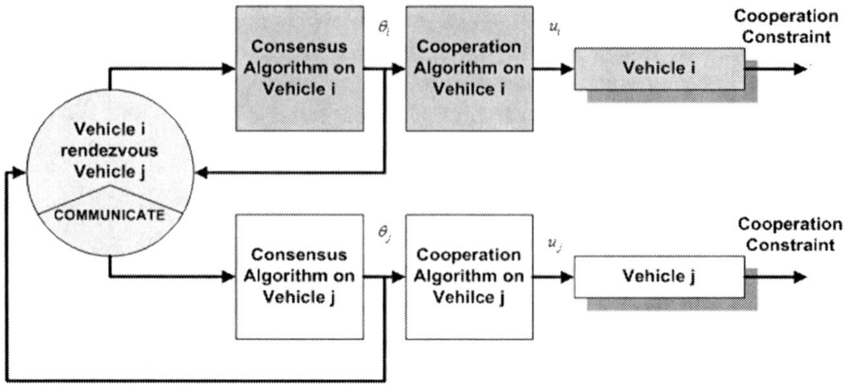

Fig. 7.3 UAV-cooperation scheme using consensus algorithms.

while the communication between the UAVs is carried according to the consensus algorithm, as shown in Figure 7.3.

7.3.2 Forest Fire Monitoring

Forest fires are catastrophic phenomena with social and economic consequences to environment [19, 20]. In order to combat forest fires effectively, early detection and continuous tracking of the perimeter of the fire is very important. Until recently the common high-tech approach to fire monitoring, is the use of orbital satellites that initially capture images of forest fires and in the sequel the fire perimeter (bounds of fire) can be extracted. Orbital satellites fail to effectively track the fire perimeter both due to: (a) their low orbital period (about ten hours and more), and (b) the low resolution of the captured images. On the other hand, fire-fighting ground-teams need to get frequently updated information about the evolution of the fire in order to schedule the fire combat mission. For these reasons, there is an urgent need to develop more effective fire surveillance technologies.

UAVs can be applied for forest fire surveillance improving the spatial and temporal resolution of satellite observations in several tasks, including building of fire risk maps, monitoring of vegetation and even generating 3D-maps with scanning sensors such as LIDARs. Cooperating with other larger systems, these Low Altitude and Short-Endurance (LASE) UAVs are playing a key role for effectively accomplishing fire monitoring missions. As technology improves, LASE UAVs emerge as a low cost alternative with advanced capability for cooperation strategies and qualitative sensing.

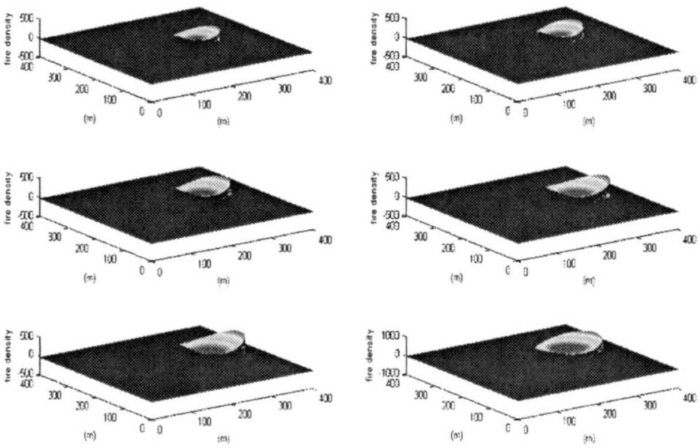

Fig. 7.4 Forest fire propagation over a 6 hour interval (hourly updates).

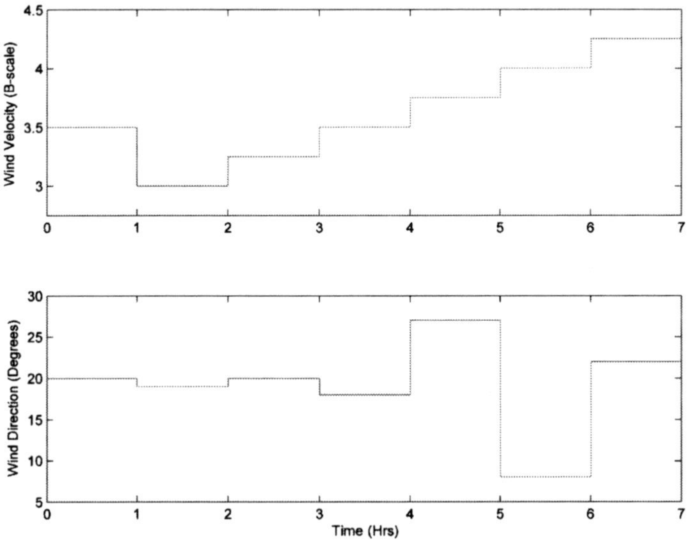

Fig. 7.5 Weather conditions for forest fire propagation.

7.3.3 Forest Fire Propagation

Wildfire modeling involves the statistical analysis of past fire events in order to create a sufficient model of spotting risks and front behavior. Simple wildfire propagation models consider fire perimeter as circles, ellipses or foliums [21]. In order to create more sufficient models, the heterogeneity of vegetation, the direction and the power of the wind and the slope steepness must be accounted for. Modern fire growth models are based on fire dynamics and evolution models [22].

Among these models for forest fire propagation, the EMBYR [23] depicts the landscape as a grid in which the area of each square cell is 2500 m^2. In this model single or multiple ignitions can occur either randomly or by specifying the coordinates of the ignited cells and a number of processes that affect the landscape. Patterns of fire spread are considered, including diffusive spread from cell to cell, the impact of wind speed and direction, ignition of non-distant cells by firebrands, and changes in the combustibility of different fuel types due to variation in fuel moisture. In the approach followed in this article the fire propagation has been modeled as a polygon with evolving perimeter depending on wind speed, direction and time. Typical snapshots of the simulation of a recent forest fire is shown in Figure 7.4, where the propagation of an evolving fire is presented over a six hour interval (with hourly increments) for low-level wind conditions (3–5 on the Beaufort scale) and almost constant wind directions, as shown in Figure 7.5.

7.4 Unmanned Quadrotor Helicopter Dynamics

For the UqH under study, shown in Figure 7.6, it is assumed that the structure is rigid and symmetrical, the center of gravity and the body fixed frame origins coincide, the propellers are rigid and the thrust and drag are proportional to the square of propeller's speed.

The dynamics of a rigid body, under external forces applied to the center of mass and expressed in the body fixed frame, are given in a Newton–Euler formulation as

$$\begin{bmatrix} mI_3 & 0 \\ 0 & J \end{bmatrix} \begin{bmatrix} \dot{V} \\ \dot{\omega} \end{bmatrix} + \begin{bmatrix} \omega \times mV \\ \omega \times J\omega \end{bmatrix} = \begin{bmatrix} F \\ \tau \end{bmatrix}, \tag{7.1}$$

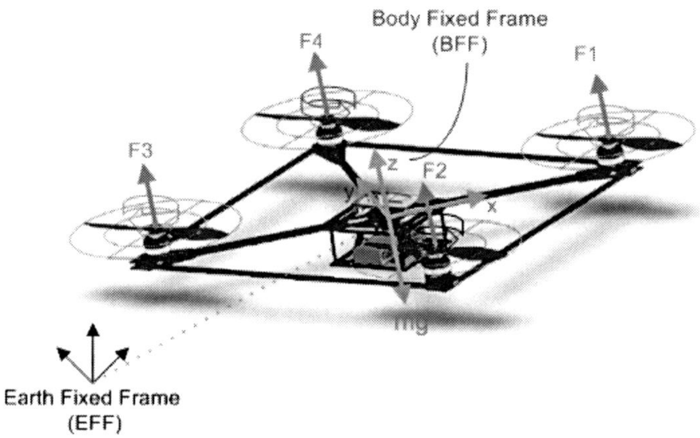

Fig. 7.6 UqH-UPATcopter's configuration frame system.

where $J \in \Re^{3\times3}$ is the inertia matrix, $m \in \Re$ is the mass of the helicopter, I_3 is the 3×3 identity matrix, V is the body linear speed vector, ω is the body angular speed and F, τ are the induced forces and torques respectively.

The aerodynamic forces and moments acting on the quadrotor during a hovering flight segment are summarized as in [24, 25].

The thrust force, T, due to vertical forces acting on all blade elements is

$$T = C_T \rho A(\Omega R_r)^2$$
$$\frac{C_T}{\sigma\alpha} = \left(\frac{1}{6} + \frac{1}{4}\mu^2\right)\theta_i - (1+\mu^2)\frac{\theta_\varphi}{8} - \frac{1}{4}\lambda, \tag{7.2}$$

where σ is the solidity ratio, λ is the inflow ratio, α is the lift slope, μ is the motor advance ratio, ρ is the air density, θ_i is the pitch incidence, θ_φ is the twist pitch, Ω is the propeller angular rate, A is the propeller disk area, R_r is the propeller radius and C_T is the thrust coefficient.

The hub force, H, due to horizontal forces acting on all the blade elements:

$$H = C_H \rho A(\Omega R_r)^2$$
$$\frac{C_H}{\sigma\alpha} = \frac{1}{4\alpha}\mu\overline{C}_d + \frac{1}{4}\lambda\mu\left(\theta_i - \frac{\theta_\varphi}{2}\right), \tag{7.3}$$

where $\overline{C}_d$ is the drag coefficient at 70% radial station and C_H is the hub force coefficient.

The drag moment, Q, due to aerodynamic forces acting on the blade elements is

$$Q = C_Q \rho A(\Omega R_r)^2 R_r$$
$$\frac{C_Q}{\sigma\alpha} = \frac{1}{8\alpha}(1+\mu^2)\overline{C}_d + \lambda\left(\frac{1}{6}\theta_i - \frac{1}{8}\theta_\varphi - \frac{1}{4}\lambda\right), \tag{7.4}$$

where C_Q is the drag coefficient.

The rolling moment, is the integration over the entire rotor of the lift of each section acting at a given radius and is formulated as

$$R_m = C_{Rm} \rho A(\Omega R_r)^2 R_r$$
$$\frac{C_{R_m}}{\sigma\alpha} = \mu\left(\frac{1}{6}\theta_i - \frac{1}{8}\theta_\varphi - \frac{1}{8}\lambda\right), \tag{7.5}$$

where C_{R_m} is the rolling moment coefficient. All the above coefficients can be calculated using experimental data, Blade Element Theory and Computational Fluid Dynamics software, while for the presented UqH, shown in Figure 7.7, these values have been determined experimentally.

Under the adoption of the coordinate frame shown in Figure 7.6 for the UPAT-copter, the 6-DOF UqH's kinematics is formulated as

Fig. 7.7 University of Patras' Quadrotor Helicopter Prototype (UPATcopter).

$$\dot{\zeta} = J_{\Theta^{\nu}}, \tag{7.6}$$

where $\zeta = [x\ y\ z\ \varphi\ \theta\ \psi]^T$ is the linear position and angular vector, ν is the generalized velocity with respect to the Body Fixed Frame (BFF) and J_Θ is the generalized inertia matrix.

The vector ν is composed of the linear and angular velocity vector as

$$\nu = [V^{BFF}, \omega^{BFF}]^{\mathrm{T}} = [u, v, w, p, q, r]^T, \tag{7.7}$$

where p, q, r correspond to the helicopter's body rates.

The generalized matrix is composed of 4 sub-matrices as

$$J_\Theta = \begin{bmatrix} R_\Theta & 0_{3\times3} \\ 0_{3\times3} & T_\Theta \end{bmatrix},$$

$$R_\Theta = \begin{bmatrix} c\psi c\theta & -s\psi c\varphi + c\psi s\theta s\varphi & s\psi s\varphi + c\psi s\theta c\varphi \\ s\psi c\theta & -c\psi c\varphi + s\psi s\theta s\varphi & -c\psi s\varphi + s\psi s\theta c\varphi \\ -s\theta & c\theta s\varphi & c\theta c\varphi \end{bmatrix},$$

$$T_\Theta = \begin{bmatrix} 1 & s\varphi t\theta & c\varphi t\theta \\ 0 & c\varphi & -s\varphi \\ 0 & s\varphi/c\theta & c\varphi/c\theta \end{bmatrix}, \tag{7.8}$$

where for notation brevity the following symbols have been used: s for sin, c for cos and t for tan.

The UqH's dynamics in a hovering maneuver (at a horizontal plane) are simplified as

$$
\begin{aligned}
J_{xx}\ddot{\varphi} &= \dot{\theta}\dot{\psi}(J_{yy} - J_{zz}) + J_r\dot{\theta}\Omega_r + l_a(T_4 - T_2) \\
&\quad - h\left(\sum_{i=1}^{4} H_{yi}\right) + (-1)^{i+1}\sum_{i=1}^{4} R_{mxi}, \\
J_{yy}\ddot{\theta} &= \dot{\varphi}\dot{\psi}(J_{zz} - J_{xx}) + J_r\dot{\varphi}\Omega_r + l_a(T_1 - T_3) \\
&\quad + h\left(\sum_{i=1}^{4} H_{xi}\right) + (-1)^{i+1}\sum_{i=1}^{4} R_{myi}, \\
J_{zz}\ddot{\psi} &= \dot{\theta}\dot{\varphi}(J_{xx} - J_{yy}) + J_r\dot{\Omega}_r + (-1)^i\sum_{i=1}^{4} Q_i \\
&\quad + l_a(H_{x2} - H_{x4}) + l_a(-H_{y1} + H_{y3}), \\
m\ddot{z} &= mg - (c\psi c\varphi)\sum_{i=1}^{4} T_i, \\
m\ddot{y} &= (-c\psi s\varphi + s\psi s\theta c\varphi)\sum_{i-1}^{4} T_1 - \sum_{i=1}^{4} H_{yi} - \frac{1}{2}C_y A_c \rho \dot{y}|\dot{y}|, \\
m\ddot{x} &= (-s\psi s\varphi + c\psi s\theta c\varphi)\sum_{i-1}^{4} T_1 - \sum_{i=1}^{4} H_{xi} - \frac{1}{2}C_x A_c \rho \dot{x}|\dot{x}|,
\end{aligned} \tag{7.9}
$$

where $J_r = J_m + J_p/4$ is the rotor inertia, J_m the motor inertial, J_p the rotor inertial, l_a the horizontal distance between the propeller and the center of gravity of the quadrotor, h the vertical distance between the propeller and the center of gravity of the quadrotor, A_C is the fuselage area, and Ω_r the overall residual propeller angular speed.

Under the assumption of a dc-motor operating as an actuator for the rotor/propeller, yields

$$
\dot{\Omega}_i = \frac{K_a}{T}V_i - K_b\Omega_i^2 - \frac{1}{T}\Omega_i, \quad i = 1, 2, 3, 4, \tag{7.10}
$$

where Ω_i is the propeller's angular rate, and V_i the applied voltage at each motor.

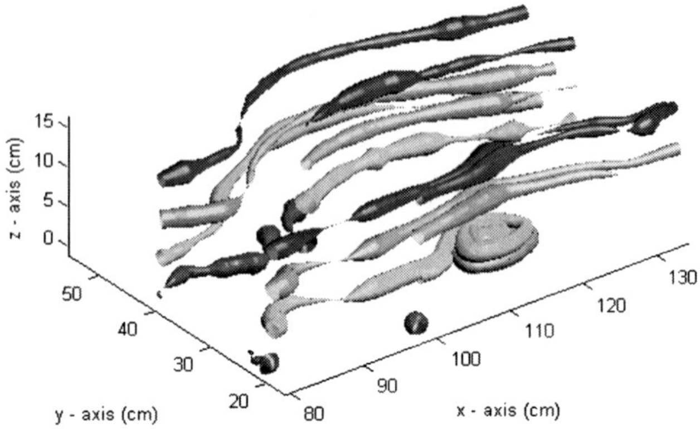

Fig. 7.8 3D-wind gust spatial velocity profile.

7.4.1 Wind Gust Modeling

A low-altitude flight over a forest-fire environment, where the temperature gradient is large, is susceptible to the forcible wind-gusts that affect the dynamics of the quadrotor. These winds gusts can be detrimental to the stability and performance of the UAV and need to be properly modeled prior to their attenuation thru the control action.

A wind gust is a complex physical phenomenon and is typically modeled using elements from stochastic fluid theory. In order to obtain a simple, yet sufficient model the following assumptions are made [26]: (a) the wind gust has a mean wind-velocity value at each axis: $V_G = [V_{GX}, V_{GY}, V_{GZ}]^T$, (b) the spatial center and the borders of the gust are fixed while the gust is attenuated around each axis by a $(1 - \cos)$ shape. Under these assumptions the gust velocities acting on the face of the airframe at each axis takes the following form:

$$\begin{aligned} V_{G,i} &= \frac{1}{2} V_{ds,i} \left(1 - \cos\left(\frac{\pi s_i}{H_i}\right)\right), \\ V_{ds,i} &= V_{r,i} F_{g,i} \left(\frac{H_i}{350}\right)^{1/6}, \quad i = X, Y, Z, \\ |V_G| &= \sqrt{V_{G,X}^2 + V_{G,Y}^2 + V_{G,Z}^2}, \end{aligned} \tag{7.11}$$

where $H_i(V_{r,i})[F_{g,i}]$ is the gust depth (reference velocity) [alleviation factor due to the aerodynamic characteristics of the airframe and the propellers], and s_i the distance between the quadrotor's center of gravity and the wind gust "focal center" along the ith axis. A typical wind-gust representation based on the aforementioned assumptions is shown in Figure 7.8.

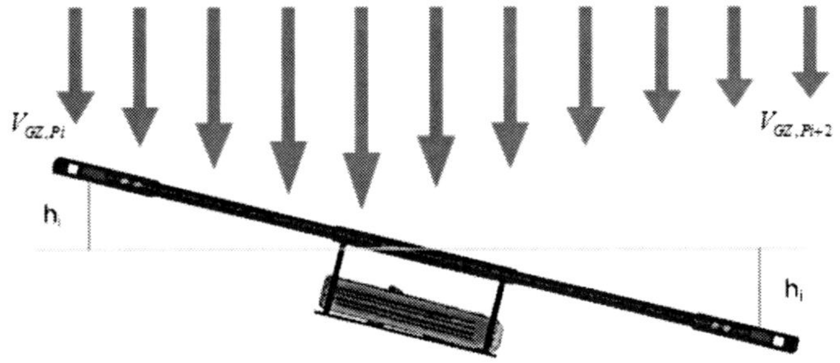

Fig. 7.9 Wind-gust differential disturbance at opposite rotors.

In order to calculate the forces that the wind gust causes to the quadrotor the following equations are used:

$$F_i = \frac{1}{2}\rho C_b V_{G,i}^2 A_{B,i} + 4\frac{1}{2}\rho C_{p,i} V_{G,i}^2 A_{P,i}, \quad i = X, Y, Z, \tag{7.12}$$

where $A_{B,i}(A_{Pi})$ is the area of the body (propellers) as seen from the $V_{G,i}$ vector. These values are calculated using the rotation matrix R_Θ. The value of drag coefficient C_b is calculated assuming that the entire area of the quadrotor is concentrated at the main rectangular part. The values of $C_{p,i}$ are calculated using computational fluid dynamics neglecting the rotation of the quadrotor.

In order to calculate the moments caused by the wind gust, the different velocity of the gust at the four different propeller positions must be taken into account as shown in Figure 7.9.

According to this the induced moments are formulated as shown below:

$$\begin{aligned}
M_A &= \frac{1}{2}\rho C_{pX} V_{GZ,P1}^2 A_{P,Z} l_a - \frac{1}{2}\rho C_{pZ} V_{GZ,P3}^2 A_{P,X} l_a,\\
M_B &= \frac{1}{2}\rho C_{pX} V_{GZ,P4}^2 A_{P,Z} l_a - \frac{1}{2}\rho C_{pZ} V_{GZ,P2}^2 A_{P,X} l_a,\\
M_C &= \frac{1}{2}\rho C_{pX} V_{GZ,P1}^2 A_{P,Z} l_a + \frac{1}{2}\rho C_{pZ} V_{GZ,P3}^2 A_{P,X} l_a\\
&\quad - \frac{1}{2}\rho C_{pX} V_{GZ,P4}^2 A_{P,Z} l_a + \frac{1}{2}\rho C_{pZ} V_{GZ,P2}^2 A_{P,X} l_a,\\
V_{GZ,Pi} &= \frac{1}{2} V_{dsZ}\left(1 - \cos\left(\frac{\pi S_Z - h_i^F}{H_Z}\right)\right), \quad i = 1, 4,\\
V_{GZ,Pj} &= \frac{1}{2} V_{dsZ}\left(1 - \cos\left(\frac{\pi S_Z - h_j^S}{H_Z}\right)\right), \quad j = 2, 3,
\end{aligned} \tag{7.13}$$

where $V_{GZ,Pi}, i = 1, \ldots, 4$ is the vertical wind-gust velocity acting on propeller i and h_i^F, h_i^S are the forward and sideward vertical distances between the opposite motors respectively.

7.4.2 Quadrotor Helicopter Dynamic Model

The overall UqH-system's dynamics correspond to a nonlinear ODE in the form of $\dot{X} = f(X, U)$, where U is the 4×1 input vector and X the 12×1 state vector corresponding to:

$$X = [\varphi, \dot{\varphi}, \theta, \dot{\theta}, \psi, \dot{\psi}, z, \dot{z}, x, \dot{x}, y, \dot{y}]^T. \tag{7.14}$$

The input vector $U = [U_1 U_2 U_3 U_4]$ is related to the propellers' angular rates as

$$\begin{aligned} U_1 &= b(\Omega_1^2 + \Omega_2^2 + \Omega_3^2 + \Omega_4^2), \\ U_2 &= b(-\Omega_2^2 + \Omega_4^2), \\ U_3 &= b(\Omega_1^2 - \Omega_3^2), \\ U_4 &= d(-\Omega_1^2 + \Omega_2^2 - \Omega_3^2 + \Omega_4^2), \end{aligned} \tag{7.15}$$

where $b(d)$ is the drag (thrust) coefficient.

Under the assumption of small angles, the vector comprising of the rate of change of the Euler angles $[\varphi, \theta, \psi]^T$ is close to the body angular velocity vector $[p, q, r]^T$ and the nonlinear mapping is

$$f_{12\times 1}(X, U) = \begin{bmatrix} \dot{\varphi}, \dot{\theta}\dot{\psi}a_1 + \dot{\theta}a_2\Omega_r + b_1U_2, \ldots \\ \dot{\theta}, \dot{\varphi}\dot{\psi}a_3 - \dot{\varphi}a_4\Omega_r + b_2U_3, \ldots \\ \dot{\psi}, \dot{\theta}\dot{\varphi}a_5 + b_3U_4, \ldots \\ \dot{z}, g - \frac{U_1 \cos\varphi \cos\theta}{m}, \dot{x}, \frac{u_xU_1}{m}, \dot{y}, \frac{u_yU_1}{m} \end{bmatrix}^T, \tag{7.16}$$

where

$$a_1 = \frac{J_{yy} - J_{zz}}{J_{xx}}, \quad a_2 = \frac{J_r}{J_{xx}}, \quad a_3 = \frac{J_{zz} - J_{xx}}{J_{yy}}, \quad a_4 = \frac{J_r}{J_{yy}}, \quad a_5 = \frac{J_{xx} - J_{yy}}{J_{zz}},$$

$$b_1 = \frac{l_a}{J_{xx}}, \quad b_2 = \frac{l_a}{J_{yy}}, \quad b_3 = \frac{1}{J_{zz}}$$

and

$$u_x = (\cos\varphi \sin\theta \cos\psi + \sin\varphi \sin\psi), \quad u_y = (\cos\varphi \sin\theta \sin\psi - \sin\varphi \cos\psi).$$

It must be noted that in this system description, the angles and their derivatives are independent of the translation components.

7.5 UqH Control Design and rendez-vous Task Formulation for Forest Fire Surveillance

7.5.1 Horizontal Trajectory Feedforward Control

Under the assumption of a flight in a plane parallel to the ground, the (x^d, y^d) position of the UqH is achieved by rolling and/or pitching the helicopter in response to the reference trajectory, as

$$x^d(k) - 2x^d(k-1) + x^d(k-2) = -\frac{T_s^2}{m}\theta^d(k)U_1^*,$$

$$y^d(k) - 2x^d(k-1) + x^d(k-2) = \frac{T_s^2}{m}\varphi^d(k)U_1^*, \tag{7.17}$$

where T_s corresponds to the time-discretization interval. Having computed the φ^d and θ^d, the horizontal motion dynamics is $m\ddot{x} = -\theta^d U_1^*$, $m\ddot{y} = \varphi^d U_1^*$.

After the arrival of the UqH at the target position, the feedforward controller lifts the helicopter at its target altitude.

7.5.2 Constrained Finite Time Optimal Attitude Control

The aforementioned position controller generates the reference values φ^d and θ^d from (xx). The CFTO-controller is responsible for regulating any perturbation $\varphi = \varphi^d + \Delta\varphi$ and $\theta = \theta^d + \Delta\theta$ from these values. This regulation should take place in lieu of any physical actuator and state constraints, and any wind-gust disturbance acting on the model.

Essentially, the CFTOC is used for the regulation of the quadrotor around its nominal angles' of operation, during the translation on a priori reference trajectory. The proposed control action is affecting only the rotations of the quadrotor around φ and while the CFTO-controller has the merit to take under consideration: a) the physical and mechanical constraints of the system, and b) the wind-gust disturbances. These factors degrade the system's performance, and are generated from the physical and mechanical characteristics of the system. During the CFTOC-synthesis phase, these factors are embedded in the system model.

Based on the system modeling in (7.16), the linearized equations for the rotations of the quadrotor are:

$$\dot{x} = Ax + Bu + W, \tag{7.18}$$

where

$$A=\begin{bmatrix}0 & 1 & 0 & 0 & 0 & 0\\ 0 & 0 & 0 & \frac{J_{yy}-J_{zz}}{2J_{xx}}\dot{\psi}^d & 0 & \frac{J_{yy}-J_{zz}}{2J_{xx}}\dot{\theta}^d\\ 0 & 0 & 0 & 1 & 0 & 0\\ 0 & \frac{J_{zz}-J_{xx}}{2J_{yy}}\dot{\psi}^d & 0 & 0 & 0 & \frac{J_{yy}-J_{xx}}{2J_{yy}}\dot{\varphi}^d\\ 0 & 0 & 0 & 0 & 0 & 1\\ 0 & \frac{J_{xx}-J_{yy}}{2J_{zz}}\dot{\theta}^d & 0 & \frac{J_{xx}-I_{yy}}{2J_{zz}}\dot{\varphi}^d & 0 & 0\end{bmatrix},\quad B=\begin{bmatrix}0 & 0 & 0 & 0\\ \frac{l_a}{J_{xx}} & 0 & 0 & \frac{J_r}{J_{xx}}\dot{\theta}^d\\ 0 & 0 & 0 & 0\\ 0 & \frac{l_a}{J_{yy}} & 0 & \frac{J_r}{J_{yy}}\dot{\theta}^d\\ 0 & 0 & 0 & 0\\ 0 & 0 & \frac{l_a}{J_{zz}} & 0\end{bmatrix},$$

$$W=[f(M_A),0,g(M_B),0,h(M_C),0]^T,\qquad x=[\Delta\varphi,\Delta\dot{\varphi},\Delta\theta,\Delta\dot{\theta},\Delta\psi,\Delta\dot{\psi}]^T$$

and

$$u=[\Delta U_1,\Delta U_2,\Delta U_3,\Delta U_4]^T,$$

where f, g, h are functions representing the disturbances in the Euler angles that are dependent of M_A, M_B, M_C. At this point we should mention that the wind gusts affect mainly the rotational movement of the quadrotor, and this the primary reason of the inclusion of the terms M_A, M_B, M_C acting as random additive disturbances. These disturbances have not been embedded in the system model as affine terms as there is no a priori information regarding their values. Depending on: (a) the partition of the x-parameter vector, (b) the reference values $\varphi^d \in [\varphi^{\min}, \varphi^{\max}] = \bigcup_{i=0}^{n-1}[\varphi^{\min} + i\Delta\varphi, \varphi^{\min} + (i+1)\Delta\varphi]$, and $\theta^d \in [\theta^{\min}, \theta^{\max}] = \bigcup_{j=0}^{m-1}[\theta^{\min} + j\Delta\theta, \theta^{\min} + (j+1)\Delta\theta]$, and c) the bounds (due to saturation) of the control inputs $U_i^{\min} \le U_i \le U_i^{\max}$, $i = 1, \dots, 4$, the aforementioned system can be cast as a switching system $\dot{x} = A_j x + B_j u + W$. Discretization with a sampling period T_s yields its discrete equivalent

$$x(k+1) = A_j^* x(k) + B_j^* u + W, \quad j = 1, \dots, L, \tag{7.19}$$

where A_j^*, B_j^* are the discrete time state space matrices, with the state and input constraints expressed in a condensed form via the following guard function:

$$\begin{bmatrix}x(k)\\ u(k)\end{bmatrix} \in P = \{H_i x + J_i u\} \le P_i, \quad i = 1, \dots, M. \tag{7.20}$$

The CFTOC ("multi-parametric") approach consists in computing the optimizer control vector $U_{k,N} = [u(k), \dots, u(k+N-1)]^T$, N being the prediction horizon, which minimizes the cost function:

$$J_{k,N} = \min_{U_{k,N}} \left\{ \|Px(k+N)\|_l + \sum_{j=0}^{N-1} (\|Ru(k+j)\|_l + \|Qx(k+j)\|_l) \right\} \tag{7.21}$$

subject to the linear system dynamics and state/input constraints defined earlier.

The cost function in (7.21) may be linear (e.g., $l\{1,\cdot\}$) or quadratic (e.g., $l = 2$) depending on the vector norm employed, $x(k+j)$ corresponds to the predicted values of the state vector thru the application of the input sequence $U_{k,N}$. Moreover Q, R and P are the full column rank weighting matrices on the corresponding optimization variables affecting the predicted states, control effort and the desired final state, respectively. The predicted final state x_N typically belongs to a predefined set $x_{k,N} \in X_s$ which is dictated by the stability and feasibility requirements.

The solution to the aforementioned stated control problem [2–4] results in a continuous piecewise affine state feedback of the following form:

$$u(k) = F_j x(k) + G_j, \;\; \text{if } x(k) \in R_j, \quad j \in \{1, \ldots, J_p\} \tag{7.22}$$

defined over convex polyhedra R_j. Rather than working with a state-regulation problem, the CFTOC-problem can also be formulated for an output set-point problem by replacing the term $Qx(k+j)$ with $Q_y(y^d - y(k+j))$.

It should be noted that the computational complexity [27] of the controller increases significantly with the number of constraints M, the prediction horizon N, and the number $L = m\,n$ of the switching systems. Essentially the number of polyhedra J_p increases almost exponentially with the previous parameters making the computation of the CFTO-control prohibitive in real-time. However, this controller can be computed in an offline manner and a lookup table can be used to derive $u(k)$ from (7.22).

7.5.3 UqHs rendez-vous for Forest Fire Surveillance

Assume a simplified forest fire surveillance mission, considered in this article, as shown in Figure 7.10. The operational goal is to track every point of the evolving fire perimeter and update the location of the fire with the least latency. The function $\delta(x,t)$ denotes the latency associated with surveillance information about the position x of the fire perimeter at time t. This function acts as a measure of the quality of the proposed cooperation strategy. For the case of static (not evolving) fire [28–30], the minimization of the latency function is reached when all UAV agents travel equal lengths between their rendez-vous.

7.5.3.1 Problem Formulation

The goal of the UqH-team, consisted of N UqHs, for the forest fire surveillance problem is to monitor and track the perimeter of the fire as it evolves and to transmit acquired information to the ground station.

Let a situation state $x_k \in X$ describing the fire perimeter coordinates and $P_c(t)$ the fire perimeter curve. The decision variable $u_k \in U$ is assumed to be the next rendez-vous point. The cooperation constraint J_{cc} is to fly every quadrotor on the

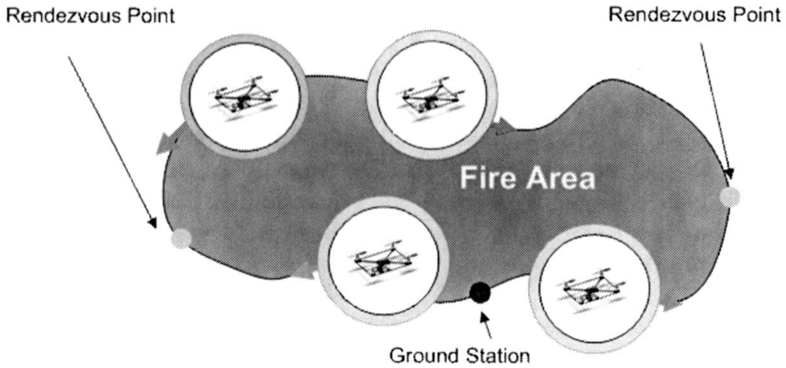

Fig. 7.10 UqH- (2 member) cooperation for forest fire surveillance.

fire perimeter P_c and the cooperation objective J_{co} is to equalize the path lengths flown by the quadrotors around the perimeter of the fire. The coordination variable θ^* is the path last flown (from the last rendez-vous). The coordination variable is a function of the situation state and the decision variable. Subsequently, the coordination function $J_{cf,i}$ is defined as the variance of the length of the path flown by the ith quadrotor, from the mean value of the length of the paths $\mu = \ell(P)/N$. The centralized cooperation scheme is formulated as

$$\ell(\theta^*) = \arg\min \left\{ \lim_{t\to\infty} \sum_{i=1}^{N} J_{cf,i}(\theta; x_i; u_i) \right\} \tag{7.23}$$

subject to

$$0 = \lim_{t\to\infty} P_c(\theta; X; U),$$

$$J_{cf,i} = \int_0^t (\ell(\theta^*) - \mu) p(\ell(\theta^*)) u_i d\tau,$$

$$\ell(\cdot) = \int_{P_c} (\cdot) d\tau.$$

A decentralized strategy which effectively copes with the forest fire perimeter surveillance problem formulated in (7.23) is presented in the sequel. The communication between several UAVs takes place only when they meet in a rendez-vous "point".

The steps of the decentralized approach are listed below:

1. Assume a generic propagating fire perimeter.
2. UAVs start their flight in sets of two and in opposite directions and with time difference equal to the time that each UAV travels a path with length $\ell(P_0^c)/N$, where P_c^0 the initial perimeter of the fire.

3. **IF** i-UAV at $P_c(\theta_i; x_k; u_i)$ rendez-vous with its j-neighbor at $P_c(\theta_j; x_k; u_i)$ **THEN**
4. Communicate the length of the paths last flown $\ell(\theta_i), \ell(\theta_j)$
5. Get new knowledge about the local part of the fire perimeter
6. Communicate knowledge between rendez-voused UAVs
7. **IF** the fire has evolved to the new perimeter P_c' then
8. Find points $x_j^i, x_i^j \in P_c'$: min $\| P_c'(\theta_i; x_k; u_i) - P_c'(\theta_j; x_k; u_i)\|_2$
9. Fly UAVs i, j to x_i^i, x_i^j
10. **END**
11. Change directions of flight
12. **ELSE**
13. Continue in current direction
14. **END**

7.6 Simulation Studies

7.6.1 Decentralized Cooperation Scenario

A set of simulation studies are presented in order to study the perfomance of the proposed decentralized cooperation strategy.

In Figures 7.11 and 7.12 selected scenes from the execution of a decentralized cooperation senario for the case of four UqHs and a time evolving fire are shown. In this simulation it must be noted that:

- The fire perimeter is assumed to be an ellipse and the evolution is modeled as a change on the ellipse foci affecting its major and minor radius. The change of the fire perimeter is occurs at time $t = 1$ hr, 2 hrs and 4 hrs respectively.
- The quadrotors take off at sets of two. The first two UqH take off at $t = 0$ and the last two depart at the time that the first couple has traveled half of the fire perimeter.
- Each UqH communicates only with the UqH with which they meet in a rendez-vous point.
- UqHs that have met in a rendez-vous point sense fire perimeter changes. If a variation of the fire perimeter is sensed then both UqH fly to the new fire perimeter at its closest point.
- Figure 7.11 (cases (a) thru (d)) and Figure 7.12 (cases (e) thru (k)) indicate the main points of the cooperation strategy: (a) shows $U_qH_{1,2}$ rendez-vous and $U_qH_{3,4}$ positions, (b) shows $U_qH_{2,4}$ and $U_qH_{1,3}$ rendez-vous, (c) shows $U_qH_{1,2}$ rendez-vous and $U_qH_{3,4}$ positions, (d) shows $U_qH_{3,4}$ rendez-vous and change fire perimeter while $U_qH_{1,2}$ still are at the previous fire perimeter. (e) shows $U_qH_{1,2}$ rendez-vous and change fire perimeter, $U_qH_{3,4}$ positions, (f) shows $U_qH_{3,4}$ rendez-vous and change fire perimeter while $U_qH_{1,2}$ still are at the

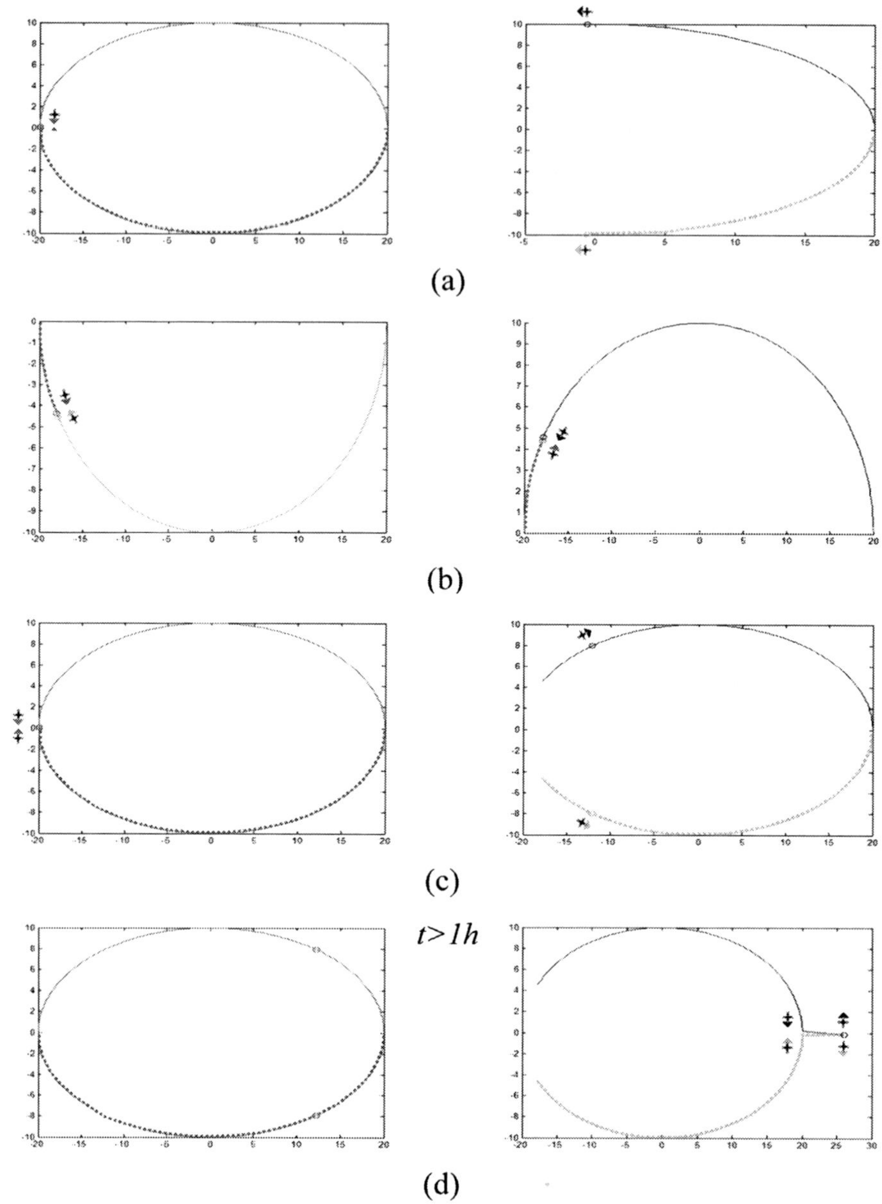

Fig. 7.11 Decentralized rendez-vous study for an evolving fire perimeter (case (a) thru (d)).

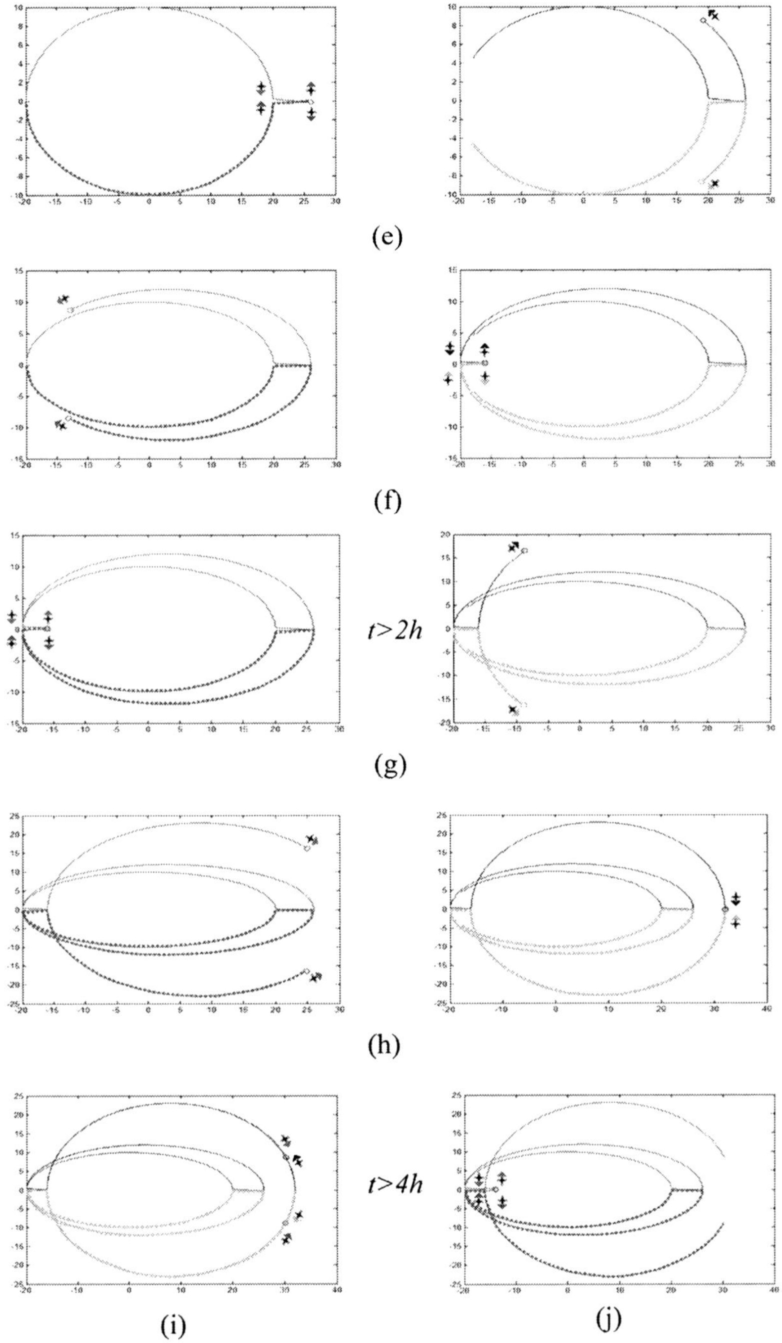

Fig. 7.12 Decentralized rendez-vous study for an evolving fire perimeter (case (e) thru (k)).

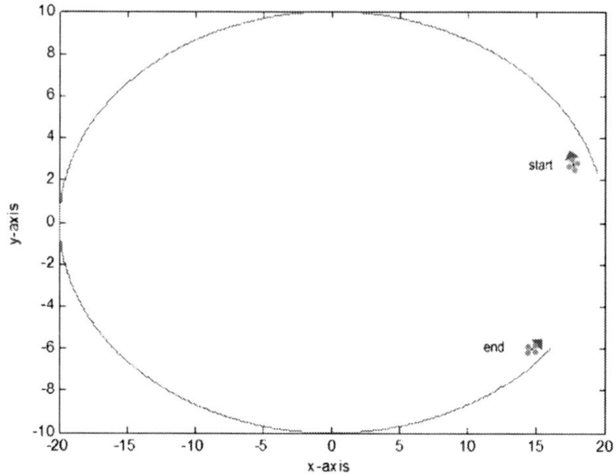

Fig. 7.13 UqH-reference path.

previous fire perimeter, (g) shows $U_qH_{1,2}$ rendez-vous and change fire perimeter, $U_qH_{3,4}$ positions, (h) shows $U_qH_{3,4}$ rendez-vous and $U_qH_{1,2}$ positions, (j) shows all UqH rendez-vous on current fire perimeter (1,3 and $U_qH_{2,4}$ rendez-vous), (k) shows $U_qH_{1,2}$ rendez-vous and change of fire perimeter.

- Overall, the team of UqHs effectively surveys the fire perimeter. After each rendez-vous point, the UqHs tend to equalize the lengths of their paths [34] while flying between their two rendez-vous points.

7.6.2 Trajectory Tracking and Wind Gust Attenuation

The proposed CFTOC-scheme is applied to the piecewise model of the UPAT-copter. Following the analysis in [7] the discrete models in (7.19) are derived for: a) $-0.05 \leq \dot{\varphi}^d, \dot{\theta}^d, \dot{\psi}^d \leq 0$, and b) $0 \leq \dot{\varphi}^d, \dot{\theta}^d, \dot{\psi}^d \leq 0.05$. The constraints on the inputs and the outputs are $0 \leq U_1 \leq 200$, $-30 \leq U_2, U_3 \leq 30$, and $-25 \leq U_4 \leq 25$, while the constraints on the states are: $-\pi/3 \leq \varphi, \theta, \psi \leq \pi/3$, and $-0.1 \leq \dot{\varphi}, \dot{\theta}, \dot{\psi} \leq 0.1$.

The tuning parameters used in the cost function of the CFTOC are $R = 100I_4$, $Q = 10^4 I_6$; the l_2-norm was selected with a prediction horizon $N = 2$. For testing the tracking performance of the applied control law a trajectory in the φ and θ plane was introduced while $\psi^d = 0$. The reference path in the xy-plane is shown in Figure 7.13. The results of this simulation are presented in Figures 7.14 and 7.15, where the desired (actual) trajectory is presented with dashed (solid) lines. A snapshot of the generated controller partition where $J_p = 3304$ based on its projection on $[\varphi, \psi, \theta]$-space is presented in Figure 7.16.

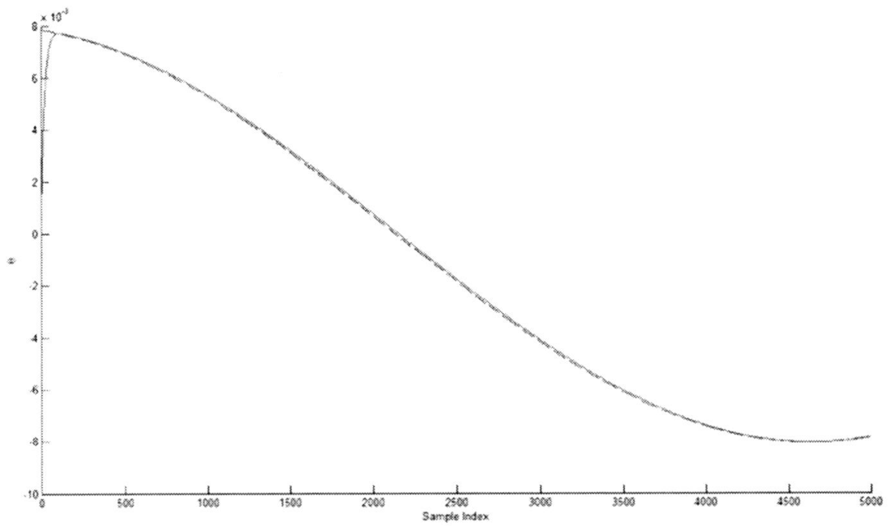

Fig. 7.14 CFTO controlled system θ-tracking response.

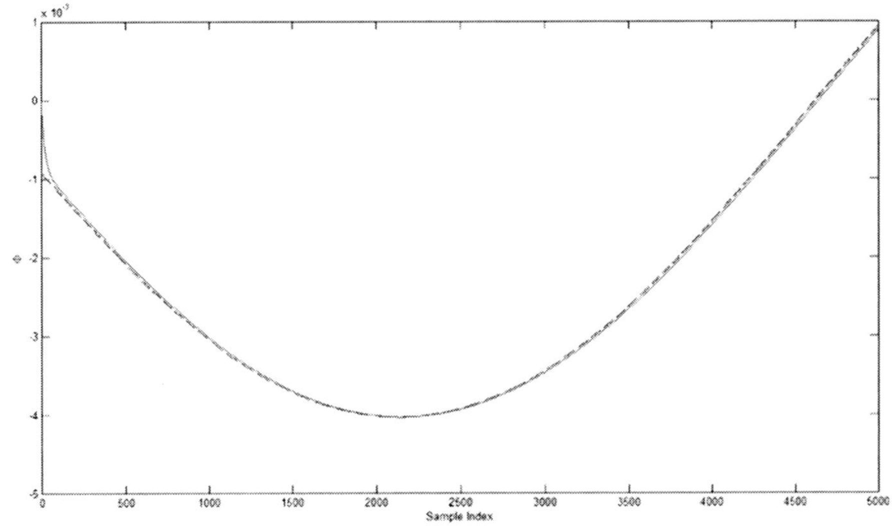

Fig. 7.15 CFTO controlled system φ-tracking response.

7.7 Conclusions

A system using Unmanned quadrotor Helicopters (UqHs) for forest fire surveillance is analyzed. The UqHs are coordinated in a decentralized manner while patrolling the perimeter of the fire. The UqHs exchange mutual information in pairs at their

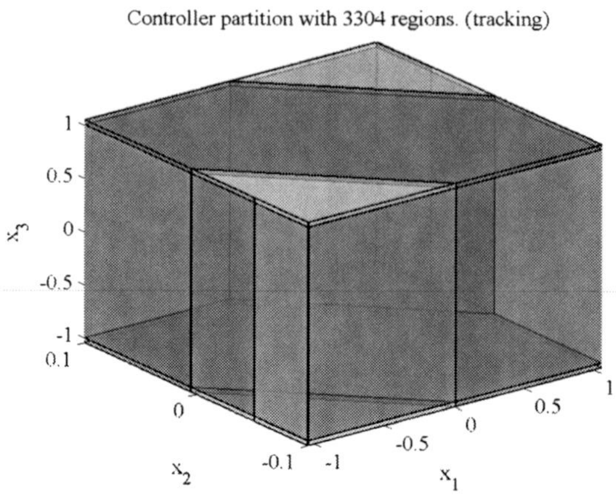

Fig. 7.16 CFTO-controller partitioning (normalized & projected).

rendez-vous points. This information concerns the reference trajectory that each UqH will follow prior to its next rendez-vous. Perturbations from this trajectory are handled via a Constrained Finite Time Optimal Controller (CFTOC) which penalizes the control effort and any state deviations over a prediction horizon. Furthermore, the CFTOC handles any state and input constraints and offers attenuation to the effects of the fire-induced wind gusts.

References

1. G.M. Saggiani and B. Teodorani, Rotary wing UAV potential applications: An analytical study through a matrix method, *Aircraft Engineering and Aerospace Technology* **76**(1), 6–14, 2004.
2. F. Borelli, M. Baotic, A. Bemporad and M. Morari, An efficient algorithm for computing the state feedback optimal control law for discrete time hybrid systems, in *Proceeding of the American Control Conference*, June 2003.
3. P. Grieder, F. Borelli, F. Torrisi and M. Morari, Computation of the constrained infinite time linear quadratic regulator, *Automatica* **40**(4), 1–708, April 2004.
4. M. Kvasnica, P. Grieder, M. Baotic and M. Morari, Multi-Parametric Toolbox (MPT), *Hybrid Systems: Computation and Control* **2993**, 448–462, 2004.
5. G. Nikolakopoulos, L. Dritsas, A. Tzes and J. Lygeros, Adaptive constrained control of uncertain ARMA-systems based on set-membership identification, in *Proceedings of the 14th IEEE Mediterranean Conference on Control and Automation (MED'06)*, June 28–30, Ancona, Italy, 2006.
6. L. Dritsas, G. Nikolakopoulos and A. Tzes, Constrained optimal control over networks with uncertain delays, in *Proceedings of the 45th IEEE Conference on Decision and Control*, San Diego, CA, USA, December 13–15, pp. 4993–4998, 2006.
7. S. Bouabdallah, A. Noth, and R. Siegwart, PID vs LQ control techniques applied to an indoor micro quadrotor, in *Proceedings 2004 IEEE/RSJ International Conference on Intelligent Robots and Systems*, Vol. 3, pp. 2451–2456, 2004.
8. A. Benallegue, A. Mokhtari and L. Fridman, Feedback linearization and high order sliding mode observer for a quadrotor UAV, in *Proceedings of International Workshop on Variable Structure Systems*, pp. 365–372, 2006.

9. S. Bouabdallah and R. Siegwart, Full control of a quadrotor, in *Proceedings of IEEE/RSJ International Conference on Intelligent Robots and Systems*, pp. 153–158, 2007.
10. G.M. Hoffmann, H. Huang, S.L. Waslander and C.J. Tomlin, Quadrotor helicopter flight dynamics and control: Theory and experiment, AIAA Guidance, Navigation and Control Conference and Exhibit, 20–23 August 2007, Hilton Head, South Carolina.
11. L. Merino et al., Cooperative fire detection using unmanned aerial vehicles, in *Proceedings of the 2005 IEEE International Conference on Robotics and Automation*, pp. 1884–1889, 2004.
12. R.W. Beard et al., Coordinated target assignment and intercept for unmanned air vehicles, *IEEE Transactions on Robotics and Automation* **18**(6), 911–922, 2002.
13. M. Alighanbari, Y. Kuwata and J.P. How, Coordination and control of multiple UAVs with timing constraints and loitering, in *Proceedings of the 2003 American Control Conference*, Vol. 6, pp. 5311–5316, 2003.
14. M. Flint, M. Polycarpou and E. Fernandez-Gaucherand, Cooperative control for multiple autonomous UAV's searching for targets, in *Proceedings of the Conference on Decision and Control*, 2002.
15. P.R. Chandler, M. Pachter and S. Rasmussen, UAV cooperative control, in *Proceedings of the 2001 American Control Conference*, Vol. 1, pp. 50–55, 2001.
16. C. Schumacher, P.R. Chandler, M. Pachter and L.S. Pachter, Optimization of air vehicles operations using mixed-integer linear programming, *Journal of the Operational Research Society*, 2007.
17. R.W. Beard and T.W. McLain, Multiple UAV cooperative search under collision avoidance and limited range communication constraints, in *Proceedings of the Conference on Decision and Control*, 2003.
18. D.B. Kingston, W. Ren and R.W. Beard, Consensus algorithms are input-to-state stable, in *Proceedings of the 2005 American Control Conference*, Vol. 3, pp. 1686–1690, 2005.
19. R. Gardner et al., Climate change, disturbances, and landscape dynamics, in *Global Change and Terrestrial Ecosystems*, International Geosphere-Biosphere Programme Book Series – Book #2, Cambridge University Press, Great Britain, pp. 149–172, 1996.
20. W. Hargrove et al., Simulating fire patterns in heterogeneous landscapes, *Ecological Modelling* **135**, 243–263, 2000.
21. Folium, Wolfram, http://mathworld.wolfram.com/Folium.html
22. T. Cebeci, J.P. Shao, F. Kafyeke and E. Laurendeau, *Computational Fluid Dynamics for Engineers*, Springer, 2005.
23. http://research.esd.ornl.gov/EMBYR/embyr.html
24. G.D. Padfield, *Helicopter Flight Dynamics*, Blackwell, Oxford, 1996.
25. G. Fay, Derivation of the aerodynamic forces for the mesicopter simulation, Internal Report, Stanford University, 2001, http://adg.stanford.edu/
26. J. Etele, Overview of wind gust modelling with application to autonomous low-level UAV control, Defense R&D Canada-Ottawa, 2006.
27. D.P. Bertsekas, *Dynamic Programming and Optimal Control*, Athena Scientific, 1995.
28. R.W. Beard and T.W. McLain, Multiple UAV cooperative search under collision avoidance and limited range communication constraints, in *Proceedings of Conference on Decision and Control*, 2003.
29. D.W. Casbeer, D.B. Kingston, R.W. Beard and T.W. McLain, Cooperative forest fire surveillance using a team of small unmanned air vehicles, *International Journal of Systems Science*, 1–18, January 2005.
30. D. Kingston, R.W. Beard and R.S. Holt, Decentralized perimeter surveillance using a team of UAVs, *IEEE Trans. on Robotics* **24**(6), December 2008.

Chapter 8
Genetic Fuzzy Rule-Based Classifiers for Land Cover Classification from Multispectral Images

D.G. Stavrakoudis, J.B. Theocharis and G.C. Zalidis

Abstract This chapter presents a Boosted Genetic Fuzzy Classifier (BGFC), for land cover classification from multispectral images. The model comprises a set of fuzzy classification rules, which resemble the reasoning employed by humans. BGFC's learning algorithm is divided into two stages. During the first stage, a number of fuzzy rules are generated in an iterative fashion, incrementally covering subspaces of the feature space, as directed by a boosting algorithm. Each rule is able to select the required features, further improving the interpretability of the obtained model. The rule base generation stage is followed by a genetic tuning stage, aiming at improving the cooperation among the fuzzy rules and, subsequently, increasing the classification performance attained after the former stage. The BGFC is tested using an IKONOS multispectral VHR image, in a lake-wetland ecosystem of international importance. For effective classification, we consider advanced feature sets, containing spectral and textural feature types. The results indicate that the proposed system is able to handle multi-dimensional feature spaces, effectively exploiting information from different feature sources.

8.1 Introduction

Over the past few decades, classification of remotely sensed images has attracted considerable research interest. The increasing need for environment protection strategies and urban development planning, along with the availability of very high

D.G. Stavrakoudis · J.B. Theocharis
Aristotle University of Thessaloniki, Department of Electrical and Computer Engineering, Division of Electronics and Computer Engineering, 54124 Thessaloniki, Greece; e-mail: jstavrak@auth.gr; theochar@eng.auth.gr

G.C. Zalidis
Laboratory of Applied Soil Science, Faculty of Agronomy, Aristotle University of Thessaloniki, 54124 Thessaloniki, Greece; e-mail: zalidis@agro.auth.gr

K.P. Valavanis (ed.), Applications of Intelligent Control to Engineering Systems, 195–221.

resolution (VHR) commercial satellite images, have boosted the demand for automated or semi-automated classification algorithms. Traditionally, statistical classifiers, like the Maximum Likelihood Classifier (MLC) [1], have been used for land cover classification tasks. However, the availability of hyperspectral satellite imagery and the exploitation of advanced features, aiming at decreasing the overlapping between the spectral signatures of the classes, have rendered the use of statistical classifiers impractical, due to the acute increase in the feature space dimensionality.

In the direction of producing reliable classification performance in high-dimensional feature spaces, various classifiers based on computational intelligence have been proposed [2]. Traditionally, neural network classifiers are suggested [3, 4], taking advantage of the enhanced mapping capabilities of neural networks. Fuzzy theory offers an attractive framework for building classifiers that are easily interpretable by humans, while at the same time modeling the imprecision encountered in nature. A hybridization of fuzzy theory with neural networks has given rise to neuro-fuzzy classifiers [5–7], which combine the mapping and learning capabilities of neural networks with the ability of fuzzy classifiers to handle imprecise data.

The main drawback of neuro-fuzzy classifier is that the interpretability of the resulting system is spoiled by their complex architecture. Undoubtedly, fuzzy rule-based systems (FRBSs) have the most intuitive representation in the context of fuzzy systems, comprising fuzzy rules that closely resemble propositions made by human beings. Recently, a few fuzzy rule-based classification systems (FRBCSs) have been proposed for land cover classification [8–11]. In these systems, fuzzy classification rules are produced and/or tuned using simulated annealing [8], self-organizing feature map (SOFM) [9], clustering algorithms such as k-means [10], as well as directly from the training data, in an explicit supervised approach [11]. However, these traditional learning techniques do not scale well with the increase in the dimensionality of the feature space.

Since the middle of the past decade, considerable research on genetic FRBCSs (GFRBCSs) has been conducted [12]. GFRBCSs exploit the enhanced search capabilities of Evolutionary Algorithms (EAs), in order to automatically extract an optimal fuzzy rule base. Feature selection schemes can easily be embedded during the genetic evolution, which is very useful when handling hyperspectral images or when considering advanced features, derived from the initial bands of a multispectral satellite image. The aforementioned merits have led to the proposal of GFRBCSs – and genetic FRBSs (GFRBSs) in general – for many and diverse application areas [13].

In this chapter, we present a GFRBCS for land cover classification of remotely sensed images. The model is developed through a two-stage learning scheme. During the first stage, an initial fuzzy rule base is derived in an incremental fashion, iteratively producing one fuzzy rule at a time. The features subspaces already covered by fuzzy rules are effectively penalized by a boosting algorithm, thus forcing new rules to focus in uncovered regions. At the same time, the boosting algorithm induces cooperation among the rules during their creation. Nonetheless, a genetic tuning stage follows the rule generation one, in order to further improve the cooperation between the rules and, thus, increase the classification performance. A local feature selection employed in the rule generation stage automatically infers the most important features in a per rule basis, allowing the effective manipulation of highly dimensional feature spaces. Therefore, advanced features can easily be considered, increasing the classification efficiency.

This chapter is organized as follows. In Section 8.2, we briefly discuss the basic concepts of GFRBCSs and Evolution Strategies (ES), which will be used in subsequent sections. A detailed description of the presented system and its learning algorithm is provided in Section 8.3. Experimental results are reported in Section 8.4, using an IKONOS multispectral image. Finally, some conclusions are summarized in Section 8.5.

8.2 Basic Concepts

The classification task consists of assigning an object defined in a certain feature space $\mathbf{x} = [x_1, \ldots, x_N] \in F^N$ to a class C_j from a predefined class set $C = \{C_1, \ldots, C_M\}$, where N and M are the dimensions of the feature and the class spaces, respectively. A FRBCS comprises a number of IF−THEN rules, where the IF part (termed *premise* or *antecedent* part) defines a fuzzy subspace in the feature space covered by the rule, and the THEN part (the *consequent* part) describes the class of the rule. The fuzzy subspace in the antecedent is formed by a conjunction of fuzzy sets, defined along each input dimension. Each fuzzy set is described by a membership function, which determines the degree to which each input value belongs to the fuzzy set. The proposed system is produced through a supervised learning scheme, which means that a set of labelled patterns $E = \{e^p = (\mathbf{x}^p, c^p), p = 1, \ldots, Q\}$ is used to produce a classifier that performs this mapping from the feature space to the class space, with the minimum possible rate of misclassifications.

8.2.1 Fuzzy Ruled-Based Classification Systems

Fuzzy classification rules can have various forms. The simplest one uses a class in the consequent, having the structure:

$$R_k\text{:}\quad \text{IF } x_1 \text{ is } A_1^k \text{ and } \cdots \text{ and } x_N \text{ is } A_N^k \text{ THEN } y \text{ is } C_j^{(k)}, \tag{8.1}$$

where $A_i^k, i = 1, \ldots, N$, are fuzzy sets defined in the i-th input dimension and y is the class $C_j^{(k)}$ to which the pattern belongs. Another form of fuzzy rules uses a class accompanied by a certainty degree in the consequent:

$$R_k\text{:}\quad \text{IF } x_1 \text{ is } A_1^k \text{ and } \cdots \text{ and } x_N \text{ is } A_N^k \text{ THEN } y \text{ is } C_j^{(k)} \text{ with } r^k, \tag{8.2}$$

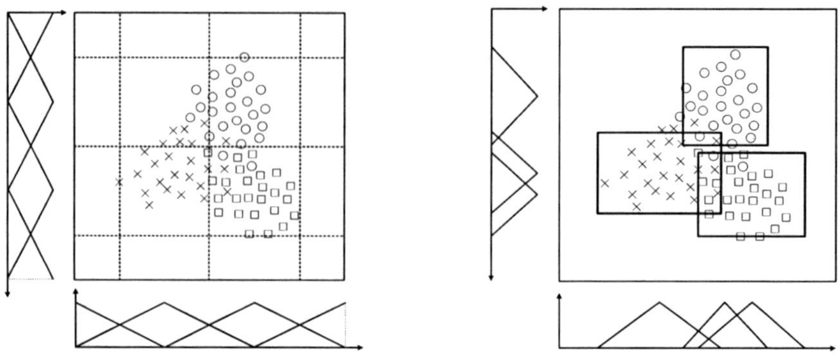

Fig. 8.1 Feature space fuzzy partitions for descriptive (left) and approximate (right) FRBCSs.

where r^k is the certainty degree of the classification in the class $C_j^{(k)}$ for a pattern belonging to the fuzzy subspace defined in the premise part of the rule. This certainty degree is usually calculated as:

$$r^k = S_j^k / S^k, \tag{8.3}$$

where S_j^k is the sum of the rule's firings for the class $C_j^{(k)}$ training patterns, and S^k is the same sum for all training patterns, irrespective of class.

Finally, the most general case is fuzzy classification rules with certainty degrees for all classes in the consequent:

$$R_k\colon \quad \text{IF } x_1 \text{ is } A_1^k \text{ and } \cdots \text{ and } x_N \text{ is } A_N^k \text{ THEN } (r_1^k, \ldots, r_M^k), \tag{8.4}$$

where r_j^k denotes the certainty degree of rule k to predict class C_j, also calculated by (8.3). The first two types of fuzzy classification rules are special cases of this type. Considering $r_j^k = 1$ and $r_h^k = 0$ if $h \neq j$ we have the first type of rules, while if $r_j^k = r^k$ and $r_h^k = 0$ if $h \neq j$ we have the second one.

Two different approaches can be considered for the formulation of the fuzzy sets in the premise part of the rule. In the first one, each input variable is partitioned into a set of possible values, represented by user-defined fuzzy sets with a linguistic meaning. The resulting system is termed *descriptive* FRBCS, because it can be described by a set of simple phrases. In the second approach, each rule is allowed to define its own fuzzy sets, rather than using linguistic terms from a global term set, resulting in the so-called *approximate* FRBCSs. Figure 8.1 illustrates the differences of the two types of fuzzy systems, for a two feature space. The advantage of a descriptive system lies in its interpretability, since the meaning of each fuzzy set is uniform for all rules. However, in order to achieve accurate results, it is often necessary to increase the granularity of the term sets in each dimension, which, in turn, radically increases the search space in high-dimensional problems. On the other hand, an approximate FRBCS is generally able to produce higher classification performance, with fewer rules. Although the latter lacks a linguistic interpretation, it can still be easily represented and interpreted visually, making it a reasonable compromise between accuracy and interpretation.

8.2.2 Fuzzy Inference

The fuzzy reasoning method determines how the rules are combined in order to infer a classification decision, given a feature vector $\mathbf{x}^p$ to be classified. The matching degree $\mu_k(\mathbf{x}^p)$ of the k-th rule is the degree of compatibility between the input vector $\mathbf{x}^p$ and the antecedent part of the rule, also known as rule firing. In the presented system, the AND operator is implemented through the algebraic product, and the matching degree is calculated as follows:

$$\mu_k(\mathbf{x}^p) = \prod_{i=1}^{N} \mu_i^k(x_i^p), \tag{8.5}$$

where $\mu_i^k(x_i^p)$ is the membership grade of the fuzzy set defined in the i-th dimension of the k-th fuzzy rule. Assuming fuzzy rules in the form of (8.2), the input pattern is finally assigned to the class:

$$C_{\max} = \underset{c=1,\ldots,M}{\arg\max} \sum_{R^k/C_j^{(k)}=c} \mu_k(\mathbf{x}^p) \cdot r^k \tag{8.6}$$

where $C_j^{(k)}$ is the class label in the consequent part of the k-th rule. In the above relation, rule aggregation is performed using the algebraic sum. This approach is known as *maximum voting scheme*, and is frequently used in FRBCSs [14–16]. Using the maximum voting scheme, all overlapping fuzzy rules contribute to the classification decision, rather than the traditional case where only the rule with the maximum matching is considered.

The methodology presented here does not impose any restriction to the membership functions describing the fuzzy sets. Therefore, in order to increase the flexibility of the resulting system, we chose to implement the fuzzy sets through double-sided Gaussian membership functions, having the following mathematical expression:

$$\mu_i^k(x_i^p) = \begin{cases} \exp\left\{-(x_i^p - m_\ell^{(k,i)})^2/(\sigma_\ell^{(k,i)})^2\right\}, & x_i^p < m_\ell^{(k,i)} \\ 1.0, & m_\ell^{(k,i)} \leq x_i^p \leq m_r^{(k,i)} \\ \exp\left\{-(x_i^p - m_r^{(k,i)})^2/(\sigma_r^{(k,i)})^2\right\}, & x_i^p > m_r^{(k,i)} \end{cases} \tag{8.7}$$

where $m_\ell^{(k,i)}$ and $m_r^{(k,i)}$, $m_\ell^{(k,i)} \leq m_r^{(k,i)}$, are the centers of the left and right Gaussian functions, and $\sigma_\ell^{(k,i)}$ and $\sigma_r^{(k,i)}$ are the respective widths.

8.2.3 Evolution Strategies

Evolutionary algorithms (EAs) are universal optimization methods, inspired from the genetic adaptation of natural evolution. There exist various implementations of EAs, with the Genetic Algorithms (GAs) being the most well-known. However, all these implementation share a more or less similar framework. In particular, an EA maintains a population of trial solutions, which evolves over time by means of genetic operators such as mutation, recombination (also known as crossover) and selection. In each step, called generation, each candidate solution is evaluated using an objective function. Fittest individuals, as defined by the objective function, have a higher probability of survival, while weak individuals are eventually removed from the population. EAs are well-known for avoiding local minima, compared to traditional supervised optimization methods.

Evolution strategies are a type of evolutionary algorithm characterized by the self-adaptation of additional strategy parameters, adapting the algorithm itself throughout the evolution, along with the object variables [17]. They are well-suited for real-valued parameter optimization problems and have been previously used in the design of GFRBSs [18–20]. Here, we concentrate on the most well-known version of ES, the so-called (μ, λ)–ES.

Each chromosome in the population is formed by a pair of real-valued vectors $(\mathbf{x}, \mathbf{s})$. The object variables $(x_1, \ldots, x_n)$ represent the candidate solution, defined in $\Re^n$. The additional vector $(s_1, \ldots, s_n)$ represents the standard deviation of the mutations, one for each object variable. The population is formed by μ individuals. At each generation, two individuals are randomly selected from the population, with uniform probability, in order to be recombined and produce two children individuals. Different recombination operators are usually employed for the object variables and the strategy parameters: the most common choice is discrete recombination for the object variables and generalized intermediate recombination for the strategy parameters. In discrete recombination, each gene of a child is randomly selected from the respective gene of one of the parents. In the generalized intermediate recombination, the two children's genes receive the same value, a linear combination of their parents' genes:

$$s_i' = a_i s_i^{(1)} + (1 - a_i) s_i^{(2)}, \tag{8.8}$$

where a_i is a uniformly distributed random number in $[0, 1]$.

After recombination, each child is mutated. The strategy parameters are the first to be mutated, through:

$$s_i' = s_i \cdot e^{\tau' \cdot N(0,1) + \tau \cdot N_i(0,1)}, \tag{8.9}$$

where $N(0, 1)$ represents a normally distributed random variable sampled once for the entire chromosome, whereas $N_i(0, 1)$ represents normal random variables sampled anew for each gene. The global factor $e^{\tau' \cdot N(0,1)}$ allows for an overall change of the mutation rate (in the object variables), while the local term $e^{\tau \cdot N_i(0,1)}$ allows for individual changes along each dimension. The step-size constants are usually set as $\tau' = (2n)^{-1/2}$ and $\tau = (2\sqrt{n})^{-1/2}$. The mutation of the object variables uses the mutated values of the strategy parameters:

$$x_i' = x_i + s_i' \cdot N_i(0, 1). \tag{8.10}$$

Recombination and mutation are repeated, until λ ($\lambda > \mu$) children individuals are produced, which are then sorted according to their fitness value. Out of these λ intermediate individuals, the μ fittest are selected to form the next parent population. Although the value of μ and λ are generally set through a trial-and-error process, a rule of thumb is that the ratio $\mu/\lambda = 1/7$ should be selected, with the $(15, 100)$–ES being the most frequently used choice [17]. Under a successful evolution in an ES run, the strategy parameters tend to zero, disallowing changes in the object variables (8.10), and thus the whole population converges to an optimal solution.

8.2.4 Genetic Fuzzy Rule-Based Classifiers

Two genetic learning approaches, adopted from the field of genetic-based machine learning systems, have been traditionally used in GFRBSs: the *Pittsburg* and the *Michigan* approaches [12]. In a Pittsburg GFRBS [21–24] each chromosome encodes an entire rule base and the individuals of the population compete among them along the evolutionary process. In the Michigan approach [16, 25, 26], also known as classifier system, each chromosome represents a single rule and the population forms the rule base. In the past few years, a new approach for learning GFRBSs has been proposed, namely, the *iterative rule learning* (IRL) methodology [27, 28]. In this approach, each chromosome represents a single rule and an evolutionary rule generation algorithm is invoked in an iterative fashion, incrementally adding new rules to the rule base. To induce the desired cooperation between the rules, since each rule is unaware of the previously obtained ones, those training examples that are sufficiently covered by the current rule set are removed from the training set, so that subsequent invocations of the rule generation algorithm will concentrate on the remaining uncovered instances. Prominent examples of GFRBSs following the IRL approach are the MOGUL [28, 29] and the SLAVE [14, 30] systems.

Following the IRL approach, rules generated at later stages may conflict with previously generated ones, as they are unaware of the removed training patterns. This is an instance of the well-known cooperation vs. competition problem, common in learning GFRBSs under the IRL approach. The MOGUL system attacks this problem by splitting the rule generation process into two stages: the first one creates an intermediate set of candidate rules, which is then post-processed in the second stage, the so-called genetic multiselection stage, in order to improve the cooperation among the rules of the final rule set. On the other hand, the SLAVE algorithm, proposed for classification tasks, learns one particular class label at a time. When all patterns of that class label have been covered, the rule generation algorithm is invoked to cover patterns of the next class label.

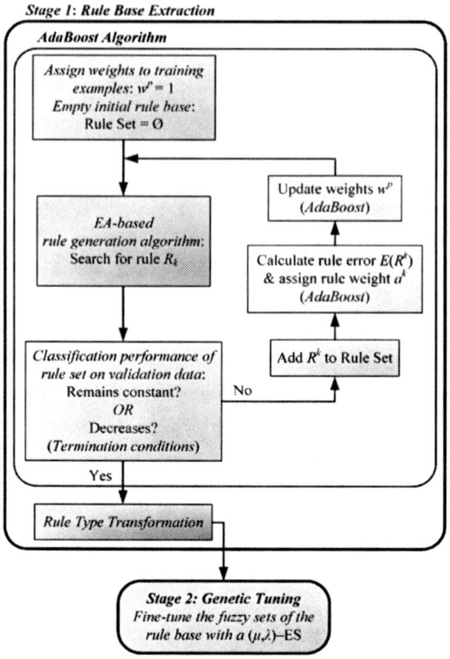

Fig. 8.2 Outline of BGFC's learning algorithm.

8.3 Boosted Genetic Fuzzy Classifier

Under the IRL approach, training examples that are covered by the current rule set are completely removed from the training set. However, this can harness the co-operation between the rules, since newly generated rules are unaware of previously derived ones. In order to cope with this problem, the use of AdaBoost has been proposed for designing GFRBCSs under the IRL approach [20, 31]. The AdaBoost algorithm [32] assigns weights to every example of the training set. Each time a new rule is generated, instead of completely removing the partially covered examples, AdaBoost effectively down-weights those examples covered by a rule, according to their matching degree. Thus, subsequent invocations to the rule generation algorithm concentrate on uncovered instances, while, at the same time, new rules are indirectly aware of the current rule set, a fact that implicitly promotes cooperation of the rules in the rule generation stage.

The BGFC proposed here is an approximate GFRBCS, designed under the IRL approach. An outline of BGFC's learning algorithm is shown in Figure 8.2. It comprises two basic stages. In the first stage, AdaBoost incrementally constructs the fuzzy rule base, through repeated invocations of a hybrid EA-based rule generation algorithm. During this step, fuzzy rules have only a class in the consequent, as in (8.1). After the whole rule base has been obtained, the type of its fuzzy classification rules is transformed into the form of (8.2), in order to increase the interpretability of the fuzzy rule base. The rule extraction algorithm also implements a local feature selection scheme, selecting only those features which best describe the current distribution of training patterns. In the second stage, the parameters of the obtained fuzzy sets are fine-tuned through a genetic tuning process, with the goal to further increase the model's classification performance. The various steps of BGFC's learning algorithm are detailed in the rest of this section.

8.3.1 AdaBoost Algorithm

AdaBoost is a boosting algorithm, commonly used to combine linear classifiers. The main idea of boosting is to combine a set of low quality classifiers (called weak hypotheses in the boosting literature) with a voting scheme, in order to generate an overall classifier that performs better than any of its constituents alone. Each fuzzy rule of a FRBCS can be considered as a simple, partial classifier, that is, as a weak hypothesis. Assuming a training set of labelled patterns $E_{trn} = \{e^p = (\mathbf{x}^p, c^p)/p = 1, \ldots, Q\}$, a weight w^p is assigned to each one of the Q training patterns, representing the relative importance of the p-th training instance. All weights are initialized with a unity value $w^p = 1$. Each time the rule generation algorithm is invoked, a new rule R_k is generated and AdaBoost computes the error $E(R_k)$ of the rule, taking into account the current weights and the matching degrees (8.5) of the patterns with the rule's antecedent:

$$E(R_k) = \frac{\sum_{k/C_j^{(k)} \neq c^p} w^p \mu_k(\mathbf{x}^p)}{\sum_{k=1}^{Q} w^p \mu_k(\mathbf{x}^p)}, \tag{8.11}$$

where $C_j^{(k)}$ is the class of the k-th rule. After the new rule generation, incorrectly classified patterns retain their former weights, while the weights of the correctly classified examples are reduced by a factor $\beta^k(\mathbf{x}^p)$:

$$w^p \leftarrow \begin{cases} w^p, & \text{if } C_j^{(k)} \neq c^p, \\ w^p \cdot \beta^k(\mathbf{x}^p), & \text{if } C_j^{(k)} = c^p. \end{cases} \tag{8.12}$$

The factor

$$\beta^k(\mathbf{x}^p) = \left(\frac{E(R_k)}{1 - E(R_k)}\right)^{\mu_k(\mathbf{x}^p)} \tag{8.13}$$

depends on the current rule's error and its matching degree with the p-th pattern. Effectively, examples that are classified correctly and match the rule antecedent are down-weighted, whereas misclassified or uncovered examples keep their original weights. Thus, when seen in a relative scope, the algorithm increases the weights of those training examples that are difficult to learn.

Under the AdaBoost philosophy, when an unseen instance $\mathbf{x}^p$ is input in the final classifier, the classification decision $C_{\max}$ is made following a weighted maximum voting scheme:

$$C_{\max} = \arg\max_{c=1,\dots,M} \sum_{R^k / C_j^{(k)} = c} a^k \cdot \mu^k(\mathbf{x}^p), \tag{8.14}$$

where

$$a^k = \log\left(\frac{1 - E(R_k)}{E(R_k)}\right) \tag{8.15}$$

is a weight showing the importance of the rule. Thus, unreliable, low performing rules have a reduced significance in the final classification decision, as compared to rules exhibiting a small error. To avoid numerical instabilities in (8.15), we constrain the error to be strictly positive, by setting:

$$E(R_k) \leftarrow \max(E(R_k), t_\ell), \tag{8.16}$$

where $t_\ell \ll 1$ is a threshold representing the smallest rule error allowed. We can set the threshold to any reasonably small value, for example $t_\ell = 0.01$.

The decision to terminate the iterative rule generation procedure depends on the rule base's performance in a separate validation set of examples. Accordingly, our initial dataset is split into three disjoint sets: the training, the validation and the checking ones. Each time a new rule is generated, the classification performance of the fuzzy rule base obtained so far is calculated in the validation set, using (8.14). The AdaBoost algorithm terminates either if this performance decreases or if it remains constant for a small number of rules (for example, four). The rule responsible for the classification performance decrement or the superfluous rules, which produced no increase in the classifier's performance, are not included in the exported fuzzy rule base.

8.3.2 Rule Generation Algorithm

The rule generation algorithm is implemented through an EA, which evolves a population of individuals in order to locate the best rule, under the current distribution of the training set. Each individual encodes a single fuzzy rule and the individuals of the population compete with each other, given the current weights w^p of the training set. The BGFC's rule generation algorithm uses a hybrid representation scheme, with each chromosome being comprised by a binary and a real-valued part, as shown in Figure 8.3. The binary part, with a length of N genes, determines whether the fuzzy clause "x_i is A_i^k" will be present in the rule. A zero value omits the respective input variable from the rule, which in fuzzy terms has the meaning of a "don't care" fuzzy set. The "don't care" fuzzy set returns a unity membership value for any input. Thus, in mathematical terms, the rule's activation is unaffected by such a fuzzy set, due to the conjunctive AND operator in the rule's premise part

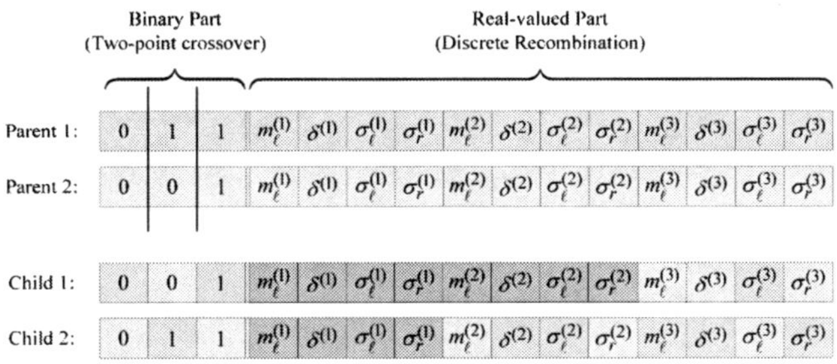

Fig. 8.3 An example of recombination between two parents, forming the two children individuals with N=3. Darker regions in the children individuals represent genes belonging to inactive feature.

(8.1)–(8.4). In this way, each rule implements its own, local feature selection process and chooses which input variables are required for the best representation of the patterns its antecedent covers. For the sake of simplicity, in the following we will refer to an input variable (feature) x_i as "active" when the fuzzy clause "x_i is A_i^k" is present in the rule and as "inactive" in the opposite case.

The real-valued part of the chromosome is formed by concatenating N four-valued parts, each encoding the parameters of the fuzzy sets pertaining in the k-th rule. Following the notation of (8.7), each fuzzy set is expressed as a four-valued part $\{m_\ell^{(k,i)}, \delta^{(k,i)}, \sigma_\ell^{(k,i)}, \sigma_r^{(k,i)}\}$, where $\delta^{(k,i)} = m_r^{(k,i)} - m_\ell^{(k,i)}$. The distance of the right center from the left is encoded in the chromosome, instead of using the absolute value of the right center directly, so that a fuzzy set remains valid as long as $\delta^{(k,i)}$ is strictly positive. The parameters of all fuzzy sets exist in the real part of the chromosome. However, if an input variable is inactive, the respective part is disregarded. Therefore, the population encodes individuals of equal size.

For the binary part, the most frequently used GA operators are considered, that is, two-point crossover and uniform mutation. The real-valued part uses the ES-specific operators presented in Section 8.2.3, which are applied only if the respective input variable in the child individual is active. Figure 8.3 depicts an example of the recombination between two parents. The binary parts are crossed first, determining the active variables of the children. Subsequently, recombination of the real-valued part is performed, in those genes that belong to active input variables.

The rest of the algorithm follows the steps of a (μ, λ)–ES, as described in Section 8.2.3. In each generation, λ individuals are generated through repeated application of the genetic operators in pairs of parents and the μ fittest are selected to form the parent population in the next generation. The rule generation algorithm terminates and returns the best rule found if the maximum number of generations has been reached, or, if the standard deviation of the μ individuals' fitness values falls below a pre-specified threshold, expressed as a percentage of its initial value. The latter condition is used to identify the convergence of the EA to a unique solution.

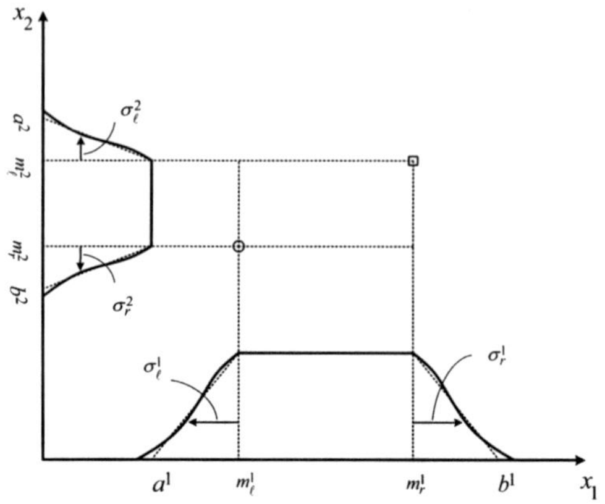

Fig. 8.4 The initialization scheme of the two-sided Gaussian fuzzy sets.

8.3.2.1 Initialization

In order to speed up the rule extraction procedure, the initial population of chromosomes is generated using the training examples directly, rather than randomly. In particular, two training patterns $\mathbf{x}^1$ and $\mathbf{x}^2$ are selected randomly, with a probability proportional to the weights w^p assigned to them. The two centers of each two-sided Gaussian membership function are selected such as the two training examples span the core of the fuzzy set (disregarding the rule index k for simplicity):

$$m_\ell^i = \min\{x_i^1, x_i^2\},$$
$$m_r^i = \max\{x_i^1, x_i^2\}, \tag{8.17}$$

as depicted in Figure 8.4. The widths of the Gaussian parts are calculated by considering an equivalent trapezoidal fuzzy set, with a support twice as wide as its core:

$$b^i - a^i = 2(m_r^i - m_\ell^i), \tag{8.18}$$

where a^i and b^i are the extreme points of the trapezoidal fuzzy set. It is then possible to approximate the initial widths of the two-sided Gaussian fuzzy set as [33]:

$$\sigma_\ell^i = \sigma_r^i = \frac{m_r^i - m_\ell^i}{2\sqrt{\pi}}. \tag{8.19}$$

The width of the core $m_r^i - m_\ell^i$ also determines the initial values of the strategy parameters $s_{m_\ell^i, m_r^i} = 0.2 \cdot (m_r^i - m_\ell^i)$ and $s_{\sigma_\ell^i, \sigma_r^i} = 0.1 \cdot (m_r^i - m_\ell^i)/\sqrt{\pi}$ in the ES algorithm. The binary part of the rule is initialized randomly.

8.3.2.2 Rule Consequent

The rule consequent $C_j^{(k)}$ can be easily calculated as the class dominating among the training instances covered by the rule antecedent:

$$C_j^{(k)} = \arg\max_{c=1,\ldots,M} \sum_{p/c^p=c} \mu_k(\mathbf{x}^p). \tag{8.20}$$

Each time the rule generation algorithm is invoked, rules describing different class labels compete with each other, taking into consideration the current weights w^p assigned to the training patterns. Assuming all classes are – roughly – equally represented in the training set, AdaBoost will eventually produce rules for all classes. However, in land cover classification tasks, it is a quite common phenomenon to have minor classes with very few examples. In such cases, it may be reasonable to fix the consequent of the first M rules (where M is the number of classes), to one class at a time, in order to assure that the obtained classifier will be able to output all class labels. Subsequently, the consequents for the rest of the rules can be calculated through (8.20). This technique was adopted for the experiments presented in Section 8.4.

8.3.2.3 Fitness Function

The goal of the rule generation algorithm is to provide rules covering a high number of examples, with as few misclassifications as possible. In order to achieve these two objectives, the EA considers a number of optimization criteria, aggregated into a single scalar fitness value. In particular, three criteria are taken into consideration in BGFC's rule generation algorithm. All of them depend on the number of positive and negative examples covered by the k-th rule, which are defined, respectively, as:

$$n^+ = \sum_{p/C_j^{(k)}=c^p} w^p \cdot \mu_k(\mathbf{x}^p), \tag{8.21}$$

and

$$n^- = \sum_{p/C_j^{(k)}\neq c^p} w^p \cdot \mu_k(\mathbf{x}^p). \tag{8.22}$$

The first criterion, called *class coverage* of the rule, is defined as the number of training instances covered by the k-th rule, compared to the overall number of training instances belonging to the rule's class:

$$f_1 = n^+ \Big/ \sum_{p/C_j^{(k)}=c^p} w^p. \tag{8.23}$$

The second criterion, referred to as the *support* of the rule, expresses the generality of the rule and is related to the fraction on training instances covered by the rule, irrespective of the class label, defined as:

$$n = n^+ \Big/ \sum_{p=1}^{Q} w^p. \tag{8.24}$$

A rule is considered general enough when it covers a significant fraction of all instances:

$$f_2 = \begin{cases} 1, & \text{if } n > k_{\text{cov}}, \\ \frac{n}{k_{\text{cov}}}, & \text{otherwise}, \end{cases} \tag{8.25}$$

with $k_{\text{cov}} \in [0.2, 0.5]$, depending on the number of classes.

Rule consistency demands a rule to cover a high number of positive examples and few negative ones. The final criterion, the so-called *k-consistency* of a rule, is defined as:

$$f_3 = \begin{cases} 0, & \text{if } k_c \cdot n^+ < n^- \\ \frac{n^+ - n^-/k_c}{n^+}, & \text{otherwise}, \end{cases} \tag{8.26}$$

with $k_c \in [0, 1]$. The EA tries to maximize all three fitness criteria, which are normalized in [0,1]. The overall fitness of each chromosome is computed by the product:

$$f = f_1 \cdot f_2 \cdot f_3. \tag{8.27}$$

Notice that all individual fitness functions depend on the current weights w^p assigned to the training instances. Thus, cooperation among the rules in the final rule base is implicitly promoted, since even examples covered by previous rules affect the creation of a new rule, but with a degree relative to their weight. In other words, the evolutionary algorithm is biased towards generating new fuzzy rules that complement well the current rule set.

8.3.3 Rule Form Transformation

The AdaBoost algorithm constructs rules with a single class label in the consequent (8.1) and assigns weights a^k to each one of them (8.15), according to their performance at the time of their creation. These weights, however, degrade the interpretability of the resulting fuzzy rule base, since they can attain large positive values, blurring the intuitive representation of the FRBCS. Moreover, their maximum value is constrained through the threshold t_ℓ (8.16), which is heuristically selected by the designer. Therefore, after all rules have been created, we disregard the rule weights a^k and transform rules to the form of (8.2) (see Figure 8.2), with the certainty degrees being calculated as:

$$r^k = \frac{\sum_{k/C_j^{(k)}=c^p} \mu_k(\mathbf{x}^p)}{\sum_{k=1}^{Q} \mu_k(\mathbf{x}^p)}. \tag{8.28}$$

The resulting system finally classifies unknown examples through (8.6). The observing reader will have noticed that the certainty degrees r^k have a similar meaning to the rule weights a^k and that (8.6) and (8.14) are mathematically equivalent. Indeed, the weights a^k are a different way of measuring the confidence of the rule. However, there are two important distinctions. First, the certainty degrees r^k are normalized in [0, 1], whereas the rule weights a^k can take any value. Second, and most importantly, the certainty degrees are only involved in deriving the soft classification decisions and, therefore, do not spoil the intuitive interpretation of the fuzzy rule base, as rule weights do.

8.3.4 Genetic Tuning

The induction of the AdaBoost algorithm in the IRL approach, presented so far, aims at implicitly promoting the cooperation between the fuzzy rules of the obtained FRBCS. As we have seen, the rule generation algorithm is responsible for tuning the fuzzy membership parameters, along with selecting the relative features for each rule, since the BGFC implements an approximate FRBCS. To that extend, the BGFC is able to obtain a high-performing, well-cooperating fuzzy rule base in a single stage. However, in difficult classification tasks, there is still a possibility that the cooperation between the rules in the final rule base obtained, enforced implicitly through the boosting scheme, is not the optimal one. Moreover, the transformation applied to the fuzzy classification rule type at the end of the rule base extraction algorithm may diminish the classification performance of the final rule base.

Therefore, a post-processing genetic tuning stage follows the rule base extraction one in BGFC's learning algorithm, as depicted in Figure 8.2. The goal of the genetic tuning stage is to optimize the cooperation, and thereby the performance, of the fuzzy rule base obtained in the previous step. Note that this stage performs only a fine tuning the parameters of the fuzzy sets in the rule base, rather than a coarse search in the feature space. In plain words, it tries to maximize the performance of a nearly optimal FRBCS, produced in the first stage. Arguably, classification problems with high signature overlapping of neighboring classes are mostly expected to benefit from this stage, since in those cases the consistency condition (8.26) discourages the creation of rules in mixed areas of the feature space.

Tuning is implemented through a real-valued (15,100)–ES, as presented in Section 8.2.3. A fuzzy set is represented by a vector of its four real-valued parameters $\{m_\ell^{(k,i)}, \sigma_\ell^{(k,i)}, m_r^{(k,i)}, \sigma_r^{(k,i)}\}$. The whole chromosome is formed by concatenating the parameters of all active fuzzy sets for all rules, in a Pittsburg approach. The parameters of the inactive features are omitted from the genetic representation, since this stage does not aim at changing the features selected for each rule. The class label

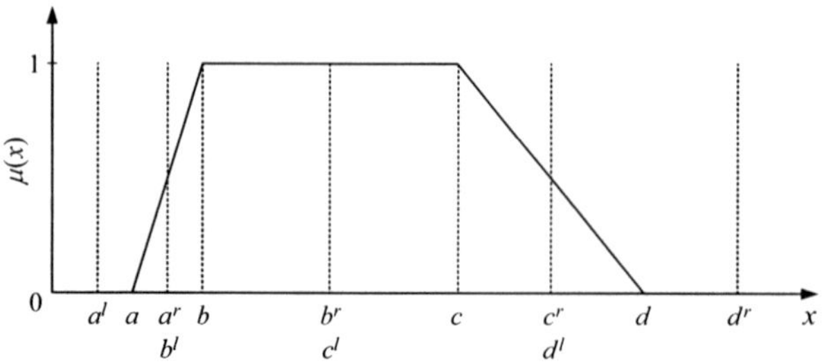

Fig. 8.5 Fixed intervals of adjustment for a trapezoidal membership function.

in the consequent part of the rule also remains fixed at the value obtained in the previous stage. Only the certainty degrees r^k are recalculated for each chromosome, through (8.28). The termination criteria are the same as those used in rule generation algorithm, that is, a maximum number of generations and a threshold in the population's fitness values standard deviation. The fitness function is based solely on the classification performance P_c, defined as the fraction of the number of correct classifications committed by the FRBCS to the total number of patterns. Here, we use both the training and the validation sets, with the fitness value determined as an average of the classification performance in the two sets:

$$f = 0.5 \cdot \left(P_c^{(trn)} + P_c^{(val)} \right). \tag{8.29}$$

Since only a fine-tuning of the fuzzy sets parameters is required in this stage, the ES performs a hard constrained optimization task. Therefore, we need to define the intervals of adjustment for each tunable parameter. Here, we chose the fixed intervals of adjustment, as proposed in [12, 29]. For the time being, assume trapezoidal membership functions, as depicted in Figure 8.5. For each parameter of the fuzzy set, a right and a left point is chosen, as the mid distance between two successive parameters of the membership function, based on the initial fuzzy sets, before the start of the evolution. For the trapezoidal function in Figure 8.5, the intervals of adjustment are defined as follows:

$$\begin{aligned}
a &\in [a^l, a^r] = [a - 0.5(b - a), a + 0.5(b - a)], \\
b &\in [b^l, b^r] = [b - 0.5(b - a), b + 0.5(c - b)], \\
c &\in [c^l, c^r] = [c - 0.5(c - b), c + 0.5(d - c)], \\
d &\in [d^l, d^r] = [d - 0.5(d - c), d + 0.5(d - c)].
\end{aligned} \tag{8.30}$$

A second source of adjustment restrictions is derived from the feature space directly. Assuming an input variable x_i, $i = 1, \ldots, N$, the maximum $x_i^{\max}$ and minimum $x_i^{\min}$ feature values encountered in the patterns of both the training and the validation sets are calculated. We therefore require for each adjustable parameter

$q = \{a, b, c, d\}$ of a fuzzy set to change in a small percentage of its input space interval:

$$q \in \left[q - p_x \cdot (x_i^{\max} - x_i^{\min}), q + p_x \cdot (x_i^{\max} - x_i^{\min})\right], \qquad (8.31)$$

where p_x is a reasonably small percentage of the input space, for example 5–10%. The final intervals of adjustment are calculated by the intersection of the respective intervals in (8.30) and (8.31). The respective intervals of adjustment for the two-sided Gaussian functions are easily derived similarly to (8.19), that is, by using the substitutions: $m_\ell^i = b$, $m_r^i = c$, $\sigma_\ell^i = (b-a)/(2\sqrt{\pi})$ and $\sigma_r^i = (d-c)/(2\sqrt{\pi})$. The initialization of the whole population is performed randomly inside its intervals of adjustment, with the exception of the first chromosome, which encodes the final rule base obtained by the rule base extraction algorithm (after the rule type transformation).

8.4 Experimental Results

Lake Koronia is located in a tectonic depression in northern Greece (40′ 41″ N, 23′ 09″ E). The lake-wetland ecosystem is surrounded by an intensively cultivated agricultural area, with maize, alfalfa, and cereals being the dominant agricultural crops. The industrial sector has recently increased, discharging untreated effluents in the lake from fabric dying, food and dairy processing activities. In recognition of its ecological importance, and to prevent further degradation, the lake-wetland system of Lakes Koronia – Volvi is protected by a number of legal and binding actions.

For classification purposes, an IKONOS bundle image, exhibiting 1 m spatial resolution in panchromatic and 4 m in multispectral (three visible and one near-infrared), is acquired covering the 134 km^2 of the study area. The image was acquired on 7 August 2005, was clear from clouds and was acquired at nadir view angle in order to minimize noise reflectance from topographic effects.

Extensive field survey was conducted to identify land cover classes, referred mainly to the agricultural and wetland area. A total number of 4100 locations were selected at regular intervals along the agricultural road network, using the stratified random sampling method. The land cover on these locations was identified by visual inspection and was afterwards labeled in 13 classes. The classification scheme included six crop types, five wetland habitats (following Annex I of Habitats Directive 92/43/EEC) and two ancillary land cover types (following the CORINE Land Cover nomenclature). Patterns were initially separated into two different sets: the training set (70%) and the testing set (30%). Due to the algorithmic requirements of the BGFC, the initial training set was further split into the training set (60%) and the validation set (40%). Two distinct zones can be identified in the whole area: the wetland zone, including the lake and its surrounding wetland vegetation, and the agricultural zone. In the wetland zone five classes were recognized (water, phragmites, tamarix, wet meadows and trees). Furthermore, we consider eight classes in the agricultural zone: six of them referring to different crop types (maize, alfalfa, cereals, orchards, vegetables and fallow) while the remaining two in other land cover types (urban areas and shrubs).

8.4.1 Features Sets

Ground cover types are usually classified using the gray level values of the bands of a multispectral image. However, prior research has proven that the use of advanced features can greatly assist land cover classification, by improving discrimination between the classes. Therefore, we derived advanced features using the initial four bands of the image, categorized into two groups, namely, textural and spectral features. The procedure employed here is similar to the one followed in [7]. In the following, a brief description of each feature set is presented; for a detailed description of the features sets and their derivation process, the interested reader can consult the aforementioned reference.

The spectral features comprise two alternative color spaces, calculated from the initial bands. The first one is the IHS transformation [34], which uses intensity (I), hue (H) and saturation (S) as the three positioned parameters, in lieu of R, G and B. Only three bands are needed to perform the transformation, so a pseudo-color RGB composite of IKONOS using channels four, three and two, respectively, was considered. The transformation produced many infeasible values (zero divisions) in saturation and, therefore, only the intensity and hue were finally used. The second color space considered is the Tasseled Cap transformation [35], which produces three data structures representing vegetation information: brightness, greenness and wetness. Overall, five spectral features (3 Tasseled Cap + Intensity + Hue) were computed for the entire image, collectively referred as *Transformed Spectral Features* (TSF).

Furthermore, two sources of textural features were considered, namely, the *Gray Level Co-occurrence Matrix* (GLCM) and the *Wavelet Features* (WF). Both feature sets provide useful information of the textural variations of the image, but from a different viewpoint. The GLCM [36] derives statistic measures from sliding windows of size $W \times W$ of the image, calculating in each window the frequency with which different gray levels occur in a specific direction θ and distance d between the pixel centers. Among 16 possible measures, four are considered to be the most important, resulting in a total of 16 textural features for the four bands. The fast wavelet transform (FWT) [37] is based on multiresolution analysis of images. A 2-level image wavelet decomposition is performed in windows of the original image, resulting in the creation of seven subimages. The total energy measure from the wavelet coefficients associated to each subimage is then calculated. Energy distributions from FWT provide detailed description of the frequency content of an image. This feature set comprises a total of 28 features (seven energy features in each one of the four bands).

Proper selection of the window size is required in order to calculate the GLCM and FWT. In addition, we need to determine the direction in the GLCM method and the wavelet basis function for the image decomposition using FWT. These parameters were calculated using a systematic approach proposed in [7]. In short, a measure of the degree of the discrimination is calculated for each window size-angle (or window size-wavelet basis function) pair. The parameters that maximize this degree were used in order to produce the features in each case. Using this technique, the optimal values for the distance d and the angle θ for the GLCM method were selected as $(21, 0°)$. For the WF, the window size and wavelet basis function selected were (7, coif4).

8.4.2 BGFC Example

In order to highlight the interpretation merits of the BGFC, we present first an example of a BGFC model. For the sake of simplicity in the representation, we assume here a model obtained for a subset of the original problem. In particular, we assume only the wetland zone and the bands and TSF features, resulting in a problem with nine features and five classes. The model is a simple one, comprising a total of nine fuzzy rules. A visual representation of the rule base is given in Figure 8.6. Each line of subplots in the figure represents a fuzzy rule, with the first line depicting the histograms of all classes in the respective features. The number on the far right part of the rule represents its certainty degree. Grayed out boxes indicate inactive features, i.e., features not considered by the fuzzy rules. Each one of the first five rules has a different consequent class, since, as mentioned in Section 8.3.2.2, the first M rules were created with a fixed consequent, where M is the number of classes ($M = 5$). The last column of subplots depicts each rule's associated class.

As a first case, we consider the fifth rule, describing the water class. A detailed view of the rule is given in the upper left part of Figure 8.7. The water class is clearly separable from the other classes in the greenness feature and, thus, the rule exhibits a certainty degree of nearly unity with only one feature. However, the water class is also separable from all other classes in the hue feature, as shown in the figure, as well as in band 4, brightness and intensity. This means that, in regard to the water class, all these features are equivalent, i.e., each one of them might be used to distinguish the water class. Having features that carry equivalent information, makes the task of global feature elimination (from the BGFC's algorithm) difficult, since the EA will select any of these features at random. In this case, the hue feature is never used by any rule. However, as the number of rules increases, the chance that a feature will never be selected becomes very small, at least in this classification problem. Arguably, if a feature is uninformative, i.e., if the feature signatures of all classes overlap to a great extent, the rule generation algorithm will never select this feature, due to the consistency condition (8.26).

The upper right part of Figure 8.7 depicts the premise parts of the first and fourth rule, describing the phragmites and trees classes, respectively. In band 2, the fuzzy

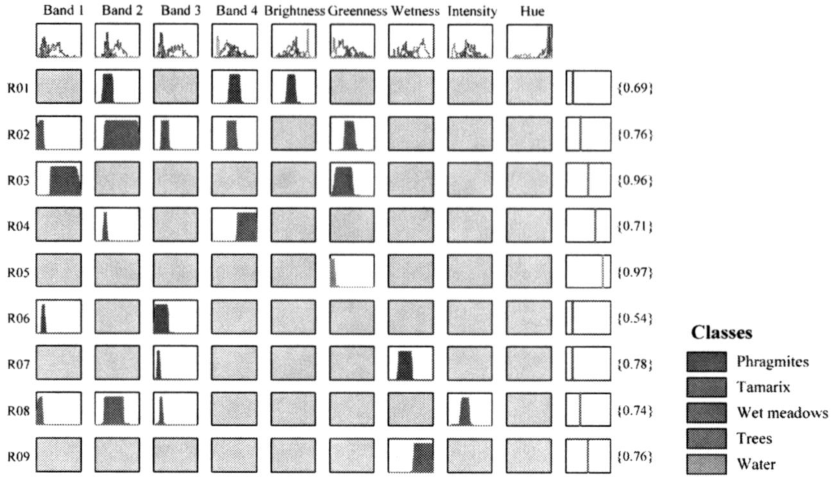

Fig. 8.6 A visual representation of the BGFC example model. The numbers in curly braces (on the far right) represent the certainty degree of each rule.

sets in the two rules overlap, but through this feature each rule is able to discriminate from the wet meadows class. The rules, however, cooperate in band 4, where the phragmites and trees classes exhibit a small overlapping. Moreover, through the fourth rule, the trees class is discriminated from the remaining classes (tamarix and water). As for the first rule, high signatures overlapping between the phragmites and tamarix classes exists, which are then separated to some extent through the brightness feature. Note that, although the two fuzzy sets in band 2 overlap, the two rules cover different subspaces of the feature space, because of the small overlapping of band's 4 fuzzy sets.

As a final case, we examine the third and ninth rule, both describing the wet meadows class. A detailed view of the rule is given in the lower part of Figure 8.7. The third rule uses band 1 to separate the wet meadows class from all other classes, and the greenness feature to avoid a few misclassifications of tamarix and tree patterns. Notice that the feature signature of the wet meadows class continues in the lower values of the feature, where it overlaps with those of other classes. The rule covers only a part of the wet meadows patterns with a high certainty degree (0.96), as dictated by the consistency condition. However, as more rules are generated, the weights assigned to the training patterns by the AdaBoost algorithm decrease, leaving higher weight values to patterns in mixed areas of the feature space. Therefore, the ninth rule tries to cover the remaining wet meadows patterns through the wetness feature, but with a lower certainty degree (0.76). This shows that the first rules obtained are the most general, while more specific ones are generated at later invocations of the rule generation algorithm. Thus, after the initial five rules (one for each class) have been generated, three out of the four remaining rules cover patterns of the major classes (phragmites and wet meadows). Moreover, one rule for the tamarix class has also been created, because this class overlaps to a high degree with

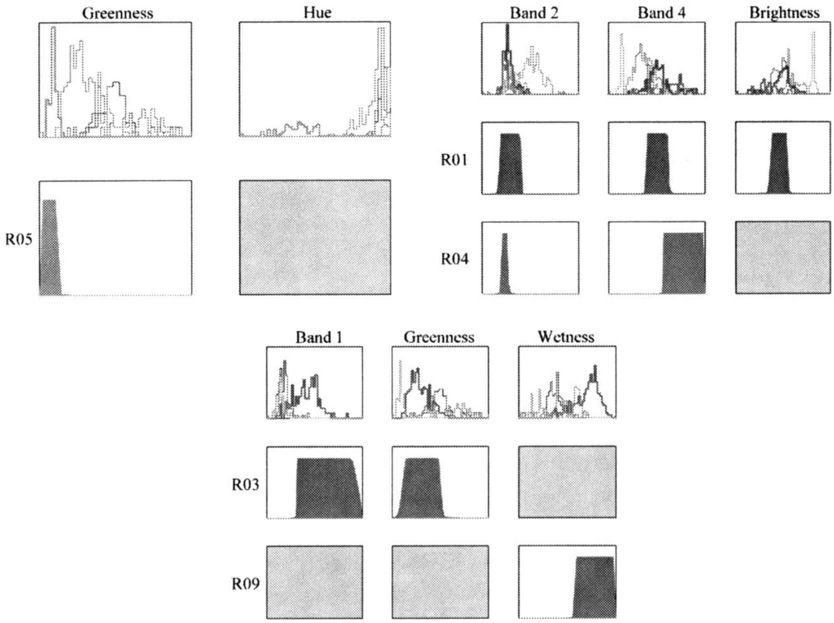

Fig. 8.7 Partial rule base views of the BGFC example model: Rule 5 (upper left), Rules 1 and 4 (upper right), and Rules 3 and 9 (lower).

all the other, making it difficult for a single rule to cover all tamarix patterns. It is worth noting, that in such cases the rule generation algorithm is forced to consider a considerably higher number of features, in an attempt to find a discriminating region of the feature space.

8.4.3 Results Obtained by BGFC

The dataset used in our experimental setup comprises a total of 53 features (bands+WF+GLCM+TSF) and 13 classes. The EA-based learning algorithm of the BGFC has stochastic characteristics, resulting in a non-deterministic final classifier structure. In order to correctly evaluate the classification performance of the BGFC, 20 independent runs of the whole algorithm were performed in the same partition of our dataset. In each case, the averaged results over these 20 runs are given, as well as the best one found. In addition, for each averaged value, the respective standard deviation is provided in parentheses, prefixed with the $\pm$ sign.

Table 8.1 summarizes the obtained results. The table presents the number of fuzzy rules and the average number of features per rule in the obtained BGFC models, as well as the percentage of correct classifications in the testing set after the rule base generation (R.G.) stage and the genetic tuning (G.T.) stage. In each case,

Table 8.1 Averaged and best results (overall accuracy and structure characteristics) of BGFC. Numbers in parentheses represent standard deviations.

	Best Result	Averaged Results
Rule number	58	55.10 (± 5.49)
Average features per rule	15.79	16.01 (± 0.68)
Classification performance after the R.G. stage (%)	70.06	69.05 (± 0.90)
Classification performance after the G.T. stage (%)	73.89	72.45 (± 0.63)

the averaged values (over 20 runs), their standard deviation values and the result obtained by the best model, in terms of accuracy, are presented.

Given the complexity of the problem, the obtained results should be considered rather satisfactory. BGFC's learning algorithm produced moderately sized rule bases, with an average number of 55 rules, comprising 16 features per rule on average. In relative terms, each rule uses only a 30% of the available features, a fact that greatly simplifies the obtained models. In no case was a global feature rejection observed, meaning that in all cases the rule base uses all features. This can be attributed to two reasons. Firstly, our dataset contains features with strong correlation, providing discriminating information between the classes from many sources, which drives the EA-based rule extraction algorithm to select any of the equivalent or nearly equivalent features. This effect was shown in the previous section, when examining the fifth rule of the example rule base. Secondly, each rule is produced independently from the previous ones. The AdaBoost algorithm provides information on previous generated rules, with respect to the patterns covered so far. However, no information on previously selected features exists, which could be evaluated to assist the global feature selection task. For both reasons, global feature rejection becomes even harder because of the moderate number of rules, which are however necessary, due to the large number of classes and features the problem comprises.

The classification performance achieved in the testing dataset is 69 and 72.5%, before and after tuning, respectively. Therefore, the first stage of the algorithm produces an overall good FRBCS, while the tuning stage optimizes the performance of the final model, by fine-tuning the decision boundaries of the classifier, ultimately gaining an additional 3.5% in classification performance. Moreover, and perhaps most importantly, the tuning stage reduces the performances standard deviation, increasing the robustness of the learning algorithm. Although the standard deviation is small for the first stage as well, tuning reduces it even more, so that the stochastic effects of the EA-based rule generation algorithm are virtually eliminated. This means that BGFC's learning algorithm is able to produce equivalent results, regardless of its initialization.

Table 8.2 shows the error confusion matrices of the best BGFC model obtained, after tuning. The classification performance of the system is 73.89%, with a Khat value of 0.71. The major part of the overestimations occurs for the alfalfa, the cereals and the wet meadows classes. This was actually expected, since these are the major classes of our dataset. We should note, however, that the selection of classes was performed considering the agricultural and the wetland zones independently.

Table 8.2 Error confusion matrix for the best BGFC model obtained, after tuning.

	a	b	c	d	e	f	g	h	i	j	k	l	m	P.A.	U.A.
a	142	4	22	1	15	2	0	0	5	0	14	4	0	70.65	67.94
b	26	185	2	3	7	27	5	0	0	0	15	0	0	88.10	68.52
c	11	0	138	2	8	0	0	0	5	2	3	2	0	77.97	80.70
d	1	0	2	33	0	0	0	1	0	0	6	0	0	63.46	76.74
e	3	0	3	1	12	1	0	0	1	0	0	0	0	22.22	57.14
f	1	4	0	0	5	21	0	0	0	1	0	0	0	35.00	65.63
g	0	0	0	0	0	0	28	0	1	0	4	0	0	54.90	84.85
h	0	2	0	0	4	0	2	61	1	0	0	0	0	98.39	87.14
i	3	2	7	4	2	1	1	0	93	8	13	0	0	82.30	69.40
j	0	0	3	0	0	0	0	0	5	40	2	0	0	67.80	80.00
k	11	13	0	4	1	8	15	0	2	3	145	0	0	70.05	71.78
l	3	0	0	2	0	0	0	0	0	5	5	50	0	89.29	76.92
m	0	0	0	2	0	0	0	0	0	0	0	0	54	100.00	96.43
Ref.	201	210	177	52	54	60	51	62	113	59	207	56	54		

a = alfalfa; b = cereals; c = maize; d = orchards; e = vegetables; f = fallow; g = shrubs; h = urban; i = phragmites; j = tamarix; k = meadows; l = trees; m = water; P.A. = producer's accuracy; U.A. = user's accuracy; Ref. = reference.

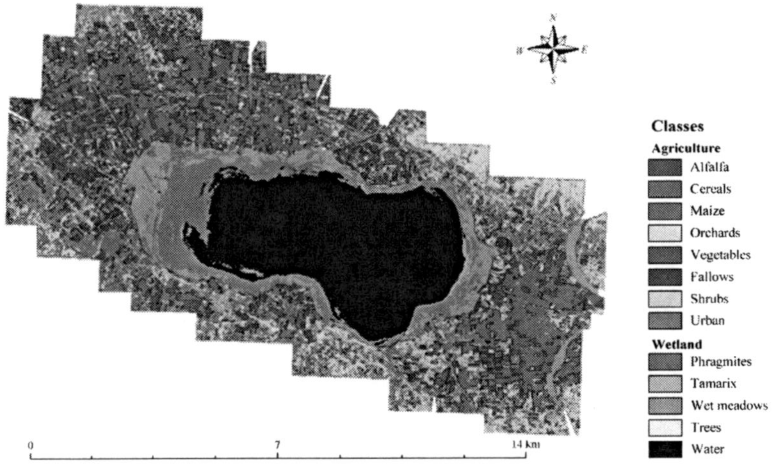

Fig. 8.8 Thematic map of IKONOS land cover classification produced by the best BGFC model obtained.

Therefore, classes like wet meadows and shrubs or trees and orchards exhibit a large degree of overlapping in the feature space. Nonetheless, we considered here the problem of the whole area, to test the BGFC in a more complex classification task.

Figure 8.8 shows the thematic map produced by the best BGFC model obtained. A detail of the map in a cultivated area north of the wetland is given in Figure 8.9.

Fig. 8.9 A detail of the land cover map produced with the best BGFC model obtained in a cultivated area north-west.

BGFC produced a thematic map with distinguishable shapes and classes for each agriculture field. Hence, a lot of parcels can be labeled using only the classified image. Moreover, a clear distinction between the cultivated areas and the wetland area is observed. Therefore, BGFC's learning algorithm is able to produce models with good quality properties.

8.4.4 Comparison with other methods

In order to validate BGFC's performance, we conducted comparative experiments with other well-known statistical and fuzzy classifiers. In particular, we considered the Maximum Likelihood Classifier (MLC) [1], the k-Nearest-Neighbor (k-NN) classifier [1], the Support Vector Machines (SVM) classifier [38], implemented with a Gaussian kernel, Quinlan's C4.5 [39], the Learning Vector Quantization (LVQ) classifier [40] and Ishibuchi's GFRBCS [16]. The last one was tested using the Keel software package [41]. For the k-NN algorithm, we tested k values in the range $[1, 20]$ and a value of $k = 3$ was selected, which maximized the algorithm's performance on the validation set. The same technique was used to determine the parameters C and σ of the SVM, the number of nodes in the LVQ models and the parameters of C4.5 and Ishibuchi's GFRBCS. The MLC does not require any parameters specification and therefore the union of the training and validation sets was used as its training set.

Table 8.3 presents the classification performance on the testing dataset for the various classifiers considered. Similarly to BGFC, Ishibuchi's GFRBCS is learned through a GA-based procedure, resulting in stochastic characteristics. Therefore, the

Table 8.3 Classification performance comparison between various methods.

Classifier	Performance (%)
MLC	50.07
k-NN	65.71
SVM	67.26
C4.5	69.76
LVQ	54.35
Ishibuchi	44.71
BGFC	72.45

average classification performance of 20 independent runs is given for this classifier in Table 8.3. The rest of the classifiers considered are deterministic ones and, therefore, only one run was required. BGFC achieved the highest performance, attaining more than 2.5% gain with regard to the second best C4.5 classifier. In the last few years, the SVM classifier is becoming a reference classifier for land cover classification tasks. Even before tuning, BGFC's performance is marginally less than that of C4.5, and still higher than any other classifier. The later fact implies that the AdaBoost algorithm is able to direct the learning algorithm in producing a well-performing classifier. There is, however, some room for further optimization, which is performed by the tuning stage, considering the whole rule base produced in the previous stage. To that extend, the tuning stage can be omitted, if the performance attained after the rule base generation stage is satisfactory.

MLC, LVQ and Ishibuchi's GFRBCS were unable to perform comparable, due to the large number of features the problem comprises. In addition to its high classification performance, BGFC produces compact fuzzy rule bases, which are easily interpretable by humans. Ishibuchi's GFRBCS produced on average 128.3 rules, which are more that two times the rules produced by BGFC in each case. Note, however, that the former classifier constructs a descriptive fuzzy rule base, which is less flexible than the approximate fuzzy rule base produced by the BGFC.

8.5 Conclusions

The BGFC is proposed for land cover classification of multispectral satellite images. The model employs fuzzy classification rules, created by an iterative EA-based boosting algorithm, one rule at a time. The local feature selection embedded in the rule generation algorithm further simplifies the, already intuitive, meaning of fuzzy rules, increasing the system's interpretability and understandability by humans. A genetic tuning stage follows the rule generation one, further increasing the performance of the obtained classifier. Comparative results with well-known classifiers in the region surrounding Lake Koronia showed that BGFC is well-suited for complex land cover classification tasks, comprising a large number of features and/or classes.

References

1. R. Duda, P. Hart and D. Stork, *Pattern Classification*, 2nd ed. Wiley, New York, 2001.
2. D. Stathakis and A. Vasilakos, Comparison of computational intelligence based classification techniques for remotely sensed optical image classification, *IEEE. Trans. Geosci. Remote Sensing* **44**(8), 2305–2318, 2008.
3. J.A. Benediktsson, P.H. Swain and O.K. Esroy, Conjugate gradient neural networks in classification of multisource and very-highdimensional remote sensing data, *Int. J. Remote Sens.* **14**, 2883–2903, 1993.
4. Z. Liu, A. Liu, C. Wang and Z. Niu, Evolving neural network using real coded genetic algorithm (GA) for multispectral image classification, *Futur. Gener. Comp. Syst.* **20**(7), 1119–1129, 2004.
5. C.-T. Lin, Y.-C. Lee and H.-C. Pu, Satellite sensor image classification using cascaded architecture of neural fuzzy network, *IEEE Trans. Geosci. Remote Sensing*, **38**(2), 1033–1043, 2000.
6. N.E. Mitrakis, C.A. Topaloglou, T.K. Alexandridis, J.B. Theocharis and G.C. Zalidis, A novel self-organizing neuro-fuzzy multilayered classifier for land cover classification of a VHR image, *Int. J. Remote Sens.* **29**(14), 4061–4087, 2008.
7. N.E. Mitrakis, C.A. Topaloglou, T.K. Alexandridis, J.B. Theocharis and G.C. Zalidis, Decision fusion of GA self-organizing neuro-fuzzy multilayered classifiers for land cover classification using textural and spectral features, *IEEE Trans. Geosci. Remote Sensing* **46**(7), 2137–2152, 2008.
8. A. Bárdossy and L. Samaniego, Fuzzy rule-based classification of remotely sensed imagery, *IEEE Trans. Geosci. Remote Sensing* **40**, 362–374, Feb. 2002.
9. A. Laha, N.R. Pal and J. Das, Land cover classification using fuzzy rules and aggregation of contextueal information through evidence theory, *IEEE Trans. Geosci. Remote Sensing* **44**(6), 1633–1641, 2006.
10. A. Ghosh, N.R. Pal and J. Das, A fuzzy rule based approach to cloud estimation, *Remote Sens. Environ.* **100**, 531–549, 2006.
11. F. Melgani, B.A.R. Al Hashemy and S.M.R. Taha, An explicit fuzzy supervised classification method for multispectral remote sensing images, *IEEE Trans. Geosci. Remote Sensing* **38**(1), 287–295, 2000.
12. O. Cordón, F. Herrera, F. Hoffmann and L. Magdalena, *Genetic Fuzzy Systems: Evolutionary Tuning and Learning of Fuzzy Knowledge Bases*, World Scientific, Singapore, 2001.
13. O. Cordón, F. Gomide, F. Herrera, F. Hoffmann and L. Magdalena, Ten years of genetic fuzzy systems: Current framework and new trends, *Fuzzy Sets Syst.* **141**, 5–31, 2004.
14. A. González and R. Pérez, SLAVE: A genetic learning system based on an iterative approach, *IEEE Trans. Fuzzy Syst.* **7**(2), 176–191, 1999.
15. A. González and R. Pérez, Completeness and consistency conditions for learning fuzzy rules, *Fuzzy Sets Syst.* **96**, 37–51, 1998.
16. H. Ishibuchi, T. Nakashima and T. Murata, Performance evaluation of fuzzy classifier systems for multidimensional pattern classification problems, *IEEE Trans. Syst. Man Cybern, Part B – Cybern.* **29**, 601–618, 1999.
17. T. Bäck, *Evolutionary Algorithms in Theory and Practice*, Oxford University Press, Oxford, 1996.
18. O. Cordón and F. Herrera, A two-stage evolutionary process for designing TSK fuzzy rule-based systems, *IEEE Trans. Syst. Man Cybern, Part B – Cybern.* **29**(6), 703–715, 1999.
19. O. Cordón and F. Herrera, Hybridizing genetic algorithms with sharing scheme and evolution strategies for designing approximate fuzzy rule-based systems, *Fuzzy Sets Syst.* **118**(2), 235–255, 2001.
20. F. Hoffmann, Combining boosting and evolutionary algorithms for learning of fuzzy classification rules, *Fuzzy Sets Syst.* **14**, 47–58, 2004.
21. P. Thrift, Fuzzy logic synthesis with genetic algorithms, in *Proc. Fourth Int. Conf. on Genetic Algorithms (ICGA'91)*, San Diego, USA, Morgan Kaufmann, Los Altos, CA, pp. 509–513, 1991.

22. J. Casillas, P. Martínez and A.D. Benítez, Learning consistent, complete and compact sets of fuzzy rules in conjunctive normal form for regression problems, *Soft Comput.*, in press, 2008.
23. S.E. Papadakis and J.B. Theocharis, A GA-based fuzzy modeling approach for generating TSK models, *Fuzzy Sets Syst.* **131**(1), 121–152, 2002.
24. H. Ishibuchi, K. Nozaki, N. Yamamoto and H. Tanaka, Selecting fuzzy if-then rules for classification problems using genetic algorithms, *IEEE Trans. Fuzzy Syst.* **3**(3), 260–270, 1995.
25. M. Valenzuela-Rendón, The fuzzy classifier system: A classifier system for continuously varying variables, in *Proc. 4th Int. Conf. Genetic Algorithms*, University of California, San Diego, July 13–16, pp. 346–353, 1991.
26. A. Parodi and P. Bonelli, A new approach to fuzzy classifier systems, in *Proc. 5th Int. Conf. Genetic Algorithms*, University of Illinois, Urbana-Champaign, July 17–21, pp. 223–230, 1993.
27. A. González and F. Herrera, Multi-stage genetic fuzzy systems based on the iterative rule learning approach, *Mathware Soft Comput.* **4**, 233–249, 1997.
28. O. Cordón and F. Herrera, A three-stage evolutionary process for learning descriptive and approximate fuzzy-logic-controller knowledge bases from examples, *Int J. Approx. Reasoning* **17**(4), 369–407, 1997.
29. O. Cordón, M.J. del Jesús, F. Herrera and M. Lozano, MOGUL: A methodology to obtain genetic fuzzy rule-based systems under the iterative rule learning approach, *Int. J. Intell. Syst.* **14**(11), 1123–1153, 1999.
30. A. González and R. Pérez, Selection of relevant features in a fuzzy genetic learning algorithm, *IEEE Trans. Syst. Man Cybern, Part B – Cybern.* **31**(3), 417–425, 2001.
31. M.J. del Jesus, F. Hoffmann, L.J. Navascués and L. Sánchez, Induction of fuzzy-rule-based classifiers with evolutionary boosting algorithms, *IEEE Trans. Fuzzy Syst.* **12**(3), 296–308, 2004.
32. Y. Freund and R. Schapire, Experiments with a new boosting algorithm, in *Proc. 13th Int. Conf. Machine Learning*, pp. 148–156, 196.
33. C.T. Lin, *Neural Fuzzy Control Systems with Structure and Parameter Learning*. World Scientific, Singapore, 1994.
34. Y. Zhang and G. Hong, An IHS and wavelet integrated approach to improve pan-sharpening visual quality of natural colour IKONOS and QuickBird images, *Information Fusion* **6**, 225–234, 2005.
35. J. H. Horne, A tasseled cap transformation for IKONOS images, in *Proc. ASPRS Annu. Conf.*, Anchorage, Alaska, 2003.
36. R.M. Haralick and L.G. Shapiro, *Robot and Computer Vision*, Vol. 1, Addison-Wesley, Reading, MA, 1992.
37. S.G. Mallat, Theory for multiresolution signal decomposition: The wavelet representation, *IEEE Trans. Pattern Anal. Mach. Intell.* **11**, 674–693, July 1989.
38. C. Cortes and V. Vapnik, Support vector networks, *Mach. Learn.* **20**, 273–297, 1995.
39. J.R. Quinlan, *C4.5: Programs for Machine Learning*, Morgan Kauffman, 1993.
40. J.C. Bezdek and L.I. Kuncheva, Nearest prototype classifier designs: An experimental study, *Int. J. Intell. Syst.* **16**(12), 1445–1473, 2001.
41. J. Alcalá-Fdez, L. Sánchez, S. García, M.J. del Jesus, S. Ventura, J.M. Garrell, J. Otero, C. Romero, J. Bacardit, V.M. Rivas, J.C. Fernández and F. Herrera, KEEL: A software tool to assess evolutionary algorithms to data mining problems, *Soft Comput.* **13**(3), 307–318, 2009. Software available online: http://www.keel.es

PART III

BIOENGINEERING/NEUROTECHNOLOGY

Chapter 9
Epileptic Seizures May Begin Hours in Advance of Clinical Onset: A Report of Five Patients*

Brian Litt, Rosana Esteller, Javier Echauz, Maryann D'Alessandro, Rachel Shor, Thomas Henry, Page Pennell, Charles Epstein, Roy Bakay, Marc Dichter and George Vachtsevanos

Abstract Research into mechanisms underlying seizure generation has traditionally focused on the seconds to at most few minutes before clinical seizure onset. In search of seizure predictors over longer periods of time, we analyzed continuous three to fourteen day EEG recordings from five consecutive patients with temporal lobe epilepsy implanted intracranially, with electrodes in and on the surface of the temporal lobes, during evaluation for epilepsy surgery. We found reliable, localized quantitative EEG changes, including increases in signal energy and brief bursts of rhythmic electrical discharges, indicating that epileptic seizures begin as a cascade of electrophysiological events that evolve over hours. Two important implications

---

Brian Litt · Marc Dichter
Department of Neurology and Bioengineering, 3 West Gates, Hospital of the University of Pennsylvania, Philadelphia, PA 19104, USA

Rosana Esteller
School of Electrical and Computer Engineering, Georgia Institute of Technology, Atlanta, GA 30332, USA and Departamento Tecnología Industrial, Universidad Simón Bolívar, 89000 Caracas, Venezuela

Javier Echauz
School of Electrical and Computer Engineering, Georgia Institute of Technology, Atlanta, GA 30332, USA and Electrical and Computer Engineering Department, University of Puerto Rico, Mayagüez, PR 00681, USA

Maryann D'Alessandro
School of Electrical and Computer Engineering, Georgia Institute of Technology, Atlanta, GA 30332, USA and Army Research Labs, Atlanta, GA, 30332, USA

Rachel Shor · George Vachtsevanos
School of Electrical and Computer Engineering Georgia Institute of Technology, Atlanta, GA 30332, USA

Thomas Henry, Page Pennell, Charles Epstein · Roy Bakay
Department of Neurology, Emory University School of Medicine, Atlanta, GA 30332, USA

* This chapter is a revised and updated version of the article as published in *Neuron*, Vol. 30, pp. 51–64, April 2001. Reprinted in part with kind permission from the publishers.

K.P. Valavanis (ed.), Applications of Intelligent Control to Engineering Systems, 225–245.

of these findings are: (1) neuroscientists investigating seizure generation must now focus on much longer time periods than were previously thought to be important to this process; and (2) an understanding of seizure precursors can be used to create implantable devices to forecast seizures and trigger intervention to prevent their electrical and clinical onset.

9.1 Introduction

Epilepsy affects 50 million people worldwide, 25% of whom have seizures that cannot be controlled by any available therapy. In most cases this is focal epilepsy, in which seizures arise from a region of abnormal brain, the epileptic focus, and spread in a stereotyped, individualized fashion. Fifty years ago Penfield and colleagues found localized abnormal discharges on the EEG prior to clinical seizures in some individuals with epilepsy [34]. They reported that seizures could be cured in these patients by removing the brain region that gave rise to this activity [10, 35]). Since that time, recognition of seizure onset on the EEG has remained the gold standard for locating epileptic foci in the brain [48].

One of the most disabling aspects of epilepsy is that seizures are unpredictable. Investigation to elucidate mechanisms underlying seizure generation have focused on seconds to minutes before clinical seizure onset. Ralston and Dichter reported preseizure EEG changes consisting of increased frequency and complexity of epileptiform discharges, commonly called spikes, which became polyphasic (developed multiple peaks), rhythmic and sustained as seizures approached [3, 4, 39]. They postulated mechanisms that initiated rapid changes in neuronal function, effected by ion channels, neurotransmitters and the cell membrane. Despite over forty years of investigation into the physiology of epilepsy, it is still not possible to explain how and over what time clinical seizures emerge from the relatively normal brain state observed between spontaneous seizures.

In recent years, routine clinical epilepsy monitoring has made it possible to study human seizures quantitatively, as they develop *in vivo*. This effort has been aided by technological advances such as digital video-EEG monitoring systems that can process data in real time, and safe, reliable electrodes for implantation directly on or in the brain (intracranial electrodes). Most work in quantitative analysis of human seizures has focused on *detecting* their electrical onset on EEG in the approximately 10 seconds that usually precede their first clinical signs (clinical onset) [12, 13, 33, 37]. As early as 1970, researchers considered methods for *predicting* epileptic seizures prior to their electrical onset, though none were implemented clinically [41, 52]. Recent research in seizure prediction exploits Chaos Theory to track non-linear parameters, such as the Principal Lyapunov Exponent and Correlation Dimension, over a few minutes prior to EEG onset of temporal lobe seizures [17, 19, 23, 24, 25, 29, 42]. These studies are somewhat limited because they concentrate on relatively few or brief data segments and the abstract parameters they explore are difficult to relate to the neurophysiology of epilepsy. These studies are important, however, as they support the idea that seizures develop over time, rather than occur as abrupt, isolated events.

Because many patients report non-specific, stereotyped symptoms hours in advance of seizures [38] we have focused on looking for quantifiable precursors of seizures far in advance of seizure onset on the EEG. We analyze continuous intracranial EEG recordings over many days and use algorithms that are sensitive to the spikes, high frequency rhythmic discharges and high amplitude slow waves that are typical of the EEG in epilepsy. This preliminary investigation in five patients is a necessary first step in identifying quantitative parameters worthy of more intensive study in a larger, prospective trial.

We chose to study patients with mesial temporal lobe epilepsy, in whom seizures arose from the amygdala, hippocampus or adjacent cortex in one or both temporal lobes, because: (1) mesial temporal lobe epilepsy frequently cannot be controlled by antiepileptic drugs, is common and is among the best understood of focal epilepsies, (2) it is common practice to record from intracranial electrodes in these patients during evaluation for epilepsy surgery, (3) the likelihood of electrodes being placed in or very close to the region generating the earliest precursors of seizures is high in these cases, and (4) electrical seizure onsets are usually well-localized on intracranial EEG in mesial temporal lobe epilepsy, have high signal-to-noise ratio, and offer the best available data for studying seizure generation in humans.

9.2 Clinical Background

We analyzed data from five consecutive patients with mesial temporal lobe epilepsy who required monitoring with intracranial electrodes during evaluation for epilepsy surgery. Subjects were the first five patients meeting these criteria admitted the Epilepsy Monitoring Unit after university approval of the research protocol. Patients included four men and one woman. Mean age was 32 years (range 21–50). The etiology of epilepsy was likely closed head injury in patient #1, although she also had febrile convulsions in early childhood, which have been associated with later development of mesial temporal lobe epilepsy. Patient #5 also had febrile convulsions, but no other risk factors for epilepsy. The other three patients did not have identified risk factors for epilepsy.

Intracranial monitoring was required in these patients because brain imaging, scalp EEG and other non-invasive studies did not definitively identify the site(s) from which seizures originated. Patient evaluations included: (1) recording seizures with scalp EEG electrodes, (2) high resolution brain magnetic resonance imaging (MRI) to look for focal lesions, (3) brain positron emission tomography scanning (PET, using [^{18}F] 2-fluoro-2-deoxyglucose), to identify regions of decreased brain metabolism usually associated with epileptic foci, and (4) neuropsychological testing, to identify focal impairment of cognitive function corresponding to brain regions that give rise to seizures. A clinical summary of the study patients, including brain imaging results and clinical outcome, are summarized in Table 9.1.

Table 9.1 Patient characteristics.

Pt	Sz Types	MRI	FDG Activity	Ictal Onset Zone(s) on Intracranial EEG and Comments	Side of Temporal Resection and Outcome	Seizure Risk Factors
1*	SP	Left hippocampal atrophy	Left temporal decrease	Left hippocampus (6/6)	Left	CHI
	CP			Incomplete resection due to language localization in ictal onset zone	Class II	FS
2	SP	Normal	Equal bilateral temporal decrease	Left hippocampus (11/13)	No surgery	None
	CP			Right hippocampus (2/13) Accumulated energy lateralized to side of onset in all seizures; all Sz propagated in < 3 s to contralateral side		
3	SP	Right frontal venous angioma	Bilateral temporal decrease	Right hippocampus (9/11)	Right	None
	CP			Left hippocampus (2/11) Accumulated energy lateralized to the right in all seizures	Class I	
4	SP	Normal	Right temporal decrease	Right hippocampus (5/5)	Right	None
	CP GTC				Class II	
5	CP	Subependymal heterotopia adjacent to left lateral ventricle	Left temporal decrease	Left hippocampus (6/10)	Left	FS
				Left temporal neocortex (3/10) Right frontal neocortex (1/10) Accumulated energy lateralized to side of onset in all seizures	Class III	

Class of outcomes refers to "Engel Classification" (Engel, 1987): Class I, free of disabling seizures; Class II, rare disabling seizures; Class III, worthwhile improvement; Class IV, no worthwhile improvement.
* Earliest EEG changes associated with seizures were localized to one or two adjacent electrode contacts.
Abbreviations: CP, complex partial; GTC, generalized tonic-clonic; SP, simple partial; CHI, closed head injury with loss of consciousness and/or amnesia; FS, febrile seizures in early childhood.

Location of seizure precursors was validated by clinical outcome after epilepsy surgery in four of the five study patients (Table 9.1). Patient #2 chose not have epilepsy surgery despite a majority of seizures arising from the right hippocampus. In this patient seizures spread from one side to the other within several seconds and brain imaging was normal, predicting a poorer chance of significant reduction in seizures after epilepsy surgery.

Each patient had one 6-contact "depth" electrode placed stereotactically in each temporal lobe, using coordinates generated from a high resolution MRI scan of the brain referenced to a coordinate frame fixed to the head of the patient for the surgery. These electrodes were placed through a burr hole in the occipital region of the skull and advanced forward so that the most anterior contact lay in the amygdala and more posterior contacts were located along the head and body of the hippocampus. Additional electrodes were placed directly on the surface of the brain (subdural electrodes) over and under the temporal lobes bilaterally through a single burr hole made in the temporal bone on each side of the skull. Three patients (#3, #4 and #5) also had subdural electrodes placed on the surface of the frontal lobes bilaterally. Figure 9.1 demonstrates placement of intracranial depth and subdural temporal electrodes in one study patient.

All EEG recordings were reviewed by four epileptologists certified by the American Board of Clinical Neurophysiology in EEG (BL, TH, PP, CE), for purposes of clinical decision-making. The first author visually reviewed the EEG for each

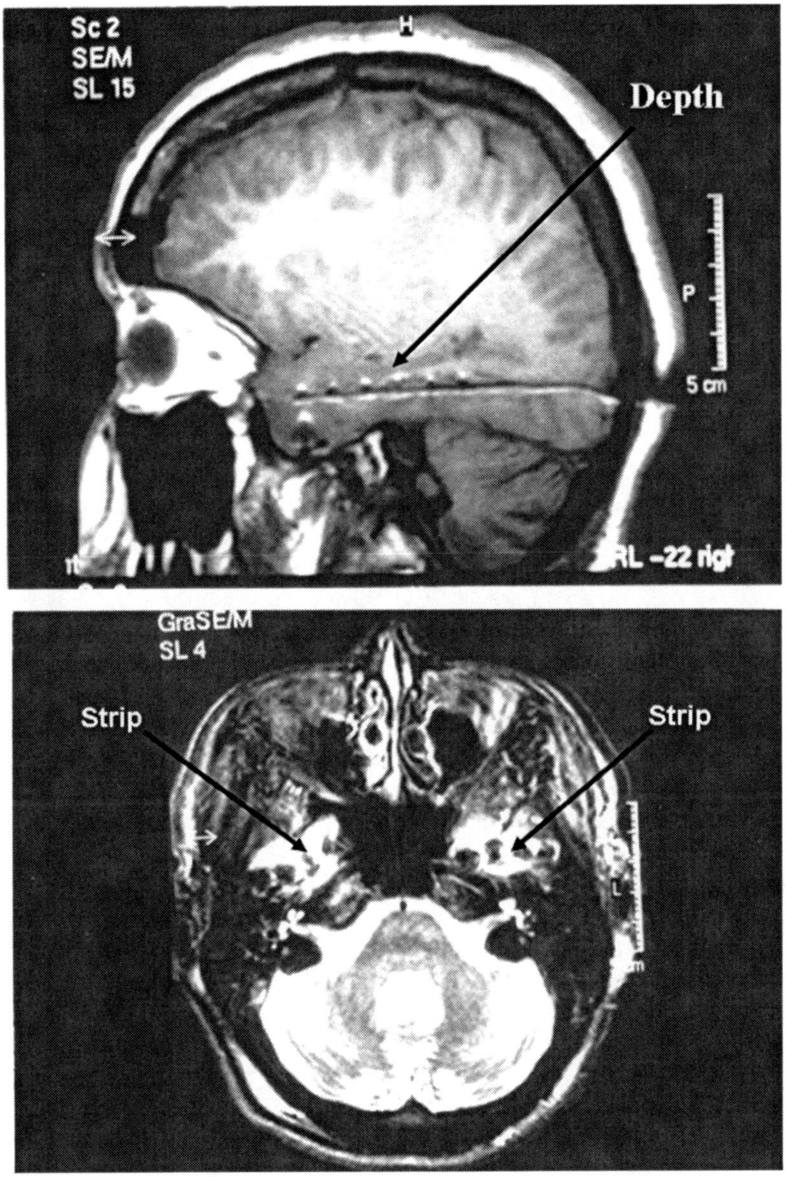

Fig. 9.1 Intracranial EEG electrode placement in patient #1, imaged on MRI scan. Subtemporal strip electrodes were placed under the temporal lobes bilaterally through burr holes through the temporal bones. Bilateral temporal depth electrodes were placed stereotaxically through burr holes in the occipital regions bilaterally.

seizure before quantitative analyses of these data, and marked each seizure at the following times: (1) the *earliest EEG change* associated with seizures: this point in time was found by identifying unequivocal seizure activity on the EEG and then moving backward in time to the point at which the first clear, sustained change from the patient's EEG baseline prior to the seizure was detected; and (2) the *unequivocal EEG onset* of seizures: this marked the time at which an EEG pattern typically associated with seizures [40] first became unquestionably clear and could be identified independent of knowing that a seizure followed. Intracranial EEG, simultaneous video and medical records were reviewed in their entirety for each patient. Important clinical activity, sleep wake cycles, times and doses of medication administered during the monitoring period were recorded and analyzed.

9.3 Results

In two patients (#1 and #4), all seizures began in one temporal lobe and earliest EEG changes were very focal, confined to one or two adjacent electrode contacts. This implies that these contacts were placed in or very close to the epileptic focus in these patients. Seizures arose independently from both temporal lobes in the remaining three patients (#2, #3, #5). Earliest EEG changes in these three patients were all poorly localized, indicating either less favorable electrode placement or that seizures arose from a more diffuse area in these individuals. In all five patients unequivocal onset of seizure activity on EEG was localized to one or two adjacent electrode contacts. Unequivocal EEG onsets of seizures were stereotyped for each patient, suggesting a reproducible, individualized pattern of initiation and spread.

Measures of energy in the EEG signal formed the core of quantitative methods used to predict seizures (see Methods). These were chosen because: (1) their calculation is simple and fast, consisting only of squaring the voltage at each point measured in the digital EEG and summing this up over a particular period of time; and (2) signal energy, as we defined it, is sensitive to EEG waveforms associated with seizures and epileptic foci in the brain, such as epileptiform spikes, rhythmic sharply contoured waveforms, and high amplitude slow (0.5–4 Hertz) activity seen after bursts of epileptiform activity.

In the first experiment a running sum of signal energy, "accumulated energy", was calculated from EEG data beginning 50 minutes prior to unequivocal seizure onset on EEG and ending 10 minutes after onset. Accumulated energy was computed at the electrode site recording the earliest or maximal (if several contacts were involved simultaneously) unequivocal EEG onset for each seizure. This was compared to two controls: (1) baseline controls: randomly chosen, one hour EEG segments recorded from the seizure focus, but separated from the beginning or end of any seizure by at least three hours of uninterrupted, artifact-free EEG, and (2) contralateral controls: data acquired during epochs leading up to seizures but from analogous electrodes in the temporal lobe opposite the seizure focus. Seizures and baseline data were compared only within the same state of consciousness, (e.g. awake vs. asleep), as baseline energy was found to be higher during sleep than wakefulness, due to normal EEG waveforms occurring during sleep (Table 9.2).

Table 9.2 Accumulated energy: number of records, classification and prediction (prior to correction for seizure clusters).

Patient Number	Mean Preseizure Declaration Time (Min)	Stdev of Pred Time (Min)	False Negatives	False Positives	Seizures		Baselines	
					Asleep	Awake	Asleep	Awake
1	23.3	15.25	1	3	1	3	6	6
2	20.78	15.77	1	2	2	2	6	4
3	17.29	10.15	0	0	1	6	5	4
4	1.3	1.83	1	0	2	3	4	5
5	21.89	11.61	0	1	7	3	4	6
All Patients	18.49	13.42	3	6	13	17	25	25

Note: Descriptions of the table entries moving from left to right are the patient identification number; mean prediction horizon, standard deviation, numbers of false positive and false negative predictions are displayed for each patient, in addition to the number of seizure and baseline epochs, and their distribution between wakefulness and sleep for each patient.

Eighty 1-hour EEG epochs were analyzed: 30 leading to seizures, and 50 baselines. Figure 9.2 displays all baseline and preseizure epochs recorded from the seizure focus in all five patients, normalized to include both sleep and awake records for all subjects. For each patient, accumulated energy during baseline periods was very similar (dashed gray lines) and accumulated energy in preseizure recordings (solid black lines) clearly rose above these baselines within 50 minutes prior to seizure onset. Table 9.2 displays prediction times calculated for each patient. This was defined as the time prior to seizure onset at which each accumulated energy curve for a particular patient crossed the level of the highest baseline accumulated energy for that patient. Using this simple threshold method of predicting seizures, data epochs were correctly classified as either preseizure or baseline initially with 89% accuracy, a sensitivity of 90% and specificity of 88%. Mean time from seizure prediction to unequivocal EEG onset of seizures by this simple algorithm was 18.5 minutes ($\pm$13.4 min). Close examination of Figure 9.2 demonstrates that accumulated energy actually rose above baselines much earlier than this time in many cases. This suggests that a method more sophisticated than a simple linear threshold will yield considerably earlier prediction horizons. These results complement reports of seizure prediction using "non-linear analysis" 2–6 minutes prior to seizure onset in 17 of 19 patients by Martinerie et al. and in 20 seizures processed by Lehnertz and Elger [24, 29].

Baseline signal energy was increased following clusters of two or more seizures within a six-hour period. When this was taken into account, four EEG epochs following clusters initially mislabeled as "pre-seizure" were correctly reclassified as baselines, improving overall accuracy to 94%. There were no clear EEG changes responsible for the two remaining false positive predictions, however, these could have occurred during prolonged bursts of energy between seizures (see below) or asymptomatic seizures hidden by gaps in recorded data found more than three hours prior to these baseline segments. There were no specific characteristics or obvious systematic errors that accounted for misclassification of three false negative epochs. The short prediction horizon for patient #4 sets him apart from the other study pa-

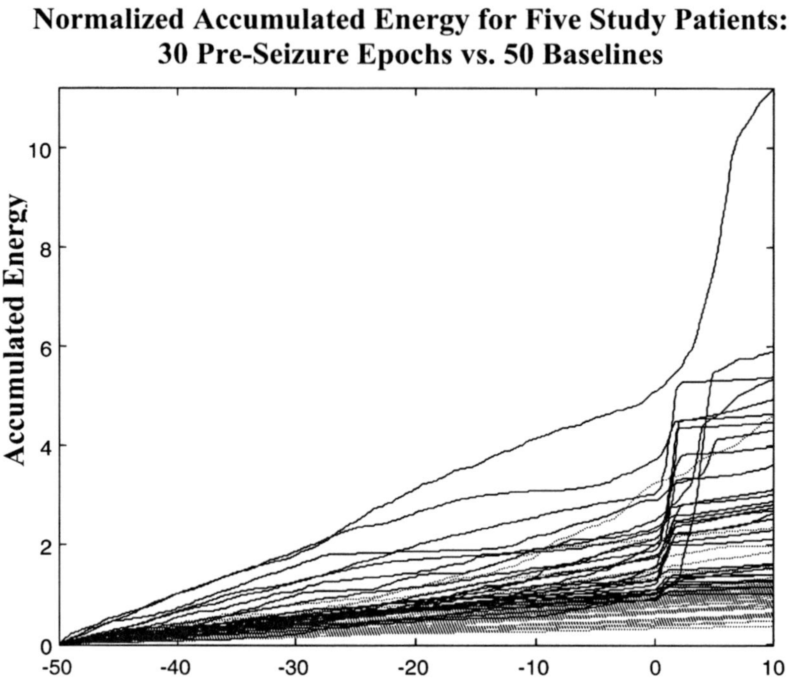

Fig. 9.2 Accumulated energy trajectories for two patients, one with seizures and baseline epochs recorded during wakefulness; the other during sleep. Baseline accumulated energy trajectories (dashed gray lines) cluster together at the bottom of the figure. Pre-seizure trajectories (black lines) deviate from baselines within 50 min prior to seizures, though a number of the pre-seizure epochs demonstrate deviation from baseline trajectories already in progress at 50 minutes prior to seizure onset, suggesting earlier pre-ictal change. Time = 0 marks unequivocal electrographic seizure onsets, marked by the first author, according to standard clinical criteria.

tients, as does the unusually short time delay between electrical and clinical seizure onset in this patient. These findings are of unclear significance.

In patients with seizure onsets confined to one temporal lobe (#1, #4), and in two of the patients with seizures arising independently from both temporal lobes (#2, #5), increased accumulated energy was found only on the same side as unequivocal EEG onset for each seizure. In patient #3, of the seven seizures that met inclusion criteria, unequivocal EEG onset was seen in the right temporal lobe for five events and on the left side in the other two. Pre-seizure changes in accumulated energy predicted onset on the right side for all seizures. These findings suggest that the epileptic focus, or region critical to generating seizures in patient #3, was likely located in the right hemisphere, perhaps near pathways allowing rapid conduction or diversion of activity to the left hippocampus under some circumstances. This hypothesis is supported by the patient's seizure-free outcome after surgical removal of the right temporal lobe.

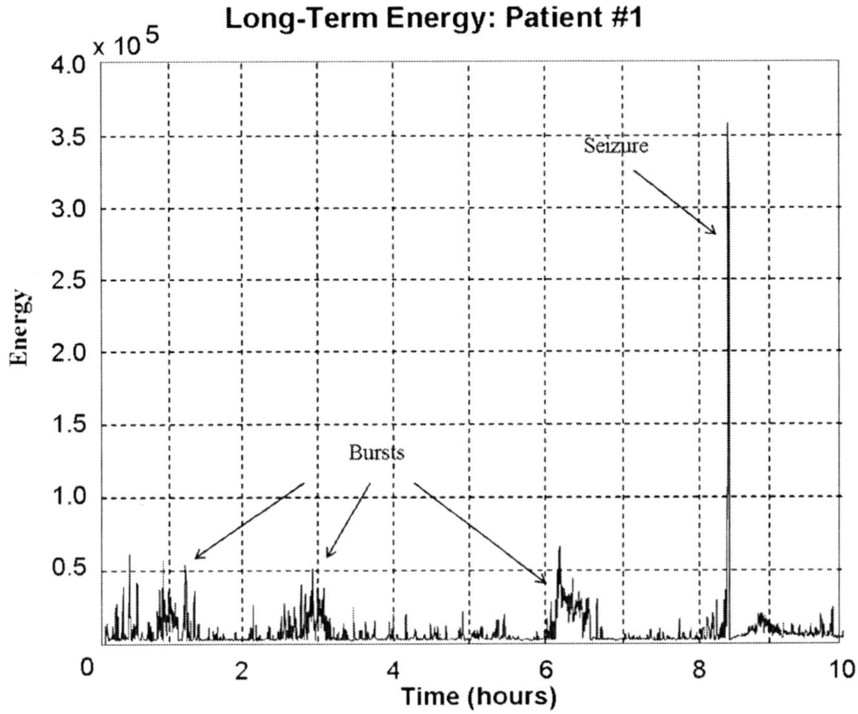

Fig. 9.3 Prolonged energy calculated from two electrodes in a bipolar montage placed proximate to the ictal onset zone prior to a single seizure in patient #1. Similar prolonged bursts of energy were seen prior to other seizures in patients #1 and #4, but not during baselines.

Visual inspection of preseizure EEG recordings did not demonstrate obvious findings that corresponded to changes in accumulated energy. The actual number of epileptiform discharges (spikes) during preseizure periods did not differ statistically from baseline periods, in agreement with Lange et al. and Katz et al. [21, 22]. Changes in accumulated energy were not confined to a particular frequency band in the EEG.

In a second experiment, signal energy in the EEG was computed over the entire hospital stay in the two patients, #1 and #4, who had highly localized earliest EEG changes prior to seizure onset. Periodic elevations in signal energy occurred in the epileptic focus throughout the hospital stay, and increased in amplitude and duration beginning approximately eight hours prior to seizures. Figure 9.3 shows a representative energy plot beginning eight hours prior to a single seizure in patient #1. The figure demonstrates several prolonged bursts of energy, sometimes lasting more than 30 minutes, as seizures approach. Bursts were defined as sustained elevations of signal energy of greater than two standard deviations above the inter-burst baseline for periods of five minutes or more in a running average of five minutes duration. Bursts were most prominent on the side of seizure onset in patients #1 and

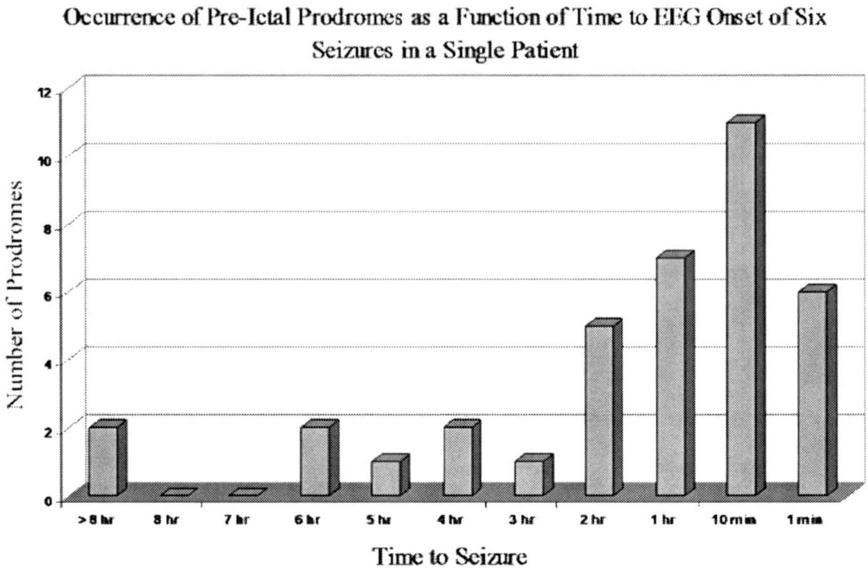

Fig. 9.4 The relationship of time of occurrence of subclinical seizures relative to unequivocal electrographic onset of seizures is displayed. As illustrated, most of these events occurred within approximately 2 hours of clinical seizures. Subclinical seizures were rare more than 3 hours removed from seizures.

#4, though lower amplitude synchronous changes in the opposite temporal lobe were also detected. Increases in signal energy were not detected at several electrode sites outside of the epileptic focus, though an exhaustive search of all electrode sites was not undertaken. Thorough analysis of these pre-ictal energy bursts has not yet been performed, thorough visual inspection revealed increasingly complex spikes accompanied by high amplitude low frequency activity (0.5–2 Hz) and loss of normal background activity in the epileptic focus channel during bursts. Prolonged bursts of signal energy occurred more commonly while patients were asleep, but were not clearly attributable to normal EEG patterns associated with sleep. Bursts of energy were less prominent and sometimes absent before seizures that clustered closely together in time, suggesting that mechanisms giving rise to clusters are likely different from those leading to individual seizures. This is an important issue for both understanding mechanisms of seizure generation and in designing treatment strategies for seizures. These findings suggest that seizure generation may begin far earlier than one hour prior to clinical seizures.

In a third observation, patient #1 demonstrated stereotyped six to twelve second bursts of very focal rhythmic activity in the left anterior hippocampus (the focus in this patient) just prior to seizures. Each "burst" began abruptly at 20–25 Hz then gradually increased in amplitude while slowing to 6–10 Hz, sometimes spreading to adjacent electrodes before either leading to seizure onset or ending in higher amplitude, irregular 1–2 Hz activity. These discharges resembled brief, highly localized seizures on EEG, but did not cause any clinical symptoms. Inspection of the patient's entire four-day intracranial EEG recording demonstrated that these electrical events clustered around clinical seizures and were extremely rare at other times. Figure 9.4 demonstrates the time of occurrence of these discharges in relation to unequivocal electrical onset of the patient's six typical seizures. The majority of these thirty-seven electrical events occurred within two hours of seizure onset. Discrete, self-limited, "subclinical seizures" in intracranial EEG recordings are reported to have important value in identifying the location of the epileptic focus in patients with mesial temporal lobe epilepsy [44, 45, 49]. Their timing, predictive value and potential role in seizure generation have not been reported previously. Of note, two of the patient's six seizures, which were excluded from the accumulated energy experiment because the EEG was briefly obscured by artifact in the three hours leading up to seizure onset, were included in this analysis.

9.3.1 Discussion

These results are important for several reasons. First, they give hints to the mechanisms underlying seizure generation in temporal lobe epilepsy and the time over which they may occur Second, they provide important insights into experimental design, data collection and processing required to conduct extensive, controlled trials in this area. Finally, they can be exploited for use in an implantable device to predict seizures and trigger abortive therapy.

9.3.2 Mechanisms

Our findings suggest that seizure generation is a multi-stage process that evolves over hours. We postulate that intermittent prolonged bursts of energy in the epileptic focus result from coherent extracellular fields in local neuronal networks that wax and wane over long periods of time. Under the influence of factors that promote seizures, for example, sleep deprivation or alcohol ingestion, increased synchronization of this activity among a larger network of cells may give rise to brief rhythmic discharges resembling highly localized subclinical seizures. As rhythmic discharges occur more frequently, tissue adjacent to the seizure focus is recruited, and a steady increase in accumulated energy occurs, heralding an oncoming seizure. This scheme agrees with previous descriptions of increasing coherence in chaotic parameters in the vicinity of the seizure focus during the preseizure period [16, 18].

There are numerous potential mechanisms that might account for these preseizure events. Prolonged bursts of energy within the epileptic focus are likely generated by hypersynchronous activation of local neuronal networks. Here EEG changes reflect synaptic currents flowing within the network, as well as intrinsic, voltage-dependent currents with longer time constants activated during localized discharges [5]. These events occur in brain regions where a pathological change has produced disrupted circuitry, enhanced recurrent excitation, and possibly enhanced synchronizing recurrent inhibition. Under normal conditions, this activity remains localized to the epileptic focus and limited in duration.

With lowered seizure threshold these bursts of self-limited activity may increase in intensity. There are known internal factors that could have this effect, such as sleep deprivation and changes in hormonal balance, but how these factors directly affect the network to initiate seizures remains speculative. Much work in this area has focused on modulation of synaptic inhibition or excitation. For example, a number of processes within the epileptic focus might promote seizures by increasing burst complexity and duration before leading to steady increases in accumulated energy. Neuronal excitability could be enhanced by slow changes in the extracellular environment, such as from slow accumulation of K^+ in extracellular spaces as buffering capacity becomes saturated [6], or slow local alkalinization [51], or slow changes in intracellular osmolarity [46]. Enhanced recurrent excitability of local networks within the focus could also effect these changes via slow potentiation of excitatory synapses, by enhanced local electrical coupling, or by the release of slowly acting neuromodulators. Increased coherence over this time period could also be promoted by transient changes in the redox state of the seizure focus, perhaps related to alterations in local perfusion, O_2 extraction or to ongoing neuronal activity, which have been shown to significantly affect NMDA receptor activity at excitatory synapses [43].

Another interesting mechanism may depend upon early increases in energy that trigger transcriptional changes in neurons within the epileptic focus. The resulting gene products may then initiate a cascade of intracellular changes leading to seizure onset in the network [26]. This idea is appealing, because communication between cell surface membrane and the nucleus may require minutes or sometimes hours to occur. In addition, long-term potentiation is known to affect gene transcription, providing one potential mechanism for electrical activity at the synapse to affect protein expression [1]. These types of changes could be analogous to those that occur in cortical neurons when memories are stored.

Although speculative, these hypotheses are testable, and could shed a new light on the physiology of seizure generation. Many of these proposed processes could be modified prior to seizure onset, leading to highly specific, targeted new therapies. Our findings, for the first time, suggest the need to evaluate these processes in broader time windows.

9.3.3 The Next Experimental Steps

One important challenge presented by these findings is how to proceed with more controlled experiments to understand our observations in humans. Because collection of human data is ethically limited to clinical necessity, this requires basic animal research. We are currently investigating seizure predictors in intracranial recordings from several animal models of human temporal lobe epilepsy, in an attempt to understand different portions of the preseizure cascade. These models include those causing widespread neuronal injury, such as systemically delivered pilocarpine and sustained electrical status epilepticus; and more focal models, such as intrahippocampal injection of kainic acid. All are problematic because they create variable lesions that are not confined to the hippocampus, and the frequency and severity of seizures associated with them is difficult to control. Genetically engineered animal models of epilepsy (i.e. knockouts) offer a controllable platform for experimentation, though most are more similar to human primary generalized epilepsy than to seizures of focal origin [2, 30, 31, 32]. Appropriate candidate genes identified in transgenic mouse epilepsy models may also be helpful in that they can be studied in primates, using gene transfer strategies [20, 53], which provide a system more homologous to humans for testing algorithms and implanted devices.

9.4 Clinical Research: Experimental Issues

Important experimental issues were raised during this study, including technical fidelity of the recordings, the effect of antiepileptic drug taper, patient state of consciousness and individual neurophysiology, which have implications for future research on seizure prediction.

All tapes were reviewed in their entirety before analysis, to screen out artifacts and occasional gaps in EEG data. Several seizures and baseline recordings were excluded from analysis due to strict inclusion criteria outlined in the methods. In retrospect these strict guidelines were justified, given our finding that clusters of seizures caused prolonged elevations in EEG signal energy that are sustained for three and perhaps up to six hours.

All study patients underwent rapid tapering of antiepileptic drugs to precipitate seizures, which is standard during presurgical evaluation. Haut et al. recently presented evidence that clustered seizures may be different from single spontaneous seizures in their appearance on EEG and site of onset [14]. Studies of seizure precursors in the absence of medication withdrawal will be important to validate prediction algorithms for application outside of the monitoring setting.

Our results demonstrate the importance of controlling for state of consciousness when evaluating seizure precursors. Accumulated energy measurements varied significantly between the awake and sleep states, and were only useful as seizure predictors when state of consciousness was taken into account. Other investigators have dealt with this issue by examining seizures only during wakefulness.

Though some of the variability in seizure predictors in our five study patients is likely due to electrode placement and variables such as medication taper, it is also likely that each of these patients has a specific neurophysiology that makes him or her unique, but that can be classified into one of a finite library of pre-ictal patterns analogous to the limited number of EEG patterns at seizure onset in temporal lobe

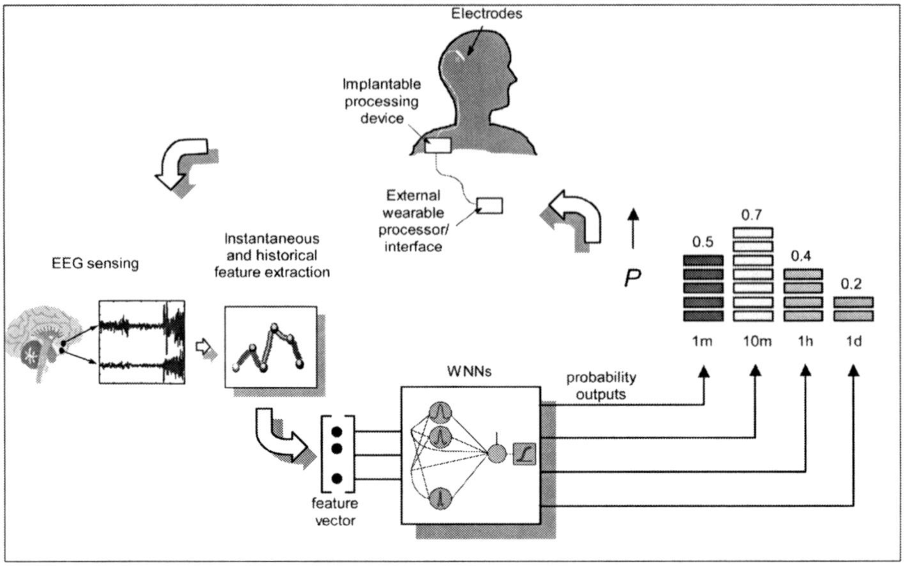

Fig. 9.5 A schematic of an implanted device for seizure prediction and intervention. EEG is recorded from electrodes implanted in the area of the seizure focus. Instantaneous features, such as signal energy and historical features, such as the time of the last subclinical seizure, are recorded. A neural network compares the behavior of these parameters to known behavior leading up to seizures for the given subject and continuously outputs the probability that a seizure will occur over four time horizons: 1 day, 1 hour, 10 minutes and 1 minute. Warnings of impending seizure or intervention in the form of electrical stimulation or infusion of drugs in the area of the seizure focus are triggered when the probability of an oncoming seizure, or the increase in probability over time exceeds a predetermined threshold value. Data processing takes place in an implanted computing unit coupled to an interactive interface worn on the subject's belt or elsewhere.

epilepsy [47]. This implies that optimal automated seizure prediction algorithms will likely be tuned to individual patients, based upon their quantitative "fingerprint" patterns on EEG.

It is important to note that all of the findings described in this pilot study refer to temporal and not extratemporal epilepsy, in which localizing seizure onset has been much more difficult. A discussion of seizure localization and prediction in extratemporal epilepsy is beyond the scope of this paper.

9.5 An Implantable Diagnostic-Therapeutic Device

Understanding mechanisms of seizure generation will ultimately produce the best therapies for poorly controlled epilepsy; however, an implantable anti-seizure device is a natural interim goal, as more definitive research progresses. Figure 9.5 displays a schematic of such an implantable device, based upon algorithms we have developed, to predict seizures and trigger treatment as they approach. The device continuously monitors features extracted from the intracranial EEG and computes the probability of an approaching seizure, in addition to checking for electrical seizure onset. A wavelet neural network is trained and optimized for each individual patient on multiple features extracted from continuous data containing both seizures and long seizure-free intervals. The patient or physician sets the probability thresholds and rate of rise that will trigger a warning or intervention.

Therapeutic interventions of several types are under investigation, including graded electrical stimulation and local drug infusion. As we envision this, when seizures approach with low probability, lower amplitude stimuli that are less likely to disrupt normal brain function, might be applied. Algorithms might include "chaos control" pacing, to disrupt rhythmic activity [11, 15, 27] or periodic low-level depolarization of a very small region. As the probability of seizure onset increases, stimulation intensity, duration and area are increased, perhaps extending subcortical structures, which may modulate seizure generation. As with all electrical interventions in brain, care must be taken to avoid stimuli that might be cause tissue injury or promote kindling of seizures in otherwise healthy tissue [28].

Infusing small amounts of antiepileptic drugs locally in the epileptic focus may also inhibit seizure generation. Candidate drugs for infusion include benzodiazepines, lidocaine, standard antiepileptic drugs (e.g. phenytoin or valproic acid), or newer antiepileptic agents. This approach has been the subject of considerable interest for the past several years [8, 50, 54]. Combination sensing electrodes/infusion cannulas are currently being investigated for this application.

9.6 Analogy to Cardiology

While the success of implants to treat cardiac arrhythmias is motivating investment into technologies to treat epilepsy, there are important differences between these applications that make intelligent devices to treat seizures more difficult to implement. Automatic implantable cardiac defibrillators serially process one channel of data, detect arrhythmias after their onset and deliver graded shocks to "re-set" the heart's electrical system. Side effects, such as pain, alteration or loss of consciousness are accepted, since the goal of therapy is to preserve life. In epilepsy, multiple channels of data must be processed in parallel. The goal of therapy is to prevent clinical symptoms, so that affected people may drive and participate in normal activities of daily living, lowering the threshold for side effects. In the brain, electrical connectivity is so widespread and conduction of seizures so rapid that delaying therapy until after electrical seizure onset may be too late to prevent clinical symptoms. EKG patterns that trigger defibrillators are the same from person to person. In epilepsy the electrical "fingerprints" of seizures are highly individualized, though relatively constant for a given person. Finally, the brain is a very dynamic organ, and intervention such as electrical stimulation can, over time, alter the way cells are connected and communicate with each other. Any device designed to treat seizures must take this into account, and allow for periodic retraining and updating of algorithms, to account for these potential changes.

9.7 Conclusion

Our results suggest that the changes in cellular and network function that lead to epileptic seizures likely develop over hours, providing exciting clues to potential mechanisms underlying seizure generation. Understanding these mechanisms should produce new, more effective and specific therapies for epilepsy. Our results also suggest that seizures in mesial temporal lobe epilepsy are predictable within time horizons that make it possible to prevent them, opening an immediate opportunity to develop intelligent, implantable therapeutic devices. As seizure precursors become better understood, it may become possible to trace the earliest precursors of seizures back to their site(s) of origin, which could be ablated by minimally invasive surgery or other techniques to replace large-scale brain resections that are now standard. Most important, these results provide new neurophysiologic evidence that may bring researchers closer to understanding how seizures are generated in the brain.

9.8 Experimental Methods

Continuous intracranial EEG and video were collected using a digital, 64-channel, 12 bit Nicolet BMS-5000 epilepsy monitoring system, and stored on videotape. Referentially recorded EEG was downloaded from tape and archived to CD-ROM for processing. EEGs were digitized at 200 Hz, and recorded after filtering through a bandpass of 0.1–100 Hz. Data were processed using a combination of MATLAB programs and custom-written C++ code for more prolonged, computationally intensive tasks on a network of four 400 MHz IBM personal computers running Windows NT and equipped with parallel processors. Hospital stays varied from three to 14 days in the five study patients, yielding roughly 50 gigabytes of raw data available for processing. Bipolar electrode montages were used to eliminate common mode artifact. A digital 60-Hz notch filter was employed to eliminate line noise. Videotapes were watched in their entirety for clinically important events, and approximation of periods of sleep and wakefulness. Finer resolution of sleep-wake cycles was achieved by examining the EEG.

In the study of accumulated energy, all preseizure epochs meeting criteria for inclusion were used. Acceptable baseline epochs were divided into hour-long segments, assigned numbers and then randomly selected for analysis. To be accepted, preseizure and baseline epochs had to be preceded by at least three hours of uninterrupted data, excluding brief artifacts lasting less than ten seconds. This excluded fourteen pre-seizure epochs from processing. A fifteenth seizure that arose outside of the temporal lobes (patient #5) was not included in the analysis. Not all excluded data epochs were rejected due to artifact. Some occurred near the beginning or end of a tape where previous or trailing data were lost, interrupted or obscured. Additionally, only one baseline epoch was allowed from a single three-hour period, to eliminate biasing results. Baseline epochs had to be recorded at least 80% during sleep or wakefulness. Target numbers for baselines were arbitrarily set at six each collected during sleep and wakefulness for each patient in this study. Table 9.2 demonstrates all usable, randomly chosen baseline segments available for processing, using the stringent selection criteria listed above.

9.8.1 Calculation of Signal Energy and Accumulated Energy

Signal energy was calculated in a sliding window of 1.25 seconds to assure signal stationarity, chosen empirically, based upon the work of Qin [36]. The instantaneous energy was obtained by squaring each value of the IEEG sequence as follows:

$$E_i[n] = x^2[n], \tag{9.1}$$

where $x[n]$ is the nth value of the intracranial EEG signal and $E_i[n]$ is the nth value of the energy.

Average energy values were computed by averaging every 250 points, equivalent to 1.25 seconds of the instantaneous energy at the given sample rate of 200 Hz. Results were computed in a moving average, overlapping consecutive calculations by 0.8 seconds (160 points) to give a higher resolution. This was accomplished by shifting the computation window in equation (9.2) by 0.45 seconds (90 points) each time.

$$E_k = \frac{1}{N}\sum x^2[n], \tag{9.2}$$

where E_k represents the kth value of the average energy over a window of 250 points and N is the total number of instantaneous energy values averaged (in this case $N = 250$).

A second level of averaging was performed to calculate the accumulated energy. Ten consecutive values of the average energy are added, divided by 10, and then added to the running tally of accumulated energy. This process is equivalent to integrating the absolute value of the EEG voltage as it is recorded on the intracranial EEG. An overlap of 50%, or five points, is used in this measure, again to improve resolution in time. Expression (9.3) summarizes how each accumulated energy value is obtained:

$$AE_m = \frac{1}{10}\sum E_k + AE_{m-1}, \tag{9.3}$$

where AE_m is the mth value of the accumulated energy and AE_{m-1} is the $m-1$th, or previous value of the accumulated energy.

9.8.2 Accuracy Calculation

Classification accuracy was calculated as the percentage of total number of correct predictions with respect to total number of records analyzed [7],

$$A = 100\left[1 - \frac{1}{N}\sum_{j=1}^{N}|y_i - \hat{y}_i|\right], \tag{9.4}$$

where A is the percentage of accurately classified data epochs, where y_i is the true classification of the ith record (1 = preseizure, 0 = baseline), $\hat{y}_i$ is the ith system's classification of the same event (1 = preseizure, 0 = baseline), N is the total number of records classified.

9.8.3 Data Handling and Processing

The large computational burden of these experiments required adopting a system of procedures to deal with the approximately 100 CDs, each containing 500 MB of digital data, processed for the above experiments. A custom toolbox of programs, written in MATLAB and C++ was developed, with an emphasis on flexibility, automation of feature extraction from raw data and tabulation of results. Two major groups of programs were developed: generic programs that were run on all data sets, including feature extraction, window length adjustment and data shifting routines; and patient-specific programs used to generate a set of individual features for each data epoch and to compile results for each trial. Batch programs were generally run overnight on the network and results were automatically saved into patient-specific work folders labeled by patient number. Of great importance, descriptive naming conventions were established for each feature generated from each patient's data, for each data epoch, for different states of consciousness and for different data classes (preseizure, baseline, control channels, etc.). These naming conventions were vital to handling the enormous amount of data processed for this study. Other issues, such as organization and control of access to the CD library of raw data, organization of files on computer hard-disk drives and assembling descriptive logs of clinical and quantitative data for each patient were labor intensive, but essential to maintaining easy access to raw and pre-processed data, calculated features, and overall results.

Acknowledgements

This work is supported by funding from the Whitaker Foundation, Epilepsy Foundation, American Epilepsy Society, University of Pennsylvania Research Foundation and National Institutes of Health grant #MH-62298RO1. Drs. Litt, Esteller,

Echauz and Vachtsevanos are founders of a small company, IntelliMedix, which supports this research. The authors express their appreciation to Steven Cranstoun M.S., Peter Crino, M.D., Ph.D., Francisco Gonzalez-Scarano, M.D., Ph.D., Steven Scherer, M.D., Ph.D. and David Raizen M.D., Ph.D. for reviewing this manuscript.

References

1. C.H. Bailey, D. Bartsch and E.R. Kandel, Toward a molecular definition of long-term memory storage, *Proc. Natl. Acad. Sci. USA* **93**, 13445–13452, 1996.
2. D.L. Burgess and J.L. Noebels, Single gene defects in mice: The role of voltage-dependent calcium channels in absence models, *Epilepsy Res.* **36**, 111–122, 1999.
3. M. Dichter and W.A. Spencer, Penicillin-induced interictal discharges from the cat hippocampus. I. Characteristics and topographical features, *Journal of Neurophysiology* **32**, 649–662, 1969.
4. M. Dichter and W.A. Spencer, Penicillin-induced interictal discharges from the cat hippocampus. II. Mechanisms underlying origin and restriction, *Journal of Neurophysiology* **32**, 663–687, 1969.
5. M.A. Dichter and G.F. Ayala, Cellular mechanisms of epilepsy: A status report, *Science* **237**, 157–164, 1987
6. M.A. Dichter, C.J. Herman, and M. Selzer, Silent cells during interictal discharges and seizures in hippocampal penicillin foci. Evidence for the role of extracellular K^+ in the transition from the interictal state to seizures, *Brain Res.* **48**, 173–183, 1972.
7. J. Echauz, S. Kim, V. Ramani and G. Vachtsevanos, Neuro-fuzzy approaches to decision making: A comparative study with an application to check authorization, *Journal of Intelligent and Fuzzy Systems* **6**, 259–278, 1998.
8. H.G. Eder, A. Stein and R.S. Fisher, Interictal and ictal activity in the rat cobalt/pilocarpine model of epilepsy decreased by local perfusion of diazepam, *Epilepsy Res.* **29**, 17–24, 1997.
9. J. Engel Jr., Outcome with respect to seizures, in *Surgical Treatment of the Epilepsies*, J. Engel Jr. (Ed.), Raven Press, New York, pp. 553–571, 1987.
10. W. Feindel, W. Penfield and H. Jasper, Localization of epileptic discharge in temporal lobe automatism, *Trans. Am. Neurol. Assoc.* **77**, 14–17, 1952.
11. J. Glanz, Do chaos-control techniques offer hope for epilepsy? [news], *Science* **265**, 1174, 1994.
12. J. Gotman, Computer analysis of the EEG in epilepsy, in *Methods of Analysis of Brain Electrical and Magnetic Signals: EEG Handbook*, A. Gevins and A. Rémond (Eds.), Elsevier, Amsterdam, pp. 171–204, 1987.
13. J. Gotman, Relationships between interictal spiking and seizures: Human and experimental evidence, *Canadian Journal of Neurologic Sciences* **18**, 573–576, 1991.
14. S.R. Haut, A.D. Legatt, C. O'Dell, S.L. Moshe and S. Shinnar, Seizure lateralization during EEG monitoring in patients with bilateral foci: The cluster effect, *Epilepsia* **38**, 937–940, 1997.
15. M.S. Heffernan, Comparative effects of microcurrent stimulation on EEG spectrum and correlation dimension, *Integr. Physiol. Behav. Sci.* **31**, 202–209, 1996.
16. L. Iasemidis, R. Gilmore, S. Roper and J. Sackellares, Dynamical interaction of the epileptogenic focus with extrafocal sites in temporal lobe epilepsy (TLE) (abstract), *Annals of Neurology* **42**, 429, 1997.
17. L. Iasemidis, L. Olson, R. Savit and J. Sackellares, Time dependencies in the occurrences of epileptic seizures: A nonlinear approach, *Epilepsy Research* **17**, 81–94, 1994.
18. L. Iasemidis, K. Pappas, R. Gilmore, S. Roper and J. Sackellares, Preictal entrainment of a critical cortical mass is a necessary condition for seizure occurrence, *Epilepsia* **37** (abstract), 1996.

19. L.D. Iasemidis, J.C. Sackellares, H.P. Zaveri and W.J. Williams, Phase space topography and the Lyapunov exponent of electrocorticograms in partial seizures, *Brain Topogr.* **2**, 187–201, 1990.
20. T. Kafri, H. van Praag, F.H. Gage and I.M. Verma, Lentiviral vectors: Regulated gene expression, *Mol. Ther.* **1**, 516–521, 2000.
21. A. Katz, D. Marks, G. McCarthy and S. Spencer, Does interictal spiking change prior to seizures? *Electroencephalography and Clinical Neurophysiology* **79**, 153–156, 1991.
22. H. Lange, J. Lieb, Jr, J. Engel and P. Crandall, Temporo-spatial patterns of pre-ictal spike activity in human temporal lobe epilepsy, *Electroencephalography and Clinical Neurophysiology* **56**, 543–555, 1983.
23. M. Le Van Quyen, C. Adam, J. Martinerie, M. Baulac, S. Clemenceau and F. Varela, Spatiotemporal characterizations of non-linear changes in intracranial activities prior to human temporal lobe seizures, *Eur. J. Neurosci.* **12**, 2124–2134, 2000.
24. K. Lehnertz and C. Elger, Can epileptic seizures be predicted? Evidence from nonlinear time series analysis of brain electrical activity, *Physical Review Letters* **80**, 5019–5022, 1998.
25. K. Lehnertz, G. Widman, R. Andrzejak, J. Arnhold and C.E. Elger, Is it possible to anticipate seizure onset by non-linear analysis of intracerebral EEG in human partial epilepsies?, *Rev. Neurol. (Paris)* **155**, 454–456, 1999.
26. F. Liang and E.G. Jones, Zif268 and Fos-like immunoreactivity in tetanus toxin-induced epilepsy: Reciprocal changes in the epileptic focus and the surround, *Brain Res.* **778**, 281–292, 1997.
27. F.H. Lopes da Silva, J.P. Pijn and W.J. Wadman, Dynamics of local neuronal networks: Control parameters and state bifurcations in epileptogenesis, *Prog. Brain Res.* **102**, 359–370, 1994.
28. E.W. Lothman and J.M. Williamson, Closely spaced recurrent hippocampal seizures elicit two types of heightened epileptogenesis: A rapidly developing, transient kindling and a slowly developing, enduring kindling, *Brain Res.* **649**, 71–84, 1994.
29. J. Martinerie, C. Adam, M.L.V. Quyen, M. Baulac, S. Clemenceau, B. Renault and F. Varela, Epileptic seizures can be anticipated by non-linear analysis, *Nature Medicine* **4**, 1173–1176, 1998.
30. J.L. Noebels, Single-gene models of epilepsy, *Adv. Neurol.* **79**, 227–238, 1999.
31. J.L. Noebels, X. Qiao, R.T. Bronson, C. Spencer and M.T. Davisson, Stargazer: A new neurological mutant on chromosome 15 in the mouse with prolonged cortical seizures, *Epilepsy Res.* **7**, 129–135, 1990. (Published erratum in *Epilepsy Res.* **11**(1), 72, March 1992.)
32. J.L. Noebels and P.A. Rutecki, Altered hippocampal network excitability in the hypernoradrenergic mutant mouse tottering, *Brain Res.* **524**, 225–230, 1990.
33. I. Osorio, M. Frei and S. Wilkinson, Real-time automated detection and quantitative analysis of seizures and short-term prediction of clinical onset, *Epilepsia* **39**, 615–627, 1998.
34. W. Penfield and T.C. Erickson, *Epilepsy and Cerebral Localization*, Charles C. Thomas, Springfield, IL, 1941.
35. W. Penfield and H. Flanigin, Surgical therapy of temporal lobe seizures, *Arch. Neurol. Psychiatr.* **64**, 491–500, 1950.
36. D. Qin, A comparison of techniques for the prediction of epileptic seizures, in *Proceedings of Eighth IEEE Symposium on Computer-Based Medical Systems*, Lubbock, TX, pp. 151–157, 1995.
37. H. Qu and J. Gotman, A patient-specific algorithm for the detection of seizure onset in long-term EEG monitoring: Possible use as a warning device, *IEEE Transactions of Biomedical Engineering* **44**, 115–122, 1997.
38. P. Rajna, et al., Hungarian multicentre epidemiologic study of the warning and initial symptoms (prodrome, aura) of epileptic seizures, *Seizure* **6**, 361–368, 1997.
39. B. Ralston, The mechanism of transition of interictal spiking foci into ictal seizure discharge, *Electroencephalography and Clinical Neurophysiology* **10**, 217–232, 1958.
40. M. Risinger, J.J. Engel, P. VanNess, T. Henry and P. Crandall, Ictal localization of temporal seizures with scalp-sphenoidal recordings, *Neurology* **39**, 1288–1293, 1989.
41. Z. Rogowski, I. Gath and E. Bental, On the prediction of epileptic seizures, *Biological Cybernetics* **42**, 9–15, 1981.

42. Y. Salant, I. Gath and O. Henriksen, Prediction of epileptic seizures from two-channel EEG, *Medical & Biological Engineering & Computing* **36**, 549–556, 1998.
43. R.M. Sanchez, C. Wang, G. Gardner, L. Orlando, D.L. Tauck, P.A. Rosenberg, E. Aizenman and F.E. Jensen, Novel role for the NMDA receptor redox modulatory site in the pathophysiology of seizures, *J. Neurosci.* **20**, 2409–2417, 2000.
44. Y. Schiller, G. Cascino, N. Busacker and F. Sharbrough, Characterization and comparison of local onset and remote propagated electrographic seizures recorded with intracranial electrodes, *Epilepsia* **39**, 380–388, 1998.
45. L.A. Schuh, T.R. Henry, D.A. Ross, B.J. Smith, K. Elisevich and I. Drury, Ictal spiking patterns recorded from temporal depth electrodes predict good outcome after anterior temporal lobectomy, *Epilepsia* **41**, 316–319, 2000.
46. P.A. Schwartzkroin, S.C. Baraban and D.W. Hochman, Osmolarity, ionic flux, and changes in brain excitability, *Epilepsy Res.* **32**, 275–285, 1998.
47. S. Spencer, P. Guimaraes, K. Katz, J. Kim and D. Spencer, Morphological patterns of seizures recorded intracranially, *Epilepsia* **33**, 537–545, 1992.
48. M. Sperling, Electrophysiology of the ictal-interictal transition in humans, in *Mechanisms of Epileptogenesis: The Transition to Seizure*, M.A. Dichter (Ed.), Plenum Press, New York, pp. 17–38, 1986.
49. M. Sperling and M. O'Connor, Auras and subclinical seizures: Characteristics and prognostic significance, *Annals of Neurology* **28**, 320–328, 1990.
50. A.G. Stein, H.G. Eder, D.E. Blum, A. Drachev and R.S. Fisher, An automated drug delivery system for focal epilepsy, *Epilepsy Res.* **39**, 103–114, 2000.
51. C.M. Tang, M. Dichter and M. Morad, Modulation of the N-methyl-D-aspartate channel by extracellular H+, *Proc. Natl. Acad. Sci. USA* **87**, 6445–6449, 1990.
52. S. Viglione, V. Ordon and F. Risch, *A Methodology for Detecting Ongoing Changes in the EEG Prior to Clinical Seizures*, McDonnel Douglas Astronautics Co., West Huntington Beach, CA, 1970.
53. L. Wang, T.C. Nichols, M.S. Read, D.A. Bellinger and I.M. Verma, Sustained expression of therapeutic level of factor IX in hemophilia B dogs by AAV-mediated gene therapy in liver, *Mol. Ther.* **1**, 154–158, 2000.
54. S. Ward and M. Rise, Techniques for treating epilepsy by brain stimulation and drug infusion, US Patent #5713923, 1998.

Chapter 10
Intelligent Control Strategies for Neurostimulation

Javier Echauz, Hiram Firpi and George Georgoulas

Abstract Neurodevices for the management of nervous system disorders have been recognized as most promising through the coming decades. Technological development is being spurred on as drugs and other standard therapies have reached diminishing returns. New data enabled by cutting-edge telemetric devices will spin off new business models, for example, in seizure rhythm management. Implantable neurostimulation devices already exist as adjunct therapy for intractable epilepsy, but paradoxically, age-old feedback control strategies remain largely unknown or underutilized in the field. In this chapter we outline strategies for intelligent feedback control of pathological oscillations. We review the state of the art in implantable devices for epilepsy and the experimental evidence for improved performance via feedback control. Then we extend an existing body of work from open-loop to continuous feedback control of phase-based models of hypersynchronization. Conversion of the results to practical devices is explored via pseudostate vector reconstruction. We conclude by outlining key components of research for continued progress in this field.

10.1 Introduction

The ability to sense, interpret, model, and modulate neurosignals together describe the foundation underlying an explosion of new, cutting-edge research that is both

Javier Echauz
JE Research, Inc., Alpharetta, GA 30022, USA; e-mail: echauz@ieee.org

Hiram Firpi
Department of Internal Medicine, 2294 CBRB, 285 Newton Road, Iowa City, IA 52242, USA; e-mail: hiram-firpi@uiowa.edu

George Georgoulas
TEI of the Ionian Islands, Computer Technology Applications in Management & Economics, 31100 Lefkada, Greece; e-mail: georgoul@teiion.gr

K.P. Valavanis (ed.), Applications of Intelligent Control to Engineering Systems, 247–263.

expanding our basic understanding of the nervous system and translating into new technologies for clinical use. At one extreme, this area focuses on charting the architecture of functional networks and neuronal subpopulations in brain to understand their function in health and disease. At the other, this new knowledge is spawning responsive implantable devices to treat human disease and new technologies to modulate or even augment normal brain function. In the last decade this research has produced prototype devices to interpret neural signals to operate robotic devices (e.g., brain-computer interfaces), to mimic the integration and neural encoding functions of sensory organs in silicon (neuromorphic engineering), responsive implantable devices to treat disease using open and closed loop control, such as in movement disorders, epilepsy, migraine, affective disorders, and rehabilitation from stroke and traumatic brain injury.

Recent efforts have focused on: (1) recording, localizing, and tracking neural signals from sensors tuned to multiple temporal and spatial scales, and (2) closed-loop control of specific networks to prevent or treat specific disease states, compensate for loss or degradation of function, or to create specific output states. To date considerable accomplishment has been attained in the first task, culminating in the successful development of algorithms for an implantable antiepileptic device currently in clinical trials. The next logical steps include work on continuous fine-grained feedback control of neural function based upon signals derived from an array of sensors implanted in animal models and humans with epilepsy. The principles outlined next form the core of technology applicable to treating a variety of disease states and medical conditions. The technology additionally has the potential for generating and controlling output in a variety of functional neuronal networks involved in processes as disparate as consciousness, attention, motor control, memory, cognition, and learning.

10.2 Implantable Neurostimulation Devices

The state of the art in implantable neurostimulation devices includes an ability to continuously monitor local field potentials of the brain in order to detect precursory or aberrant epileptiform activity with high resolution and certainty. Paradoxically, during the interventional stimulation period immediately following any such event, current systems revert to a plain open-loop or very coarse-grained type of electrical stimulation. Predicated by the success of responsive therapies for the heart (demand pacing, cardioversion, and defibrillation), responsive therapies for the brain [1–3] have been designed to deliver, directly or peripherally, preprogrammed waveforms upon detection of discrete events. The waveforms are predefined to be biphasic rectangular or experimental sinusoidal pulse trains with a combinatorial explosion of stimulation parameters to choose from (channel, amplitude, frequency, duty cycle, train duration, number of therapies in a row, etc.). If a computer could evaluate each combination in this parameter space at a rate of one per second for a given patient, it would still take millions of years to exhaustively search the space, without any guarantee that an optimal setting of these open-loop parameters would yield maximum achievable efficacy. Manual selection of these programmable parameters has yielded at best a 43% responder rate ($\geq$ 50% seizure frequency reduction) in clinical trials so far [4].

Similar to cardiac devices, it is also possible using current neurostimulation technology to automatically "adapt" the onset time of stimulation and the number of successive therapies to the sensed EEG during the epileptiform event period, in an attempt to increase efficacy, although this remains largely underutilized owing to a lack of basic studies or clinical experience. Our research recently made connections between real-time prestimulation EEG features and their subsequent acute stimulation efficacy (labeled as "successful" vs. "unsuccessful"). For example, the instantaneous phase of intracranial EEG at stimulation onset time is a statistically significant predictor of efficacy [5]. This information could be exploited by coupling a Hilbert instantaneous phase extractor to the synchonized stimulation start time capability of current devices. It is also in agreement with population models of neural desynchronization such as in the phase resetting of noisy coupled phase oscillators [6–10]. However, even under the idealized conditions of these models, the theory predicts that other stimulation variables – current intensity and duration of constant pulse – in addition to exact timing, would need precise and frequent tuning to approach maximum efficacy. Experience with implanted pulse generators so far suggests that the spatial configuration of stimuli may also require periodic tuning over time. In summary, present day's programmable responsive neurostimulators have efficacy bounded low by open-loop control, leave this efficacy largely to random chance, and are impossible to truly optimize in practice given the sheer number of programmable parameters.

In contrast to the cardiac-inspired approach, we favor a fine-grained feedback control design in which a stimulation value is determined on the fly, at every discrete time instant, based on the evolving state variables estimated from sensed EEG during an epileptiform event. The resulting actuating "waveforms" can incorporate current density, charge balance, actuation rate of change, energy expenditure, or other safety and device constraints, but their precise shape is not known a priori because they are a function of the particular dynamics of the system upon which they act. One algorithm (with its few parameters) replaces the combinatorial explosion of programmable waveform parameters. To understand the potential significance of this kind of feedback control paradigm, consider that suppressing seizures with preprogrammed waveforms is analogous to balancing a broomstick on the palm of one's hand with eyes closed, whereas continuous-time feedback allows the eyes to be open. We believe a prescription for seizure freedom in difficult-to-treat epilepsy is loosely speaking 5% open-loop (e.g., DBS, VNS, TMS) or triggered open-loop control (e.g., responsive neurostimulation RNS; all available today), but 95% fine-grained closed-loop feedback control, adaptive control (online adaptation of parameters), and intelligent control (online learning of hierarchical structure and parameters).

10.3 Experimental Evidence for Feedback Devices

In the context of treating disorders of the nervous system, experimental evidence for the effectiveness of continuous feedback control is limited but encouraging. The idea of continuous-time closed-loop feedback control to electrically suppress seizures has been around since at least 1973 with Saul Liss' U.S. patent "Apparatus for monitoring and counteracting excess brain electrical energy to prevent epileptic seizures and the like" [11]. Professor Chkhenkeli's research started yielding clinical experience with coarse-grained feedback (manually triggered DBS) since 1976 in Tbilisi [12]. Experiments showing injection of feedback currents into brain slices have been described by Nakagawa and Durand [13], Schiff et al. [14], and Lian et al. [15]. The same concept, applied to directional cancellation of the electric field, was shown by Gluckman et al. [16]. Systematic experiments by Mogul et al. have explored several PID controls to suppress penicillin-induced seizures in anesthetized rats, using EEG recorded in one electrode as "error" signal and stimulating with the other bipolar electrode, where sporadic suppressions have been displayed [17].

In silico experiments and more theoretical efforts include ensembles simulated with several neuron models that were suppressed by raw feedback of the network's "mean field" [18]. Control theorist Kostas Tsakalis, and professors Iasemidis et al. have also suggested the use of robust control (structure designed for worst case that handles model uncertainty) and have applied pulse trains to an oscillation model using coupled Rössler neurons [19].

In the above "cancellation" experiments, suppression is achieved by counteracting the amplitude of an individual neuron's output (or its polarization field leading to that output), or feeding back EEG amplitude directly. Unfortunately, this may be equivalent to "seizure suppression by brain death". What we aim to control is a property of the collective epileptic network, not of individual neurons/glia. As will be illustrated next, it is possible to disrupt synchronization by attacking the phase rather than the amplitude distribution of a network. "Chaos anticontrol" is more along these lines, since it seeks to repeatedly kick a system out of "periodicity". However, its converse, "chaos control", which was demonstrated in physical systems such as metallic ribbons, electronic circuits, and lasers, has yet to show practical utility for a biological system as well understood as the heart. In Slutzky's replication of brain slice experiments, a shadow of doubt is cast on earlier publications by distinguishing between stimulated vs. natural interburst intervals in the plots, and suggesting that demand pacing and not chaos control may have been at work [20].

10.4 Computer Simulation of Pathologic Oscillations and Feedback Therapy

Computer simulation can accelerate the development of therapeutic devices. Simulations can be run millions of times in order to explore network dynamics, the effect of parameter changes in the system, the effect of therapeutic intervention, etc., without the cost or physical harm involved in experimental setups. Focusing on small details of the nervous system has produced some spectacular successes, but does not provide much information on processes such as memory or seizures, many of whose characteristics are emergent and are not understandable below the level of ensembles of neurons. When investigating the dynamics of epilepsy, the most appropriate level of description would seem to be a population or group of neurons [21] (in contrast to the level of compartmentalized models of neural networks). Other authors conclude that effective therapy should be aimed at desynchronizing populations rather than naively at the concept of excitation/inhibition balance [22]. To test how epileptic activity is "sustained", measurements have shown zero net phase lag, suggesting that seizures behave more like coupled oscillators than reentrant loops or than projections from a discrete pacemaker [23].

Promising population-based computer simulations of seizure-like activity and therapeutic interventions can be based on the models of Suffczyński [24, 25], Wendling et al. [26], and Tass [6–10]. One attractive aspect of the Suffczyński model is spontaneity of seizure-like activity, which is not directly available in Wendling's. Additional dynamical equations to drive the variation of the gain parameters would be needed (a transition to seizure can be constructed by manually changing these gains). A disadvantage of both Suffczyński's and Wendling's models is that external inputs are either not emphasized or are accommodated in terms of average pulse densities – positive quantities that do not model the bipolar nature of neurostimulation current. The Tass model discussed next specifically includes the effect of stimulation on network dynamics. Like Wendling's, the Tass model is perpetually in a given qualitative state of oscillation unless driven out of it by additional means. For Tass, this is accomplished by the influence of therapeutic intervention.

Tass has developed an infinite-dimensional (via partial differential equations) model of desynchronization in the framework of stochastic phase resetting [6]. Although initially motivated in the context of Parkinsonian tremor, the principles are fundamental and should be broadly applicable to other biological oscillations, including epileptic seizures. A phase oscillator continually sweeps angle from 0 to 2π and repeats the cycle ad infinitum. In isolation, it will do this at a constant angular speed Ω. Interconnected, phases can wax and wane to try to synchronize. The times of spiking/bursting of a neuron are represented each time a rotator up-crosses the $+x$-axis in the complex plane (amplitude information is discarded here). Think of the network as an infinite population of rotators whose collective synchronization depends on mutual coupling, noise, and stimulation resetting influences, and whose dynamics are given by a partial differential equation derived in the asymptotic limit from Winfree's, Kuramoto's, and other prior models of oscillatory behavior.

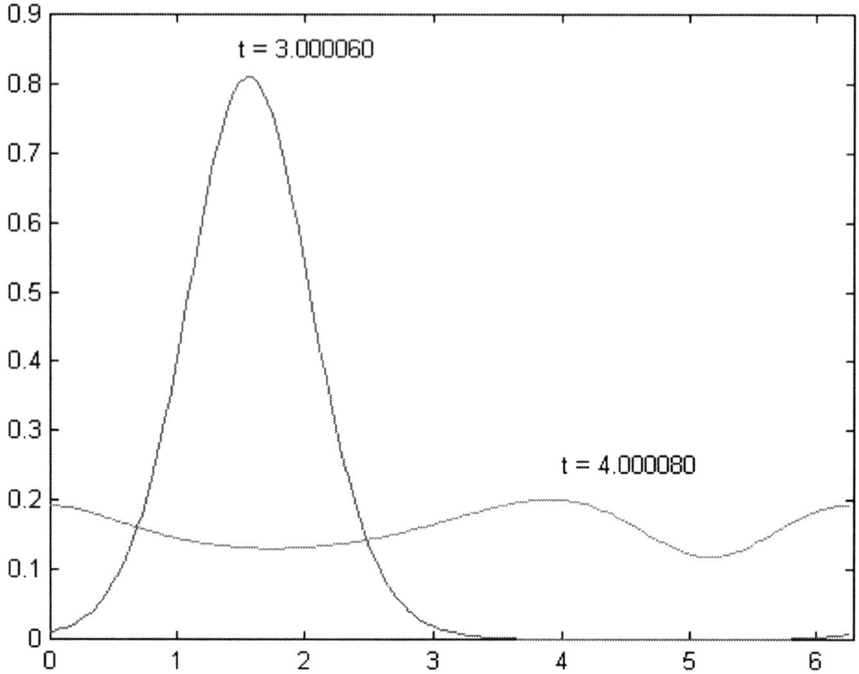

Fig. 10.1 Two snapshots in time of the distribution of phases in a network that is synchronized at $t = 3$ and desynchronized at $t = 4$. Horizontal axis is phase angle from 0 to 2π.

The model output is the average number density $n(\psi, t)$ – a time-varying p.d.f. of phases (think of it as the "fraction" of neurons firing at phase ψ at a given time t). It can be shown that this density is a kind of average of all marginal distributions of an underlying joint density $p(\psi_1, \psi_2, \ldots)$ of the rotators. For numerical stability, the evolution of $n(\psi, t)$ is solved as ordinary differential equations in the Fourier series domain using Euler or Runge–Kutta with 100 modes, so the infinite-dimensional state becomes a vector with 100 state variables (the Fourier coefficients of the p.d.f.). Given an initial p.d.f. $n(\psi, 0)$, depending on bifurcation parameters, the network will be either incoherent or attracted to a limit cycle which we associate to "seizure". In the synchronized state, the whole network acts like one giant oscillator [7]. At any given time, the network will show at least one preferred range of phases in the form of peakedness in the p.d.f. Extreme synchronization would be an impulse at a single phase, while complete desynchronization is the uniform distribution.

Figure 10.1 shows the p.d.f.s at $t = 3$ and $t = 4$ corresponding to Figure 10.2c. Instead of a constant, we made stimulation intensity a function of time to allow for arbitrary control laws. In the simulations, the influence of stimulation had a single Fourier mode: $I(t) \cos \psi$ (effect is singly-periodic with phase, and overall synchronization or desynchronization is encouraged or discouraged depending on timing and current $I(t)$); rotators "self-oscillate" with eigenfrequency $\Omega = 2\pi$;

Gaussian noise (which dampens synchronization) had variance $Q = 0.4$; influence of mutual coupling also had a single Fourier mode $-K \sin \phi$, with coupling constant $K = 2\pi$; and initial phase distribution was a circularly periodic Gaussian with central location $\pi/2$ and variance 0.6.

10.5 Extension to Continuous Feedback Control

We have derived fine-grained feedback control laws (formulas translating states into control actions) that automatically desynchronize or hypersynchronize, at will, the Tass model of neural synchronization. Tass first showed that this model can be desynchronized, though requiring cumbersome phase-resetting calibrations, using open-loop schemes: single pulse, periodic pulse train, double pulse, bipolar double pulse, and coarse-grained schemes: "demand-controlled" timing and duration [6–10], and recently, calibration-free methods using multiple sites and delays feedback coordination [27].

Let $y(t)$ be the first Fourier "mode" of $n(\psi, t)$, here the complex coefficient associated with the fundamental frequency when the snapshot p.d.f. $n(\psi, t)$ is decomposed into its Fourier Series (F.S.). The perfect desynchronized state is attained when this p.d.f. is uniform (no undulation) which has first mode zero (whole F.S. equals just the constant $(2\pi)^{-1}$). Thus, we make $y(t)$ the controlled variable. A proportional control law can be motivated noting that the network gets closer to desired state by driving the "error" signal

$$\begin{aligned} e(t) &= (y(t) - y_{\text{desired}}) \\ &= (y(t) - [0 + j0]) \\ &= (\text{Re}\{y(t)\} - 0) + j(\text{Im}\{y(t)\} - 0) \end{aligned} \tag{10.1}$$

to zero, and for almost every point in the complex plane, the sum

$$I(t) = \text{Re}\{y(t)\} + \text{Im}\{y(t)\} \tag{10.2}$$

provides a signed "corrective" signal. This real-valued law still leaves a null space where desynchronization is not guaranteed, however, in general, addition of integral control eliminates steady state error, while derivative control improves the transient response. Remarkably, regulating just the first Fourier mode of $n(\psi, t)$ to zero drives all other (infinitely many) modes of the phase distribution to zero, which is an illustration of Haken's enslaving principle (at a phase transition and during further evolution, a few order parameters dominate all others [28]).

In Figure 10.2, we compare the firing density $n(0, t)$ of the network using a biphasic pulse train stimulation with typically chosen parameters (Figure 10.2a) versus a proportional closed-loop control law whose sole parameter is a gain (Figure 10.2c), both applied for 1 second between times $t = 3$ and $t = 4$. The cor-

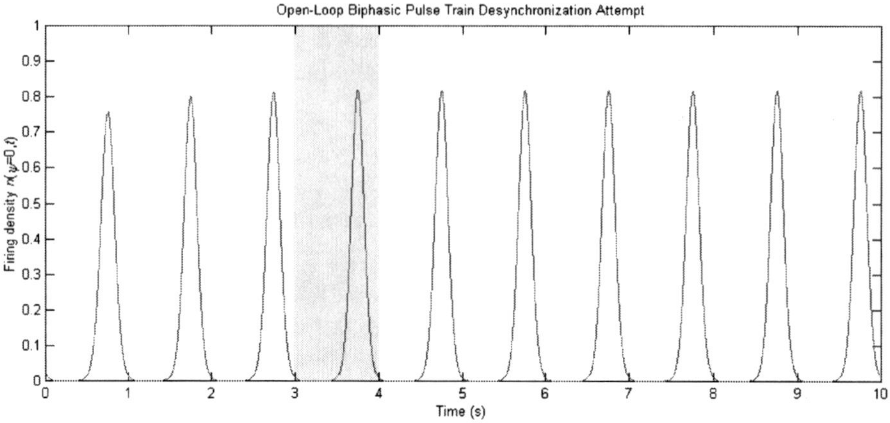

Fig. 10.2a Open-loop biphasic pulse train control was applied between times $t = 3$ and $t = 4$. This did not randomly hit during any susceptibility period and there is no effect on firing density, which continues to oscillate in its synchronized state.

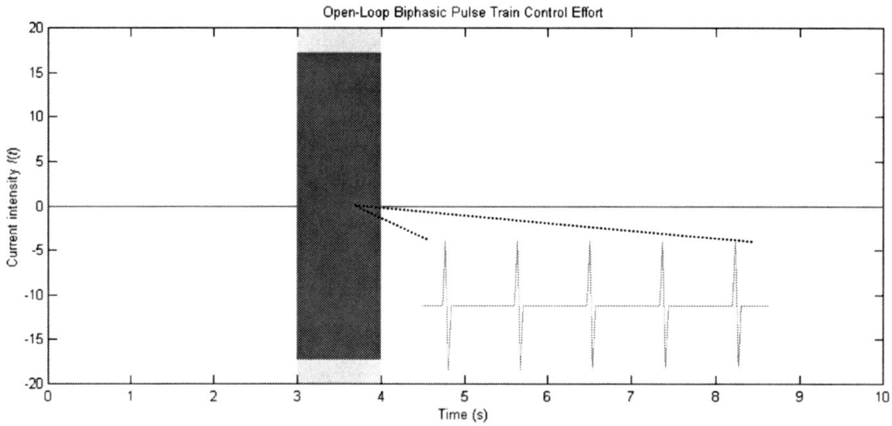

Fig. 10.2b Biphasic pulse train control effort has amplitude 17.15 (to make total energy expended equal to that of the control signal in (d)), 200 Hz frequency, and 5% duty cycle per phase. In human trials, the burst duration can be shorter but can be reinitiated several times within the same seizure by automatic seizure detection.

responding stimulation currents, i.e., the control effort signals $I(t)$, are shown in Figures 10.2b and 10.2d. The example illustrates the principle of applying a continuously updated, precisely aimed stimulation according to network output, as opposed to blind open-loop pulses. The current amplitude over-/undershoots are presumably within safe limits.

In practice, the amount of battery drain for the feedback device would be orders of magnitude less than that of chronic open-loop stimulation, since only quickly-reacting, small-amplitude, fleeting perturbations are required in the feed-

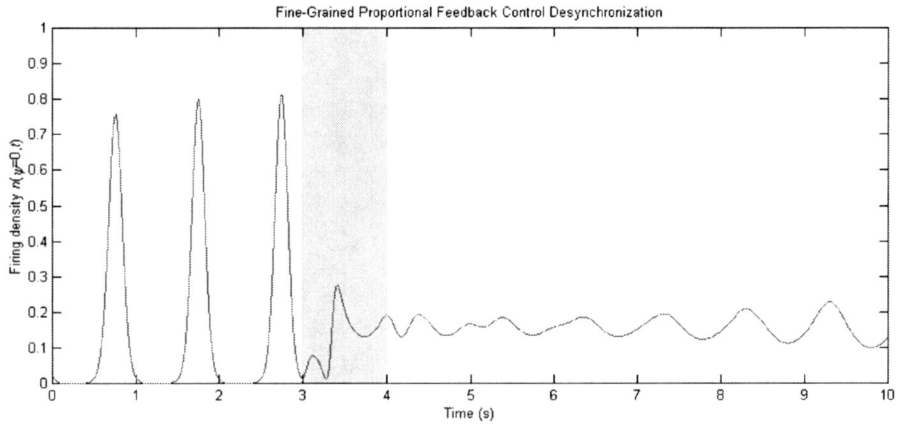

Fig. 10.2c Fine-grained closed-loop control was applied between times $t = 3$ and $t = 4$. Firing density approaches its maximally desynchronized state value of $1/(2\pi) = 0.1592$ (uniform distribution of phases). Exact timing is irrelevant in achieving this here. In a few more seconds after stimulation stops, the network will spiral back to its synchronized-state limit cycle.

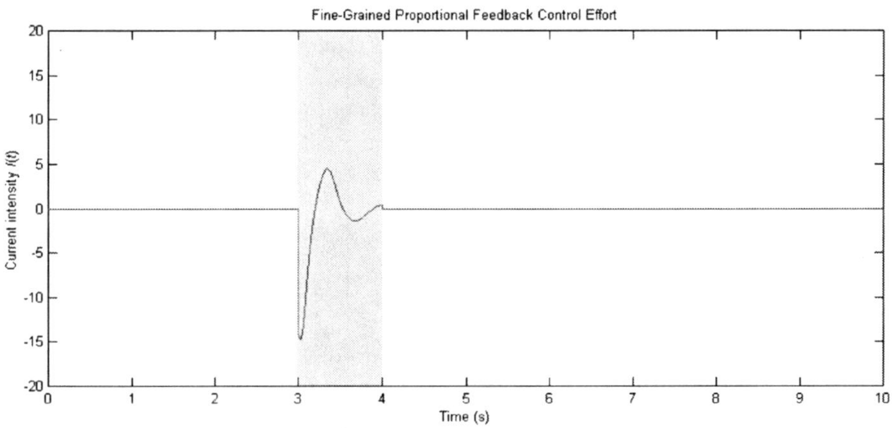

Fig. 10.2d Control effort $I(t) = K_p(\mathrm{Re}\{y(t)\} + \mathrm{Im}\{y(t)\})$, where $K_p = 100$ and $y(t) =$ 1st Fourier mode of $n(\psi, t)$ in the phase axis, cannot be any of the prescribed waveforms found in current devices. The time-varying intensity ("shape") is automatically determined from a measure of network state. Furthermore, the gain K_p is not a critical parameter in achieving desynchronization; for a wide range it only changes undershoot and settling time.

back scheme to maintain the system within a "non-seizure envelope". A key aspect of this method is that the signal being fed back to the controller is a measure of the distribution of phases in the network, not simply the firing density or EEG directly as in all prior studies of feedback control in this context.

Other control laws desynchonize the network even faster; for example, we find that the Tass system of equations is feedback linearizable (nonlinearity can be can-

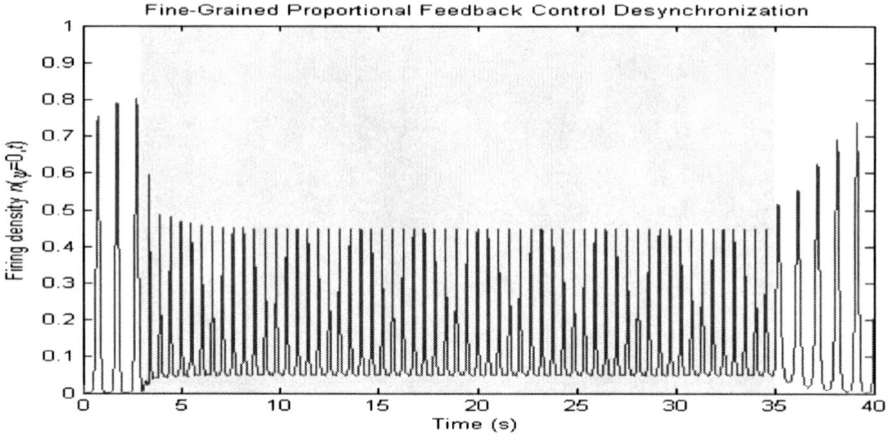

Fig. 10.3 Prolonged attempt at synchronization ignoring information from the imaginary component of the proportional control law showing failure.

celled and poles placed at any desired location). Let x be the 100-dimensional state vector containing the truncated F.S. coefficients (each associated with modes). The dynamics of the first mode x_1 $(= y(t))$, which is a function of only the first two modes x_1 and x_2, can be arranged into linear and nonlinear components

$$\dot{x}_1 = (r_1^{(1)} + k^{(3)} + k^{(4)})x_1 + r_1^{(2)}u + k_1^{(1)}x_1^* x_2 + k_1^{(2)}x_2 u^*, \tag{10.3}$$

where $r_1^{(1)} = K/(4\pi)$, $k^{(3)} = -j\Omega$, $k^{(4)} = -Q/2$, $r_1^{(2)} = -j/(2\pi)$, $k_1^{(1)} = K/2$, $k_1^{(2)} = -j$, and $u = I(t)/2$ which is the F.S. of the single-mode stimulation influence $I(t)\cos\psi$ (reduces to this single coefficient). Then, to obtain linear dynamics in the closed-loop having say, a pole at -5, we would build the control law as

$$u = \frac{-5x_1 - (r_1^{(1)} + k^{(3)} + k^{(4)})x_1 - k_1^{(1)}x_1^* x_2}{r_1^{(2)} + k_1^{(2)}x_2} \tag{10.4}$$

which when plugged into (10.3) leaves the dynamics $\dot{x}_1 = -5x_1$ (converging fast to zero). However, this strategy leads to a complex-valued current $I(t)$, which has no physical meaning in this context. Ignoring the imaginary component here, or in the real-valued imaginary component of Equation (10.2), fails as shown in Figure 10.3.

Figures 10.4a and 10.4b show the firing density and control effort respectively after applying a Particle Swarm Optimization (PSO) with 50 particles to a PID resulting in parameters $K_p = 107.4411$, $K_i = 12.9379$, and $K_d = 0.1001$. This type of control shows more effectiveness than the untuned proportional feedback of Figures 10.2c and 10.2d in two ways. First the steady-state amplitude of firing density at the end of 10 seconds is smaller, meaning that stimulation hit the network closer to its maximal desynchronized state, and second, the control effort achieved both

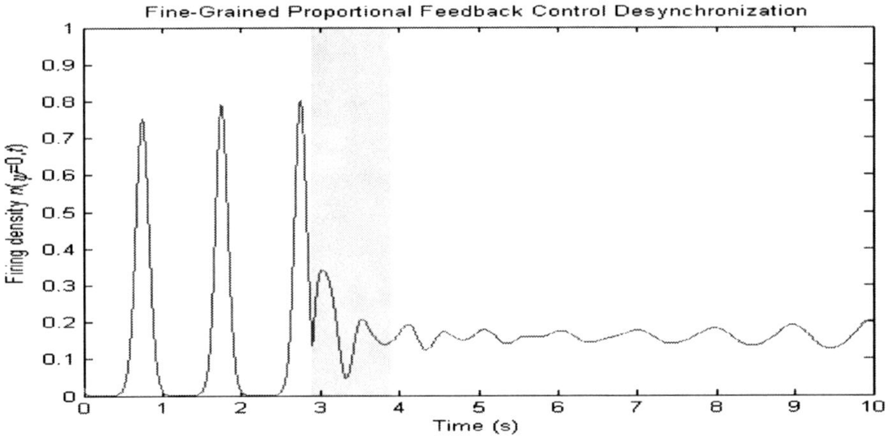

Fig. 10.4a Firing density for PSO-tuned PID controller.

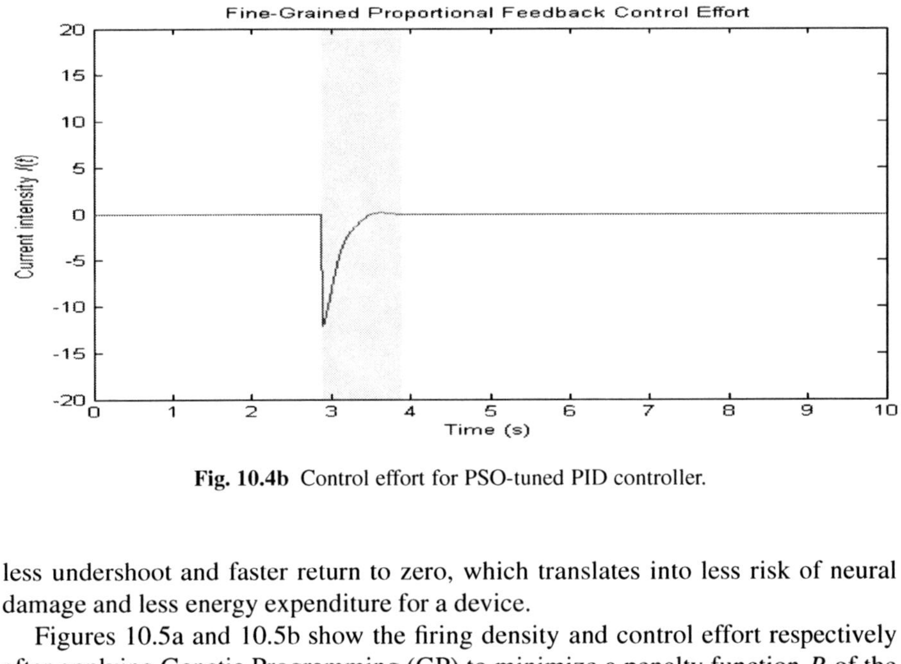

Fig. 10.4b Control effort for PSO-tuned PID controller.

less undershoot and faster return to zero, which translates into less risk of neural damage and less energy expenditure for a device.

Figures 10.5a and 10.5b show the firing density and control effort respectively after applying Genetic Programming (GP) to minimize a penalty function P of the form

$$P = \sum_{t_k=3}^{4} \left| \eta(0, t_k) - \frac{1}{2\pi} \right|. \tag{10.5}$$

This function is a discrete version of the integral of absolute deviations between the firing density and the maximally desynchronized state. Since the summation is being applied only during the stimulation period, GP returns a solution scoring $P =$

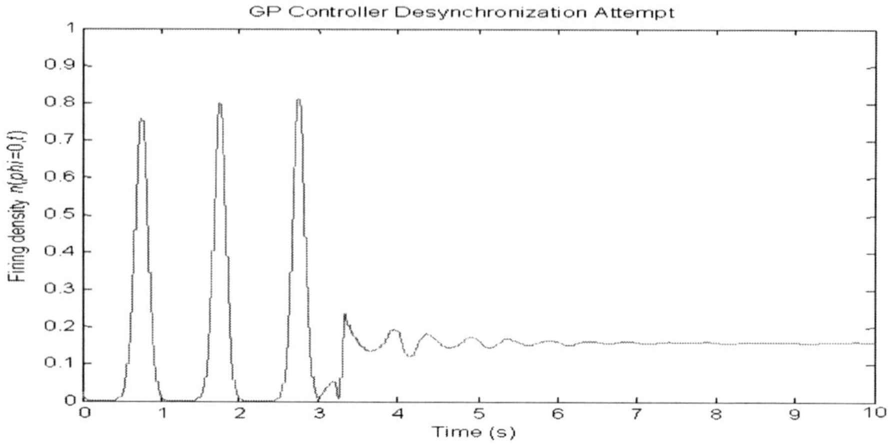

Fig. 10.5a Desynchronized signal using an aggressive GP-based controller.

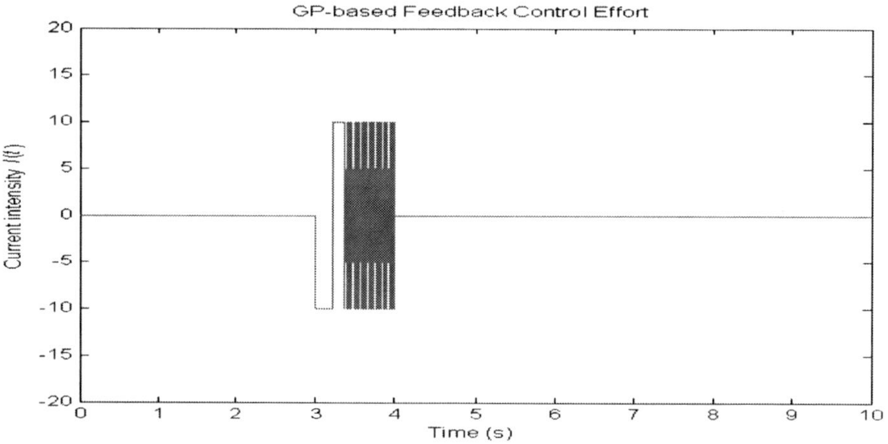

Fig. 10.5b Stimulus generated by the GP controller with bang-bang behavior.

269.84 that tries to desynchronize as soon as possible before stimulation is turned off, thus leading to a bang-bang type of control. We observe remarkable steady-state performance at the expense of potentially harmful saturated stimulations.

Like often the case in natural biological evolution, GP tends to find notoriously nonparsimonious solutions that nevertheless "just work". For example, GP-controllers that minimize the penalty function in terms of the first Fourier mode of $n(\psi, t)$ can display control laws of the form

$$I_{\mathrm{GP}} = \log_2\left(\left|\frac{\sqrt{|\mathrm{Im}\,(x_1(k))|}\,|\mathrm{Im}\,(x_1(k))|}{\mathrm{Re}\,(x_1(k)) + \mathrm{Im}\,(x_1(k))}\right|\right) - \sin\left((\sin(U(t_1, t_2)))^2\right), \quad (10.6)$$

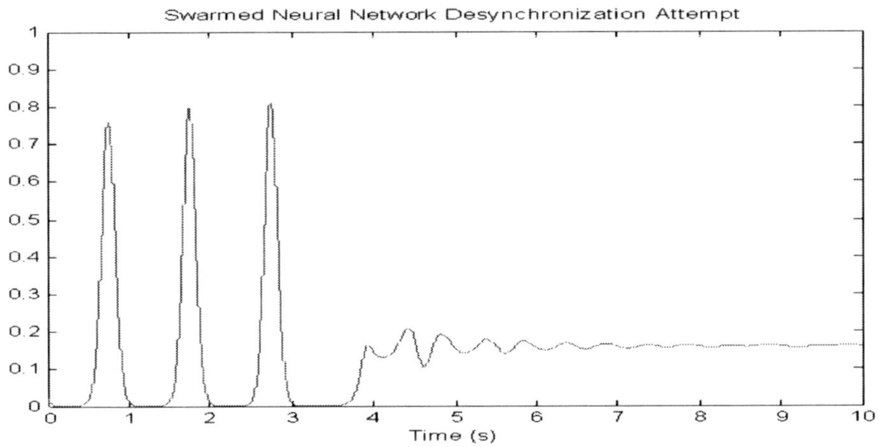

Fig. 10.6a Desynchronized signal using a swarmed neural network controller.

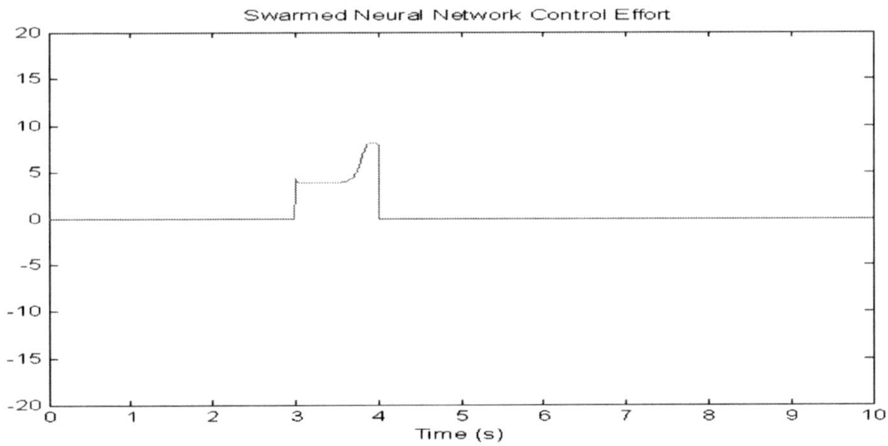

Fig. 10.6b Stimulus generated by the swarmed neural network controller.

where U is a uniformly-distributed random number. Non-parsimony, and the complexity of the output and input stimulus waveforms, can also be explicitly encoded into the penalty function for the GP to optimize.

Figures 10.6a and 10.6b show the firing density and control effort respectively after applying PSO to an artificial neural network minimize a penalty function of the form

$$P = \sum_{t_k=3}^{10} \left(\eta(0, t_k) - \frac{1}{2\pi} \right)^2 . \tag{10.7}$$

This function displays smoother characteristics and results in equally excellent steady-state response. The stimulus however is monopolar, whereas bipolar would have been preferable from a safety standpoint.

10.5.1 Extension to Practical Control via Pseudostate Vector

Further extensions of this work show that the network can be desynchronized even without knowledge of the internal state variable (1st Fourier mode), which is crucial in a practical implementation, by applying delay embedding and extracting modes from a sliding-window FFT, at expense of additional delay in achieving the desired state. Figures 10.7a and 10.7b show the firing density and control effort respectively after applying this principle of state reconstruction from measurable signals. The stimulation time must be extended in order to attain comparable steady-state performance.

In the simulation, the proportional control gain is $K_p = 300$, the Euler step is $2(10)^{-4}$, i.e., sampling frequency is 5 KHz, and the length of the FFT window is 256 points, sliding every 1 point. The key ingredient is that in applying Equation (10.2), instead of $y(t)$ which is an internal (and inaccessible) state variable of the network, we use an estimate $\hat{y}$ obtained strictly from the measurable firing density. Thus the discrete-time control law is

$$I(k) = \text{Re}\,\{\hat{y}(k-1)\} + \text{Im}\,\{(k-1)\}, \qquad (10.8)$$

where $\hat{y}$ is the Fourier coefficient associated with the fundamental (second element of FFT vector) when applying FFT to a window of tapped delays of the firing density:

$$\text{FFT}\{[n(0,k-255) \quad n(0,k-254) \quad \ldots \quad (n(0,k)]\}. \qquad (10.9)$$

A problem that was not addressed in this formulation is that of over- and undershoot exceeding safety limits in the stimulation signal as seen in Figure 10.7b.

10.6 Conclusion

Open-loop controllers (VNS, DBS, TMS, drugs) are sensorless and simply deliver preprogrammed control action. Triggered open-loop controllers (RNS) use sensing only to know when to start/stop delivery of preprogrammed control action (these have come to be known as "closed-loop" brain stimulators in some of the literature). Feedback controllers (also termed continuous or fine-grained here) use sensing to continually correct the control action. Trainable controllers can be tuned via external re-programming of parameters (a coarse form of feedback can occur only during these training occasions). Adaptive controllers continually change their fixed-structure parameters based on feedback signals. Learning controllers subsume

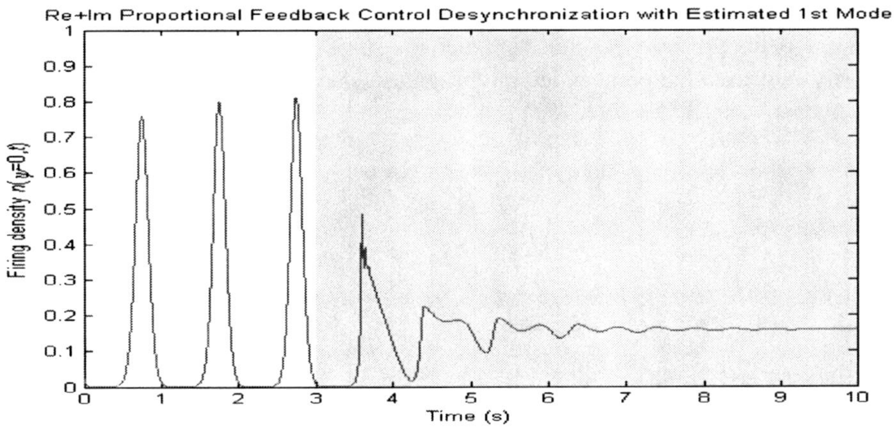

Fig. 10.7a Proportional feedback control with $K_p = 300$ and 1st Fourier mode estimated via delay embedding of the measurable firing density signal.

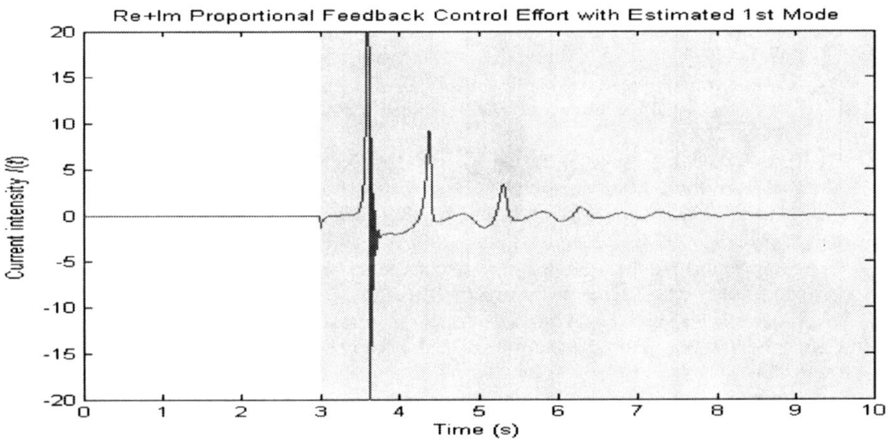

Fig. 10.7b Stimulus generated by the controller estimated via delay embedding.

all trainable and adaptive controllers, and can further automatically vary their structure.

Clearly, all the currently available therapies are trainable, but their sensorless control action hardly qualifies them as intelligent. Thus, intelligent controllers are *learning controllers that employ fine-grained feedback*. The fine-grained feedback aspect of controllers discussed here is not far from current neurostimulation technology since pulse-width or pulse-density modulation of existing pulse trains can produce the desired variable amplitudes. The identification of a fixed model of in vivo seizures reflecting dynamics of measurable outputs under stimulation inputs may prove to be sufficient for further progress in this field, given the stereotypical nature of many seizures, in which case the control problem would be reduced to model-predictive or generalized-predictive control (MPC/GPC), widely used in industry, and for which embedded targets and medical device applications are becoming increasingly possible [29].

References

1. S. Ward and M. Rise, Techniques for treating epilepsy by brain stimulation and drug infusion, U.S. Patent 5,713,923, March 13, 1996.
2. R. Fischell, D. Fischell and A. Upton, System for treatment of neurological disorders, U.S. Patent 6,016,449, October 27, 1997.
3. J. Zabara, Neurocybernetic prosthesis, U.S. Patent 4,702,254, December 30, 1985.
4. G. Worrell, R. Wharen, R. Goodman et al., Safety and evidence for efficacy of an implantable responsive neurostimulator (RNS®) for the treatment of medically intractable partial onset epilepsy in adults, *Epilepsia* **46**(s8), 226, 2005.
5. P. Betterton, personal communication, 2005.
6. P.A. Tass, *Phase Resetting in Medicine and Biology: Stochastic Modelling and Data Analysis*, Springer-Verlag, Berlin, 1999.
7. P.A. Tass, Desynchronizing double-pulse phase resetting and application to deep brain stimulation, *Biological Cybernetics* **85**, 343–354, 2001.
8. P.A. Tass, Effective desynchronization with bipolar double-pulse stimulation, *Physics Review E* **66**, 036226, 2002.
9. P.A. Tass, A model of desynchronizing deep brain stimulation with a demand-controlled coordinated reset of neural subpopulations, *Biological Cybernetics* **89**, 81–88, 2003.
10. P.A. Tass and M. Majtanik, Long-term anti-kindling effects of desynchronizing brain stimulation: A theoretical study, *Biological Cybernetics* **94**(1), 58–66, 2006.
11. S. Liss, Apparatus for monitoring and counteracting excess brain electrical energy to prevent epileptic seizures and the like, U.S. Patent 3,850,161, April 9, 1973.
12. S.A. Chkhenkeli, Direct deep brain stimulation: A first step towards the feedback control of seizures, in *Epilepsy as a Dynamic Disease*, J. Milton and P. Jung (Eds.), Springer-Verlag, Berlin, 2003.
13. M. Nakagawa and D. Durand, Suppression of spontaneous epileptiform activity with applied currents, *Brain Research* **567**(2), 241–247, 1991.
14. S.J. Schiff, K. Jerger, D.H. Duong et al., Controlling chaos in the brain, *Nature* **370**, 615–620, 1994.
15. J. Lian, J. Shuai and D. Durand, Control of phase synchronization of neuronal activity in the rat hippocampus, *Journal of Neural Engineering* **1**, 46–54, 2004.
16. B.J. Gluckman, H. Nguyen, S.L. Weinstein et al., Adaptive electric field control of epileptic seizures, *The Journal of Neuroscience* **21**(2), 590–600, 2001.
17. D.J. Mogul, Y. Li and M.E. Colpan, Using electrical stimulation and control feedback to modulate seizure activity in rat hippocampus, *Epilepsia* **46**(s8), 331, 2005.
18. M. Rosenblum and A. Pikovsky, Delayed feedback control of collective synchrony: An approach to suppression of pathological brain rhythms, *Physical Review E* **70**(041904), 1–11, 2004.
19. K. Tsakalis, Prediction and control of epileptic seizures: Coupled oscillator models. Presentation, February 2005.
20. M.W. Slutzky, P. Cvitanovic and D.J. Mogul, Manipulating epileptiform bursting in the rat hippocampus using chaos control and adaptive techniques, *IEEE Transactions on Biomedical Engineering* **5**(5), 559–570, 2003.

21. R. Larter, R. Worth and B. Speelman, Nonlinear dynamics in biochemical and biophysical systems: From enzyme kinetics to epilepsy, in *Self-Organized Biological Dynamics and Nonlinear Control: Toward Understanding Complexity, Chaos and Emergent Function in Living Systems*, J. Walleczek (Ed.), Cambridge University Press, Port Chester, NY, p. 51, 2001.
22. J.S. Ebersole and J. Milton, The electroencephalogram (EEG): A measure of neural synchrony, in *Epilepsy as a Dynamic Disease*, Springer-Verlag, Berlin, 2003.
23. J. Jefferys, Models and mechanisms of experimental epilepsies, *Epilepsia* **44**(s12), 44–50, 2003.
24. P. Suffczyński, S. Kalitzin and F. Lopes da Silva, Dynamics of non-convulsive epileptic phenomena modeled by a bistable neuronal network, *Neuroscience* **126**(2), 467–484, 2004.
25. P. Suffczyński, S. Kalitzin and F. Lopes da Silva, A lumped model of thalamic oscillations, in *Proceedings of Computational Neuroscience Meeting*, Brugge, Belgium, 2000.
26. F. Wendling, F. Bartolomei, J.J. Bellanger et al., Epileptic fast activities can be explained by a model of impaired GABAergic dendritic inhibition, *European Journal of Neuroscience* **15**(9), 1499–1508, 2002.
27. C. Hauptmann, O. Popovych and P. Tass, Effectively desynchronizing deep brain stimulation based on a coordinated delayed feedback stimulation via several sites: A computational study, *Biological Cybernetics* **93**(6), 463–470, 2005.
28. H. Haken, *Advanced Synergetics*, Springer, Berlin, 1983.
29. L.G. Bleris et al., Towards embedded model predictive control for system-on-a-chip applications, *Journal of Process Control* **16**, 255–264, 2006.

PART IV

INTELLIGENT CONTROL SYSTEMS

Chapter 11
Software Technology for Implementing Reusable, Distributed Control Systems*

Bonnie S. Heck, Linda M. Wills and George J. Vachtsevanos

Control systems such as those used in satellites, spacecraft, automobiles, chemical processing plants, and manufacturing rely heavily on the software that is used to implement them. In fact, engineers at Boeing Co. and Honeywell Inc. have estimated that 60–80% of the development of a complex control system is the software development effort, while only 20–40% is the actual control system design. The software effort includes programming individual software modules as well as writing code for communications between comnponents. Much of the software development time is spent on making the software stable and reliable, such as tracing possible hazard conditions and inserting software fault protection methods. When the operation of a control system is highly critical due to human safety factors or the high cost of failure in damaged capital or products, the software designers must expend extra effort to validate and verify their software before it can be released. In flight-critical operations, validation and verification are part of the flight certification process.

Many new software technologies exist that can facilitate the development and deployment of complex control systems. For example, component-based architectures [1, 2] promote code reuse, thereby decreasing development and validation time. Moreover, these architectures encourage flexible "plug-and-play" extensibility and evolution of systems. Distributed object computing allows heterogeneous components to interoperate across diverse platforms and network protocols [3, 4]. Open standards for software mean that products from different vendors can interoperate transparently to the user. New advances are being made to enable dynamic reconfiguration and evolution of systems while they are still running [5, 6]. New commu-

---

Bonnie S. Heck · Linda M. Wills · George J. Vachtsevanos
School of Electrical and Computer Engineering, Georgia Institute of Technology,
Atlanta, GA 30332-0250, USA; e-mail: gjv@ece.gatech.edu

* Reprinted, with permission, from *IEEE Control Systems Magazine*, February 2003, pp. 21–35.

nication technologies are being developed to allow networked embedded devices to connect to each other and to self-organize [7].

Historically, there has been a disconnect between the control engineers who design the controls and the software engineers who implement them. A thorough understanding of the methodologies used in both disciplines would make the overall process more cohesive. This article gives a tutorial overview of specific new software technologies that are useful for implementing and reusing complex control systems. The focus is on distributed controls, which utilize multiple processors, particularly those that need to be reconfigured either online or offline. This article first discusses technologies for component-based design and reuse of complex control systems, including object-oriented approaches, patterns, and frameworks. It then describes distributed computing and middleware technology for supporting the composition and interaction among control system components that are distributed. The article also presents advances in real-time distributed computing that are needed to support the real-time demands of control systems in coordinating distributed interactions. Finally, it surveys the current technology that is commercially available and on the research horizon for building reusable distributed control systems. The new software technologies discussed in this article can enhance the performance of these existing commercial systems and may form the basis for the next generation of distributed control systems.

11.1 Component-Based Architectures

Consider the engineering term *component*, which means a part or a subsystem in a complex engineering system. Typically, components are designed to be generic enough to be used in multiple different applications. An engineer might design a system by composing various components, ideally commercial-off-the-shelf (COTS). Similarly, software engineers strive to build software components [1, 2], which are modules that can be used in several applications and can be composed together to build larger applications. Some may be available as COTS. Components that interconnect across a network are called *distributed components*.

An advantage of software components is that they promote code reuse across many different applications, thereby reducing development and validation time. To construct components with this property, a designer must first assess a range of different applications and then determine generic patterns, which are processing needs and computational solutions to them that are similar across all of the applications [8]. The designer then creates software components implementing the common solutions. Typically, the components can be specialized further in the individual applications.

The software architecture of an entire control system may be component based. As an example, consider the components in a standard hierarchical control system: sensors, low-level control algorithms (such as stabilizing controllers in aircraft or set-point controllers in process control), high-level control algorithms (such as supervisory control or fault-tolerant control), and actuators. Many of these components have both hardware and software elements (e.g., a hardware sensor usually has software drivers used to transmit measurements to a processor). Similarly, there are standard control algorithms, which could be used in many applications (such as PID controllers, Kalman filters, fuzzy-logic-based controllers, neural network controllers, etc.), that could be coded as software components that are generic and yet can be tuned to a specific application.

Component-based design of control systems has several advantages over the current practice of design, especially for streamlining the transition from simulation to implementation. In particular, it makes sense to build a simulation that contains separate software modules for each component, including one that represents the plant dynamics. Setting and abiding by standards for the interfaces between the components facilitates rapid prototyping from the simulation stage to the implementation stage. For example, the code for all the controller and signal processing/filtering components would be the same from the simulation to the implementation, while the component representing the vehicle dynamics model would be replaced with the actual vehicle. Similarly, the component(s) representing sensor models would be replaced with the actual sensor drivers and hardware. Ideally, the signal processing/filtering component would not know if the information it is getting comes from the simulation code or from the sensor itself. Therefore, it would not need to change if the simulated sensor were replaced with the actual physical sensor.

As a promising future direction for control system design, suppose software components that model physical devices (such as sensors and actuators) are available from different vendors for their products. If these components adhere to specific and uniform standards that govern their interfaces, then they may be fully compatible in simulations. Standard control algorithms may also be available that adhere to these standards. A designer selects likely components, sets the control parameters, and integrates the components with a simulation of the plant. This type of design process is beneficial in several ways: the initial simulation is a more accurate representation of the final implementation, since it includes all the components and the component interactions; the designer can test the performance of sensors or actuators of different types in simulation *before* making a final selection; and the final system integration is streamlined since the software components can be replaced with the actual hardware components and their software drivers.

11.1.1 Object-Oriented Approach to Reusable Control Systems

Software components may be objects (such as a Java application or a C++ program), or they may be non-object-ori-ented modules (such as C code or legacy Fortran code). Objects contain attributes, called instance variables, representing an object's state, and operations, called methods, that access and manipulate the state. An object's state can only be accessed and changed by using the object's methods. The methods form a standard interface for an object that allows other objects to interact

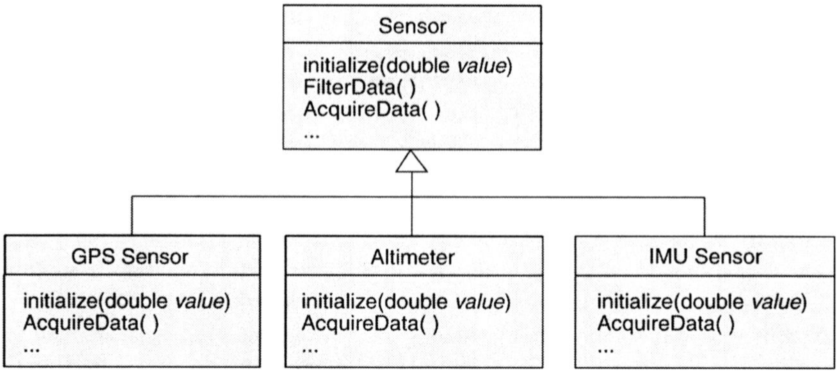

Fig. 11.1 Example class hierarchy for sensor objects.

with it. This is a property called encapsulation, which is useful for implementing components. The inner workings of the object are kept hidden from the outside world; only the interface is visible. The programmer can make changes and upgrades to the implementation of an object's methods without requiring changes to how other objects are implemented as long as the interface remains the same. In object-oriented languages, an object is defined by specifying a class, which defines the instance variables and methods of the objects that are instances of the class. For example, in the design of a control system, a sensor class may be defined as shown in Figure 1, which has methods for initialization, data acquisition, and filtering. These are functions that are commonly used in interacting with a sensor object. In Figure 11.1, three additional classes are defined for specialized types of sensors: global positioning system (GPS) sensors, altimeters, and inertial measurement unit (IMU) sensors. These are related as subclasses to the Sensor class, meaning that they share the interface defined for the Sensor class, but they specialize it in some way. They may specialize it by providing a more specific implementation for some of its methods or by adding additional instance variables or methods specific to the subclass. Any number of object instances can be created for a class. Our example may be used, for instance, in a system in which there are two altimeters with differing ranges, each an instance of the Altimeter class.

The relationships between the classes are graphically shown in Figure 11.1 using a *unified modeling language* (UML) class diagram. The UML [9] is an object-oriented modeling language that is becoming a de facto standard in the software industry. The UML provides a set of commonly used graphical notations for modeling various aspects of a software system, including requirements (e.g., use-case diagrams), temporal behavior (e.g., interaction sequencing and state charts), and implementation (e.g., class diagrams). In the class diagram, each rectangle represents a class whose name is at the top and whose methods are listed in the bottom portion of the rectangle. Instance variables, if any, are included in an additional middle section of the rectangle.

The use of subclasses in our example illustrates two key concepts in object-oriented programming that facilitate reuse: inheritance and polymorphism. When a subclass, such as `GPS_Sensor`, is defined, it automatically inherits the methods defined in the parent class `Sensor`. This means that attributes and operations common to a set of subclasses can be defined once and shared efficiently across the subclasses. An example is the `FilterData` method, which may be implemented in the `Sensor` class as a common first-order filter to reduce noise in the sensor data. When each of the subclasses is defined, the `FilterData` method comes for free in that it does not need to be redefined for each subclass. The method is defined once in the parent Sensor class and reused by the subclasses.

For flexibility, it is important to be able to tailor or specialize subclasses so that they can extend or adapt the functionality of the parent class. In particular, subclasses can be tailored by adding new methods and instance variables and/or overriding the methods defined in the parent class. Thus, a method in the parent class may be implemented differently in the different subclasses. For example, the specifics of initialization may differ from sensor to sensor. So the `initialize` method may need to be defined and implemented differently in each of the subclasses, as shown in Figure 11.1. Nevertheless, objects that interact with any sensor object are able to invoke the `initialize` method on any type of sensor and the correct implementation of the initialization will be transparently used. This is known as polymorphism, since the method can take on multiple different forms for the same function. The key benefit for reuse is that code that interfaces with the sensor class, such as a navigation component, does not need to be rewritten when a new sensor subclass is added. As long as the new sensor class satisfies the interface, it can be used just like any other sensor. Objects with the same interface can be substituted for one another at run time, a concept also referred to as "substitutability". Inheritance and polymorphism promote reuse by pulling out the generic parts of the code into classes while allowing customizations to be made in subclasses. Other specific examples of object-oriented control system software can be found in [10].

Because of the advantages of object-oriented programming, especially that of encapsulation, it is common practice to wrap code around a non-object-oriented (procedural) component to make the component appear to be an object to the rest of the world. This allows legacy procedural code to be reused in object-oriented systems.

Although object-orientation can increase the reusability of a control system, as the software evolves, it can be vulnerable to problems arising from the undisciplined use of implementation inheritance. If a parent, or "base", class is changed, it might have unexpected consequences for independently created subclasses that reused aspects of the base class. This is called the fragile base class problem [2]. Implementation dependencies due to subclassing can limit flexibility and reusability in new domains, since any part of the inherited implementation that is not applicable in the new domain could require changes to the base class or restructuring of the class hierarchy [8]. One way to avoid these problems is to constrain the design to inherit only from abstract classes that have no method implementations (only interfaces). This is done in object-oriented frameworks, which are described in the next section. Another common approach is to use object composition instead of class inheritance to reuse functionality. Objects are reused by assembling them into more complex objects; at run time, objects obtain references to other objects and forward (or "delegate") messages to them to perform supporting functionality. This requires that the interfaces of the objects be well defined and general enough that each ŞpartŤ object can be composed with many others. The advantage is that the objects can be changed and replaced independently without affecting the objects with which it is composed as long as the new object adheres to the interface. In practice, the components typically available to a designer are never exactly what is needed, so it is common to use inheritance along with object composition to allow flexibility and maximize reuse [8].

See the following Web sites for more information.

OSACA
www.mel.nist.gov/proj/interop.htm

Object Management Group (OMG)
www.omg.com

QNX
www.qnx.com

VxWorks
www.windriver.com

Real-Time CORBA (TAO)
www.cs.wustl.edu/ schmidt/TAO.html

Mathworks
www.mathworks.com

RTI
www.rti.com

Sun Microsystems Java
java.sun.com

11.1.2 Composing Components: Enhancing Reuse with Frameworks

Most commercial tool environments for building control systems make available repositories of prebuilt and preverified reusable components (such as common types of PID controllers, filters, and matrix computations) for efficient construction of systems. This is an important type of reuse. However, there are tremendous opportunities for reuse at an architectural level as well. For example, common controller configurations are encountered and used time and time again (e.g., the common sensor-controller-actua-tor feedback loop, feedforward loops, and common evaluation mechanisms that compare expected and actual environmental factors). Similarly, common patterns [8] for constructing dynamically reconfigurable systems are starting to emerge, particularly in the hybrid controls area. For example, hybrid systems often contain a common pattern for gradually transitioning from one control strategy to another [11, 12]. It may encapsulate a strategy dictating how quickly one algorithm can be switched for another or whether a redundant component needs to work concurrently with the component it is replacing before the swap occurs to allow the new component to Şcome up to speedŤ or transfer state. If generalized and made available for reuse, this pattern can be specialized with user-specified decision logic for managing the reconfiguration and applying the appropriate blending strategies to smooth the transients during the transition.

What is reusable in these cases are the "skeletons" of the design – how the components are composed or integrated. For example, the generic part can be implemented by an object-oriented framework [13], which is a set of abstract classes. An abstract class has a standard interface but contains no concrete object instances or method implementations. These abstract classes form a framework that can be applied to a particular application by adding concrete subclasses that specialize the abstract classes and by plugging in application-specific method implementations. Reuse through object-oriented frameworks has several advantages, such as reducing development costs and design errors across multiple similar applications, increasing understandability and maintainability of the systems, and facilitating system evolution. A key idea behind designing for reuse using frameworks is that the designer explicitly identifies the places where the design is anticipated to change in the future. These have been called hot spots [14] and plug-in points [15] to express the idea that the design is built to withstand evolution in predefined ways. By performing a domain analysis [16] of a family of related control system designs, it is possible to identify the common, generic, reusable structure of each mechanism and to "encapsulate the variation" [8] that is expected. The places where the code can be varied (or specialized to a specific application) are localized to the concrete methods that the application engineer is expected to write and plug into the framework. Examples of control-specific frameworks include the AOCS Framework (developed for satellite control systems) [17], Realtime Framework (developed for robotics and automation) [18], NEXUS (for robotics) [19], and OSACA (for automation) [71].

One of the most critical active areas of research in com-ponent-based reuse is how to compose components in systems that have stringent resource and dependability constraints, such as high performance, reliability, availability, consistency, survivability, timeliness, and security [20]. The loose coupling that is needed for flexible reuse and reconfigurability favors using composition mechanisms that decouple interaction. This may be done through asynchronous or anonymous communication mechanisms, such as using a shared data space in a blackboard architecture or using a publish/subscribe model (described in the next section) [21]. At the same time, it is critical that resource and dependability constraints be enforced in the use of these interaction mechanisms. How to specify and verify these constraints is an important research problem. Failure to deal with these constraints in the composition of components (not just enforcing them on individual components) can lead to unpredictable behavior and resource conflicts, such as frame overruns [21].

Recently, a broader and more powerful notion of frameworks has been defined [22] to encompass not only components and how they are composed but also the set of constraints in their interaction and the benefits resulting from these constraints. A framework in this definition specifies rigorously the types of components integrated, the knowledge they have of one another, protocol mechanisms, and the vocabulary used in component interaction (e.g., possible message types). The advantage of this richer notion of frameworks is that it may enable distributed control system architectures to self-adapt dynamically and robustly as changes in the networked environment occur and as new components are plugged into the system at run time. This provides a rich context in which to specify timing and resource constraints so that they can be enforced as an integral part of the design process.

Design for reuse by factoring out similarities across applications and identifying generic components is essential in reducing the complexity and cost of developing control systems. In today's embedded controls applications and pervasive computing environments, it is often the case that these components are distributed (both geographically across different processors and logically across different processes in a multitasking environment). Distributed computing software technology has been developed to support the composition of and interaction among components that are distributed. This technology is described next.

11.2 Distributed Computing Overview

Computer systems are typically set up using one of four different organizations. The system may have one very large centralized machine (such as a mainframe) with smart terminals that can be used to access the data or functions on that machine. While most companies and universities employed this organization during the 1970s and 1980s, the use of centralized processing has fallen off dramatically. An alternative approach to centralized processing is that of completely decentralized processing, where a large number of small units (such as desktop machines) independently perform operations with very little interaction. This provides a great deal of flexibility for individual users but makes data sharing difficult. A third possible organization is hierarchical, which integrates relatively similar small-scale units through a centralized machine, often called a server, that performs many higher-level functions. This organization is commonly employed in university computer clusters. It is also used in Web applications where the Web server is the centralized machine and the clients are the remote machines that log into the Web server to access data.

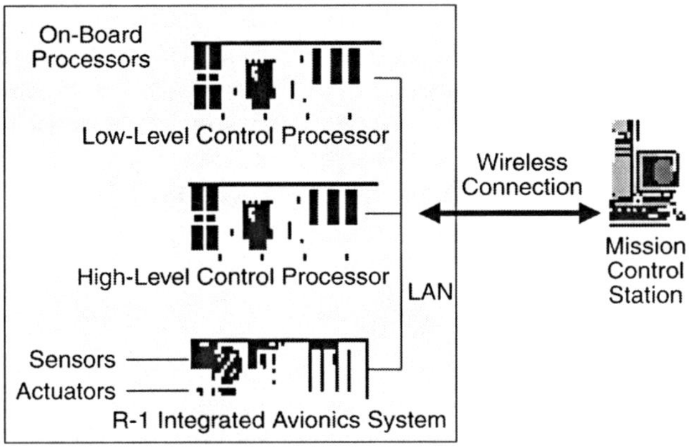

Fig. 11.2 Distributed computing architecture for an uninhabited aerial vehicle.

The fourth organization for computer systems is distributed. In this organization, each of the computers performs functions vital to the overall operations and is integrated with the other machines in the system [23]. It is differentiated from decentralized processing in that the machines are more tightly coupled to complete higher-level tasks. It is a flatter architecture than the hierarchical since the higher-level functions are performed by the individual processors and not by a high-level, centralized machine. Distributed processing can be considered a form of parallel processing, except that the processing can be distributed over separate computers, not just separate processors. An example from the business world is one in which a company has one large-scale computer for marketing, one for inventory (including shipping and receiving), and one for accounting. Clearly, all of these machines would need to perform distinct operations, and yet they need to be integrated to share data. A customer placing an order on the system may not realize that he/she is using three different computers that are working together to execute the order. Distributed computing is more robust than using centralized or hierarchical architectures since one machine can go down without necessarily disrupting the operations being performed on the other machines. Also, upgrades can be performed in a more modular fashion and thus will be less disruptive and done in smaller chunks. Distributed computing has gained acceptance, primarily due to advances in communication networks that make the transfer of data fast and reliable and due to advances in creating middleware, which hides low-level details about network programming. As a result, distributed architectures are now the most common for computer systems in businesses, universities, and government.

Distributed computing is ideal for complex control systems as well. These systems typically have multiple processors, with each performing distinct functions that need to be integrated. An example of distributed computing for the controls of an uninhabited aerial vehicle (UAV) is shown in Figure 11.2. There may be multiple processors on board the vehicle: one for avionics (containing sensor and actuator drivers and sensor processing), one for low-level control function (such as stability augmentation), and one for high-level control functions (such as trajectory planning and fault detection). The mission control processor is located remotely, either on the ground or in another vehicle (perhaps the mother ship). The onboard processors are connected via a LAN, while the remote processor is connected via a wireless connection (such as Ethernet for close range or serial for longer range). See [24] for details of an example architecture for UAV flight controls.

11.2.1 Network Communications

Integrating software components (or, more generally, application software) over a network requires extensive knowledge about data transfer between software applications and across networks. Consider first the task of setting up a communications network. In hardware, the ports and the communication lines must be connected. In software, the sockets (which are software analogs of ports) must be configured, a software interface for opening and closing sockets must be written, and the communication protocols must be set, including routing, the method to break up the data into packets, and error checking. Ethernet is a common network solution for high-speed, high-volume communication among a large number of nodes (that is, machines hooked to the network). A problem with using Ethernet for control applications is that the data transfer rate and the size of packets are not deterministic [25]. As a result, there is no guarantee of arrival time for data packets and no guarantee of information flow rate. If a control system closes the loop over a network, then this nondeterminism gives rise to stochastically varying time delays. Or if high-priority calls (such as fault isolation and reconfiguration strategies) were delayed over the network, then the system's health may be at risk. Protocols are being developed to attack some of these issues [26].

An alternative to Ethernet is to use a LAN with protocols that are better suited to smaller, more frequent data packet transfers found in control applications. Often called fieldbus systems, these control networks include ASI, BITBUS, CAN, DeviceNet, Hart, IEC-Fieldbus, InterBus-S, LonWorkds, Pi-NET, Profibus, Rackbus, Secros, SDS, TON, and WordFIP [27]. CAN (control area network) is used widely in automotive applications and is an event-driven protocol; that is, it assumes that data packets are sent in a sporadic manner, and it has a decision structure for deciding the route and the priority for transmitting different packets. In contrast, time-driven protocols (TTPs) set specific schedules for transmitting different data and nominally repeat these schedules on a periodic basis. In a comparison, Kopetz [28] determined that time-triggered protocols are well suited for safety-critical systems with hard real-time dependability constraints (by making worst-case timing assumptions), and CAN is more suitable for soft real-time systems where average-case performance and flexibility are the primary design considerations. A comparison of Ethernet (with CSMA/CD), a CAN, and a token ring network for control applications is given in [29].

Numerous communication protocols exist for dedicated controls-related LANs, as mentioned earlier; however, there is a need to develop an Internet-based protocol that is well suited for controls applications. Although a great deal of work is being done to develop protocols tailored for multimedia applications over the Internet, the same is not true for Internet-based protocols for control systems. For example, asynchronous transfer mode (ATM), which uses fixed data packets, making it more deterministic than protocols used with Ethernet networks [25], was first established to be efficient with streaming video and audio as well as standard data communications. Other work in the multimedia community aimed at making streaming audio and video reliable and steady has produced communication solutions such as voice over IP or Firewire. Some of these multimedia solutions may work their way into controls applications, but there could be work aimed at making a WAN protocol designed specifically for control systems. In fact, this suggestion was made at the recent NSF Workshop on Future Directions in Control Systems [30]. As hinted by this brief description of network protocols, programming at the network level requires detailed knowledge about networks that typical control engineers lack. Learning enough to implement even a simple control law takes considerable time. This issue is being addressed by the use of distributed computing middleware, as described next.

11.2.2 Middleware

Middleware is software that provides a substrate through which software components can communicate with each other [2]. It sits between the operating system and the application software and transparently handles many mundane low-level details generally required for data transfer between applications and over the network. As such, the use of middleware for integrating a networked control system would substantially decrease the complexity of programming, since the low-level network details are hidden from the controls engineer.

Middleware can be used when the application software is colocated on the same processor (local) or on a distributed computing system linking many processors across a network (remote). For example, consider the middleware used on a Microsoft Windows-based desktop machine that facilitates the interaction of different software applications running in different windows. As long as vendors adhere to certain standards, their application software is generally compatible with other software running on the platform. The programs can be run at the same time, and data can be passed from one program to another. A popular middleware used for desktop units is Microsoft's Component Object Model (COM) [23]. Distributed Component Object Model (DCOM) is a version of this middleware used to integrate application software running on different Windows machines and connected via a network.

Application-specific middleware can provide good examples of code reuse and show the benefits of compo-nent-based architectures. For example, avionics systems have numerous sensors and measurement devices that need to be integrated. Rather

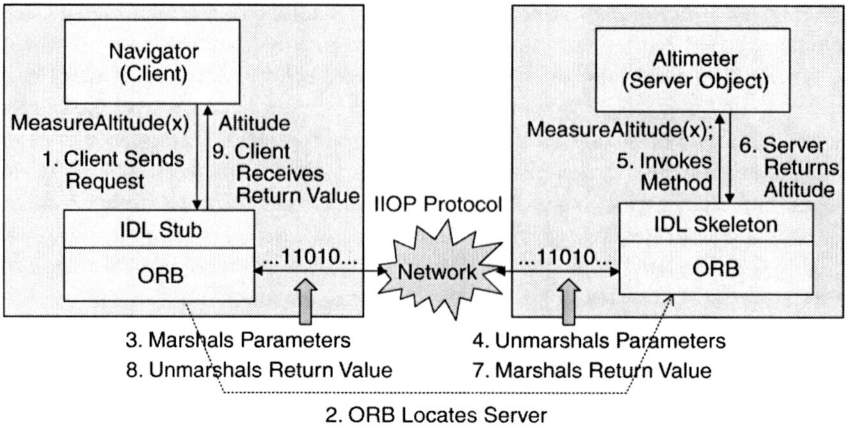

Fig. 11.3 Distributed communication through the CORBA standard.

than build completely specialized avionics packages for each aircraft, the code is separated into parts that are common among all the aircraft and parts that need to be specialized for different types of aircraft. The communication software is something that can be made common across all applications. For this reason, Boeing has developed a middleware that can provide a common communication substrate for all of their measurement devices [31]. Thus, design of a new avionic system starts with this middleware, which is then customized by adding components for the particular devices used for a particular aircraft. The overall software development time is reduced dramatically by reusing code across different applications.

To be effective for use in distributed computing, middle-ware should have certain properties: it must be abel to automate common network programming tasks, it must have a standard interface so that all application software that uses the standard can be integrated easily, and it must support interoperability in heterogeneous environments. Interoperability allows application programs that are written in different languages or for different target operating system and hardware platforms to work together seamlessly. The ultimate goal for control systems would be to use middle-ware that has plug-and-play capability. For the sake of adaptability, it would be helpful to support run-time reconfiguration of the software components (such as updating or changing algorithms online in real time).

While COM (and DCOM) are popular for desktop computing applications, they lack the interoperability needed for integrating software between machines that are not Windows based. Data transfer between different machine platforms is not trivial; for example, the memory byte ordering is different between Windows-based platforms and Unix platforms, and word size differences on 32-bit versus 64-bit processor architectures cause incompatibilities. Two other common types of middleware are CORBA (common object request broker architecture) and Java-based middleware.

CORBA is a software standard developed by a consortium called the Object Management Group (OMG). As of this writing, there are 800 member companies of OMG that develop and adopt products that abide by the standards set forth by OMG. A basic feature of CORBA is the object request broker (ORB), which handles remote method calls. As illustrated in Figure 11.3, when an object (such as a navigation system) calls a method (e.g., MeasureAltitude) of another object (such as a remote sensor) distributed elsewhere on the network, the ORB intercepts the call and directs it. The client object does not need to know the location of the remote server object, a principle known as location transparency that greatly simplifies the programming of distributed applications. The way this works is that a programmer specifies an interface for each object using a standard interface definition language (IDL). These interfaces are compiled into what are referred to as client IDL "stubs", which act as proxies for the actual objects, and server object IDL "skeletons". All components are registered with and interact through the ORB. When a client object invokes a method (e.g., MeasureAltitude), it does so as if it is making a method call to a local object, but it is actually invoking it on a proxy that is the IDL stub. The method call goes through the ORB, which locates the server object (e.g., the Altimeter). If the server object is remote, the ORB needs to send the request to the remote object's ORB over the network. This involves marshalling the parameters, which means translating data values from their local representations to a common network protocol format, such as the Internet Inter-ORB Protocol (IIOP) standard defined by OMG. On the server object side, the parameters are demarshalled and the method invocation is passed to the IDL skeleton, which invokes the method on the actual server object. The server returns the requested value in a similar fashion through the ORB. Note that the client does not have to be aware of where the server object is located, its programming language, its operating system, or any other system aspects that are not part of an object's interface.

The use of CORBA in control engineering is rather new; see [31–33] for details on the development of a platform that is based on a CORBA substrate. However, some of the high-level features of CORBA make it very attractive for future control engineering products used to integrate systems. Its main benefits are that it has a well-accepted standard for component interfaces, it allows for interoperability of software components written in a range of different languages (including Java and C++) and/or running on different platforms, and it handles all the network communication protocols to seamlessly interconnect multiple components within and across control systems. Components can also be integrated that are not object oriented (such as legacy code) by wrapping the code in a layer that provides an objected-oriented interface. Services are continually being added to the ORB structure that extend the capabilities of CORBA.

The third common way of integrating components is with Java-based technologies. Java is an object-oriented language that is platform independent, so it has good promise for use in middleware development. Java is based on a layered architecture, with new open-source APIs being developed on a continuing basis. Some of the APIs that enable Java to be used for distributed applications are the Enterprise Java Beans API, the Remote Method Invocation (RMI) API, the RMI over IIOP API, and the Java ORB API. Java Beans is a way to make software components called Beans out of standard Java applets or applications (that is, unlike applets or applications, which cannot interact with one another, Beans can be composed together to form a larger application). The Enterprise version allows these Beans to exist on different machines and to be connected across a network. The RMI API provides the means for distributed objects to interface through remote method calls (similar to remote procedure calls but specialized to Java objects). The RMI IIOP API provides the remote method call capability using a CORBA ORB. The Java ORB API provides a Java-based CORBA ORB. It should be noted that Java components can interact with non-Java components (such as those written in C) through the use of the Java Native Interface (JNI). JNI works by providing a standard for writing Java code to wrap around non-Java code; the combined code then can be treated as Java object.

Although many of these Java APIs provide services useful in middleware, more complete packages are available such as the Jini Technology, J2EE (Java 2 Platform Enterprise Edition), and J2ME (Java 2 Platform Micro Edition). Jini is a software architecture for integrating hardware over a network. The real strength of Jini is the discovery service, whereby new devices can be added to the network at run time and be recognized and registered. J2EE is a component-based technology useful for integrating business applications. Both Jini and J2EE have the option of performing remote method calls using the Java RMI technology, CORBA services, or the Extensible Markup Language (XML) [34]. Thus, these technologies can be used as complementary technologies to CORBA. J2ME is an optimized Java virtual machine meant for running on small embedded platforms, and it uses the Java RMI technology. There is a strong continuing effort in the Java community to build support for real-time applications of Java, such as the Real-Time Specification for Java [35] and the requirements for Real-Time Java [36]. For example, a version of J2ME is available for VxWorks, a commercial real-time operating system developed by Wind River Systems. Since this field changes rapidly, the reader is encouraged to visit the Sun Microsystems Java Web site (java.sun.com) for updates.

A competing technology for J2EE is Microsoft's .NET [38], which is a framework that supports Internet-based connectivity between machines by providing the communications middleware for doing remote procedure calls (as well as other services). For communications, .NET relies on SOAP (simple object access protocol), an XML-over-HTTP protocol [37]. There is some question as to the suitability of HTTP for real-time applications with the additional overhead of XML [38]. Another drawback is .NET's reliance on the Microsoft Windows operating system.

Middleware is generally constructed to provide communication between application software that adheres to one of the following communication models:

- *Point-to-point* (or *peer-to-peer*): In this model, all communication is one-to-one between pairs of entities. A common example for this type of communication is a telephone call between two parties.
- *Client-server*: Client-server is a many-to-one communication model where many clients can be connected simultaneously to one server. The server acts as a centralized source of data.

- *Publish-subscribe* (or *producer-consumer*): The pub-lish-subscribe model allows many components to interact with one another in a many-to-many architecture by introducing an intermediate level between the components that mediates the communications. Components can be designated as publishers, which produce data, and subscribers, which require data as input. An example of a publisher is a sensor that produces measurements. An example of a subscriber is a control algorithm that requires the measurements from the sensor. In a complex control system, there may be many publishers and many subscribers. For example, a process control system may have distributed components using many processors, or at least many separate algorithms all using similar data, and numerous sensors producing data.

As an example of these models, consider the architecture for implementing a control system for an uninhabited aerial vehicle, as shown in Figure 11.4. The control components are the software modules that perform the tasks of mission planning, target tracking, obstacle avoidance, and attitude control. The sensor components consist of the device drivers as well as the sensor hardware. In this case, the sensors used are a camera for vision information, a forward-looking infrared (FLIR) camera for nighttime vision, a GPS sensor, and an IMU. The architecture shown in Figure 11.4a uses a point-to-point model where the communication is between pairs of components. As a result, there is a tight coupling between components that is achieved by hard-coding calls between control components and sensor components. Such tight coupling of software components has been standard practice in control implementation due to the efficiency gained at run time, but it makes upgrades and evolution of software difficult. For example, suppose the FLIR sensor is added at a later time than the other three sensors. Then adding this sensor requires modifying three different control components, which all require image data.

Now consider the same example implemented using a publish-subscribe middleware as shown in Figure 11.4b.

Here the middleware provides a communication abstraction called an event channel that acts as a mediator between components. The vision component registers with the event channel as a publisher of image data, while the mission planning, target tracking, and obstacle avoidance components are subscribers to that data. This removes the need to hard-code a call from a specific subscriber to a specific source of data in order for components to interact. With the event channel, the interconnection structure between components is much more flexible. Subscribers and publishers register with the event channel to receive (or provide) data at a given rate. Subscribers get the data from whatever publishers are currently registered to provide that data. If a new sensor (e.g., an FLIR) is added, then it can replace the old sensor by registering the new sensor driver with the event channel as a publisher. Other publishers of the same type of data may be removed by changing their registration. No other changes to the interconnection structure between components are necessary.

Multiple publishers of a particular type of data can be registered. The event channel must have strategies for determining which data to send to the subscribers of that data. One strategy used by a real-time publish-subscribe middleware, NDDS, is to give each publisher a weight, and the middleware forwards whichever data set

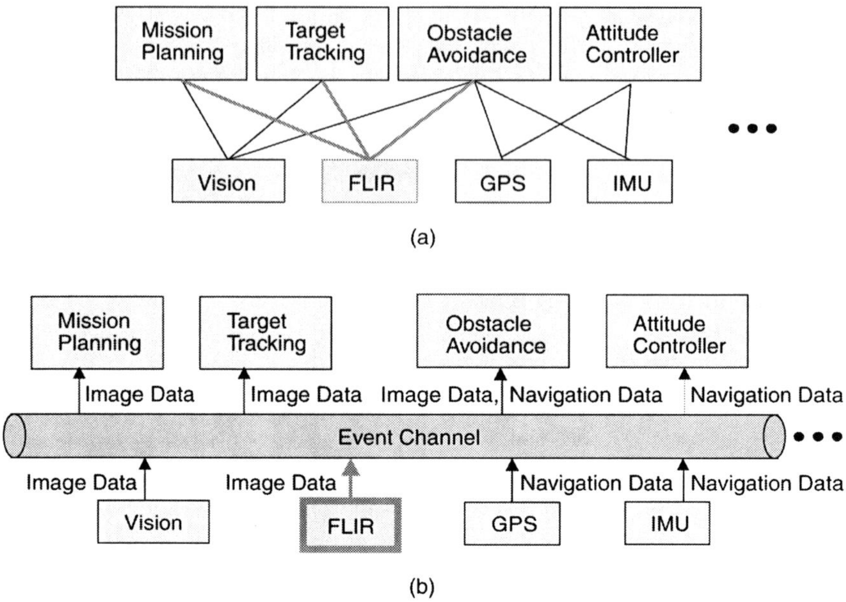

Fig. 11.4 Decoupling components through the publish-subscribe communication model. (a) In the point-to-point model, adding an FLIR component requires many changes. (b) In the publish-subscribe model, adding an FLIR component requires local change.

comes from the publisher with the highest weight [39, 40]. (The weights can be reset in real time.) Other strategies exist for determining an average of the data from the different publishers [41]. Thus, the event channel is a means to facilitate communication between components in a flexible enough manner that is easily adjusted to accommodate evolution of the system software. CORBA and DCOM use a client-server model, but an event service has been layered on the CORBA specification that emulates a publish-subscribe model.

In summary, distributed computing middleware provides the ability to automate common network programming tasks, provides a standard interface for component integration, and facilitates interoperability across different machine platforms and networks. Complex control systems that require distributed interaction among diverse reusable components can benefit from this type of technology. However, it is also critical that the interactions among components in these systems be coordinated in real time. The next section discusses new software technology that supports real-time interactions among distributed components.

11.3 Real-Time Computing

The essence of real-time systems is that they are able to respond to external stimuli within a certain predictable period of time. Building real-time computing systems is challenging due to requirements for reliability and efficiency, as well as for predictability in the interaction among components. Real-time operating systems (RTOSs) such as QNX and VxWorks facilitate real-time behavior by scheduling processes to meet the timing constraints imposed by the application.

Control systems are among the most demanding of real-time applications. There are constraints on the allowable time delays in the feedback loop (due to latency and jitter in computation and in communication), as well as the speed of response to an external input such as changing environmental conditions or detected faulted conditions. If the timing constraints are not met, the system may become unstable. This differs from multimedia applications, for example, where the user would at worst be irritated by the poor quality of an audio transmission or a jerky video or animation sequence.

These stringent timing constraints are particularly challenging to meet when used in a multitasking environment. Multitasking is useful in implementing multirate control and for allowing concurrent component execution in a complex system, such as enabling a fault detection module to run concurrently with a nominal control module. Since the multitasked processes must share CPU time, the corresponding tasks must be scheduled. Typically, such processes are scheduled statically assuming periodic operation, say τ ms per computation cycle. In this strategy, the worst-case execution time (WCET) is calculated for each process. That amount of time is allotted for that process during the t ms computation cycle (or frame). Since WCET is very conservative, this strategy can result in a small utilization of the CPU, thereby wasting computing resources. The classic technique for static real-time scheduling commonly used in real-time operating systems is rate monotonic scheduling (RMS) [42]. It is a fixed-priority scheduling technique in that it assigns each process a fixed priority based on how frequently the process needs to run (i.e., its rate). At run time, the static scheduler always chooses the process with the highest priority to run at any given point, preempting the running process if necessary [43].

Scheduling may be performed based on event triggers or on time triggers. Event triggers are external stimuli, such as operator inputs, that introduce interrupts which are handled as soon as they occur, based on priority. Loop rates can be enforced in this manner by using a hardware timer to supply periodic interrupts that are handled by sampling the sensor data and computing the corresponding control signal. The processing must be completed before the next timer interrupt; this time period is often described as a frame or as a computation cycle. Other interrupts, such as an operator input, may occur during any frame and must be handled using predefined logic on priorities. An alternative architecture is one based on time triggers rather than external event triggers. In this case, the frame is subdivided into time periods for different tasks to be completed based on the WCET. The tasks are scheduled at the beginning of each frame. External interrupts are handled by waiting until the beginning of a frame and then scheduling the corresponding handling process to be completed during the next frame, making the process more predictable than those produced by event triggers. This is an advantage for periodic processes such as control systems that have strict rate requirements. In a comparison of time-triggered architectures to event-triggered architectures, Kopetz [44] concluded that event-triggered architectures are better suited to nonperiodic processes that require flexibility in structure (e.g., to accommodate reconfiguration of the control), whereas time-triggered architectures are better suited to periodic tasks that have strict requirements on their rates (including prediction). In particular, time-triggered architectures may be less efficient in that they schedule processes based on WCET. However, if there are many interrupts, event-triggered architectures become less efficient because of the added overhead needed for interrupt handling. If events are very sensitive to latency (such as reaction to faulted conditions), then event-triggered architectures may be preferable. The alternative is to make the frames in time-triggered architectures smaller to shorten the time to react to interrupts, but this reduces efficiency due to the overhead in scheduling needed at the beginning of each frame.

To improve processor utilization, several researchers employ dynamic scheduling algorithms [3, 45, 46] for use in real-time systems. Dynamic scheduling allows run-time changes in the control system architecture while maintaining real-time operation. Since a fixed schedule is not used, higher utilization of resources can be achieved when nonperiodic tasks are used. While dynamic scheduling is common on desktop machines, the timing constraints for real-time systems are more challenging. The concept of quality of service (QoS) has been borrowed from network communications and applied to dynamic scheduling for real-time systems. In this application, QoS resource requirements are defined for each task. These typically include parameters such as update rates for different signals, priorities in case two events occur simultaneously (such as a fault occurrence flag having higher priority than a nominal control update signal), WCET, and typical execution time (i.e., the amount of time a task usually takes). The dynamic scheduling algorithm uses these parameters to determine the order in which to start tasks so that they have the computational resources they need to complete within their deadlines or at their required rates.

Figure 11.5 shows examples of different schedules created using two popular dynamic scheduling techniques: earliest deadline first (EDF) and minimum laxity first (MLF). In EDF, the task with the earliest deadline is scheduled first [47]. In MLF, the task with the least "laxity" is scheduled to run first, where laxity is the difference in the time remaining before a task's deadline and the task's execution time [48].

Run-time reconfiguration of a control system may require changing the QoS parameters in real time so that tasks that might have previously been running at a lower priority than others can be placed at a higher priority to respond to a failure or some unexpected condition. For example, an aerial vehicle's navigation component may need data at a higher rate, quality, and priority during an extreme maneuver than during other types of maneuvers, such as hovering. QoS parameters may also need to be adjusted due to changes in the system resources, such as the addition of new processes, and due to computer system failures.

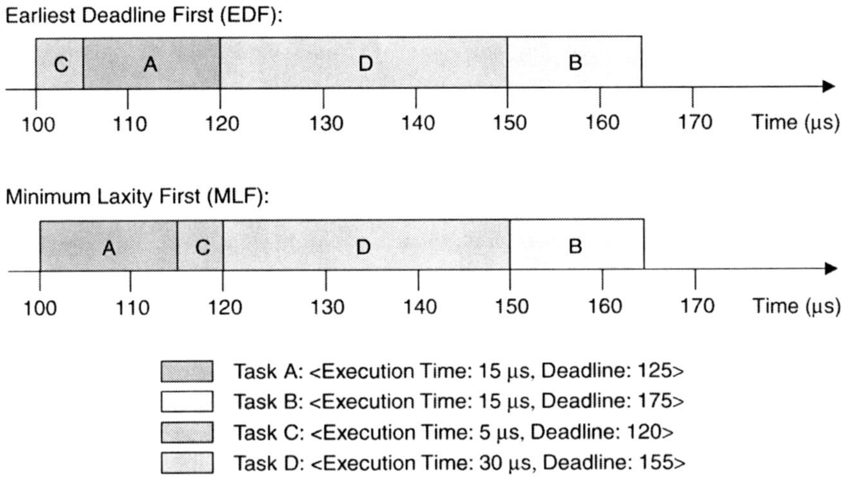

Fig. 11.5 Example schedules from two well-known dynamic scheduling algorithms.

To support these needs, new dynamic scheduling algorithms [4, 45, 46] and adaptive resource management techniques [49–52] are being developed. Adaptive resource management allocates processor and network resources based on changing client demands and based on changes in available system resources. Feedback mechanisms have been employed by several researchers to adjust scheduling of resources to ensure that the actual QoS pa-rameters are maintained within acceptable QoS parameter regions; see, for example, the control-based middleware developed by Li and Nahrstedt [53] and Honeywell's adaptive resource management system (RT-ARM) [49, 50]. Some of these methods have been integrated by Boeing as part of the open control platform (under the DARPA Software-En-abled Control program) [31, 32]. By using dynamic scheduling algorithms in the event channel scheduler, run-time changes can be made in QoS requirements for events by allowing QoS parameter changes to be specified and enforced at run time.

Combining dynamic scheduling and adaptive resource management enables a particularly valuable class of algorithms, called anytime algorithms or imprecise computation [54, 55]. These are continually running algorithms that can produce results at any time but continually refine their results as time passes and the system state changes. Anytime algorithms are valuable when dealing with variable amounts of available computational resources. A feature of these algorithms is that they monotonically converge on a solution. If they are given much processing time, they converge to a very accurate solution. Given a smaller amount of processing time and/or resources, they provide a more crude, but acceptable, answer. This contrasts with other types of algorithms that would not be able to produce any result if their execution time is reduced. Examples of controls applications in which these algorithms could be used are route planning and failure prediction, particularly those using iterative methods such as genetic algorithms or methods that gradually improve as they learn, such as neural-net-work-based techniques.

An alternative approach, but one that also uses adaptive resource management, is to allow the scheduler to switch components to ones that are less demanding when it determines that there are insufficient resources to use the nominal component. Honeywell employed this technique in conjunction with RT-ARM for multiagent coordination [54].

Honeywell developed a multiresolution trajectory planning algorithm with varying levels of computational needs. The algorithm is useful in its most demanding form, which gives the best results, under nominal conditions, while the least computationally demanding form, which gives only crude results, is used when there are diminished resources.

When the processing is distributed over multiple machines, the influences on real-time performance become much more complex. For example, in determining the latency of a remote method call made across a network, the communication latency must be added to the processing latencies inherent in the participating processors. Therefore, the network communications QoS must be coordinated with the individual processor QoS. Static scheduling, which is done a priori using worst-time execution time and worst-case transport delays, is common practice. The time-triggered architecture must rely on the notion of a global clock when applied to a distributed system, so that the timers on all the processors are synchronized to within a certain small error of one another [44]. Dynamic scheduling for distributed processes is currently an open research topic due to the complexity of interactions between the communication scheduling and the processor scheduling; see, for example, [56, 57].

Methods exist to reduce communication latency in distributed systems. In particular, there are numerous methods for scheduling communications to reduce latency. For example, a dynamic scheduling algorithm for communications is proposed in [58]. Another method that may be used to reduce latency in a real-time distributed system is replication [44, 47], which is normally used to provide fault tolerance through redundancy in a computational system but provides efficiency advantages as well. Replication refers to the practice of creating copies (or "replicas") of objects at multiple computational nodes. These nodes must be synchronized with respect to time and data to maintain consistency between objects and their replicas. If a replica of an object is available locally, then a remote method call may be redirected to the local replica, thereby avoiding the communication delay. There is a tradeoff, of course, between maintaining replica consistency and reducing communication delays in that the use of replication may result in too much extraneous network traffic if the replica update rate is too high compared to the rate at which the replica is accessed.

A specification for Real-Time CORBA [4] has been released by OMG that defines standard features for managing processor utilization, network, and memory resources. These standard features support end-to-end predictable behavior for fixed-priority operations (i.e., operations with prespecified static priorities) between CORBA clients and servers. By defining standard policies and capabilities for resource management, the Real-Time CORBA specification helps ensure QoS predictability for fixed-priority real-time applications. An open-source implementation of Real-Time CORBA, called TAO, is publicly available at http://www.cs.wustl.edu/~schmidt/TAO.html.

11.4 Commercial Products and Market Trends

Commercial software tools for control systems are aimed at either design and analysis or at implementation. The most common design and analysis tool is MATLAB from The MathWorks. MATLAB is often used in conjunction with the Simulink package to provide a block-diagram-based graphical user interface along with some expanded simulation capabilities. To ease system implementation, the Real-Time Workshop toolbox can be used to generate C code automatically from a Simulink block diagram. The MathWorks is working toward making the resulting code more efficient, in terms of both processing time and size (an important feature for embedded processors). Another MathWorks product, Stateflow, which is well suited to hybrid systems, also has an autocode generator that can be purchased as an option. Unfortunately, MathWorks products tend to generate monolithic code rather than component-based code. This makes it more difficult to validate or update the code. As an alternative, designers can use MATLAB to generate individual portions of the code, which can then be made into components. For example, the designer can obtain the code for a low-level controller operating under nominal conditions, as well as separate code for a controller operating under faulted conditions. The designer can then integrate these separate components using middleware.

Distributed control systems have been available commercially for many years. The majority of these systems consist of industrial-grade processors or PLCs connected to hardware devices by dedicated fieldbus networks (CAN, DeviceNet, etc.). The applications are in process control, environmental control, automotive control, mechatronic systems, and power systems, to name a few. This market is dominated by a few large corporations such as Rockwell Automation (including the subsidiary Allen-Bradley), Honeywell, Siemens, Invensys, and ABB, which supply 97.5% of the market [59]. These control suppliers provide the entire system: the measurement and actuator devices, the processors, the network interfaces and cables, and service. Very few third-party vendors (suppliers of specialized components such as sensors) have broken into this market because the interfaces to these networked systems are proprietary.

The desire for open standards has prompted the increased use of PC-based controllers, which has become the largest growth area in the distributed control system market [59]. PC-based controllers are cheaper than the indus-trial-grade processors and have a more open architecture. This open architecture means that third-party vendors are able to supply more of the components, resulting in large control houses having a smaller market share for PC-based control (75% as compared to 97.5% in the standard distributed control system market). Most control houses now offer interface cards for PCs to connect to their standard fieldbus networks.

Another trend in the market is the move toward nonproprietary communication protocols such as TCP/IP; in particular, much work is being done to develop Ethernet-based networked control systems. For example, two commercial middleware products for controls systems are specifically for networks running TCP/IP: products from Advanced Realtime Control Systems, Inc. (ARCS) [60] and NDDS from Real-Time Innovations, Inc. (RTI) [39, 40]. ARCS products use Java-based technologies interfaced with Simulink for code generation. NDDS, which uses a real-time pub-lish-and-subscribe communication model, performs well for relatively fast periodic events, as is needed, for example, by a low-level stability augmentation controller used in flight controls. RTI also markets ControlShell, a graphical tool that allows integration of control systems in an ob-ject-oriented fashion. Consult the RTI Web site (www.rti.com) for several white papers that further describe the use of the publish-and-subscribe communication model in controls, including [40]. In answer to this move in the market, many of the large control houses that make dedicated, proprietary networked systems are adding interfaces from their devices to Ethernet-based systems so that people can create interconnected networks such as a fieldbus network system to a Web-based system. This works well in connecting a local control operation to a plantwide monitoring system such as one based on SCADA devices [61].

11.5 Research Tools on the Horizon

A few notable research tools are emerging that support modeling and construction of distributed, heterogeneous systems. One such system is Ptolemy II, which focuses on modeling embedded systems that rely on a combination of technologies or, more formally, "models of computation" that define how components execute and interact with each other [62, 63]. For example, differential equations can be used to model analog circuits, and finite state machines can be used to model control logic. Ptolemy II allows multiple models of computation to be used to model hierarchical compositions of hetereogeneous components. This is necessary in designing complex, hybrid embedded systems that include, for example, mixed analog/digital technologies or continuous adaptive controllers whose activation is governed by higher-level discrete coordination logic. Ptolemy supports a systematic and principled approach to designing real-world embedded systems that are compositions of heterogeneous components. Since the focus is on the interactions among components, communication protocols and timing properties/constraints become primary design considerations, which is critical in developing resource-con-strained systems.

While Ptolemy focuses on modeling the interaction semantics governing compositions of diverse components, a complementary system, called Cactus [64–66], focuses on providing customizable middleware services for distributed systems that implement useful interaction styles. Cactus allows the QoS attributes (including fault tolerance and timeliness) of these services to be tailored to the requirements of a given application. The Cactus system synthesizes middleware software as a layer between the operating system and the application. The approach is to provide a set of "micro-protocol" software modules, each of which implements a well-defined service property such as atomicity, access control, or replication or retransmission for fault-tolerant communication. Desired, customized services are then created by composing micro-protocols (if their logical properties are compatible) to address a given application's specific requirements. What are reused in this case are fundamental fine-grained distributed computing abstractions that can be combined to form configurable middleware services, customizable to a given application.

Giotto [67, 68] is a tool-supported methodology for developing embedded control systems, particularly for safety-critical applications that have hard real-time constraints. It provides a programming model based on the time-triggered paradigm [69]. A key innovation is that Giotto supports developing a system model of functionality and timing that is independent of the real-time platform (i.e., separate from platform-specific computation and communication scheduling). Giotto provides an abstraction layer that lies between the high-level mathematical model of the control design and the platform-specific executable code. Giotto first converts a mathematical design model to a high-level embedded program (termed an embedded software model). This software model specifies the interactions (including timing requirements) between software processes, as well as the timing requirements for the software with respect to the physical system being controlled. Giotto's retargetable compiler then converts this software model into executable code that meets the scheduling and communication constraints when run on a specific target machine. The positioning of the Giotto software model at an intermediate layer provides a place for validation and verification of software that is easier to manage than at the plat-form-specific code level.

Another notable system is the generic modeling environment (GME), which supports reuse of system modeling concepts themselves [70]. GME is a design environment that provides a set of generic modeling and visualization concepts that are common to a wide range of domains. These can be used to efficiently create a modeling environment for a specific domain (e.g., the signal processing domain or reconfigurable control systems) by customizing the generic modeling concepts to support a given domain language. The customized GME models are then used to automatically synthesize applications in the given domain.

11.6 Summary

The software engineering concepts introduced in this article make implementing control systems more flexible so that they can be dynamically reconfigured with ease and can be upgraded or adapted in a flexible manner. The desire to reduce software development and validation time has led to design procedures that reuse architectural patterns as well as software components. Common architectural patterns are captured in frameworks that identify common compositions of components that have proven to be valuable in a particular domain. In addition, new distributed and real-time computing technologies are emerging to help orchestrate the distributed interaction of the diverse components in today's complex control systems. The application of these software technologies to controls systems is relatively new but has tremendous potential for growth. Some of the leading applications that motivate research on this topic are robotics, automation, and flight control, where the increasing complexity of the design process and the need for rapid run-time changes to the design pose challenges that can only be met by these new technologies.

Acknowledgments

This work is supported under the DARPA Software-Enabled Control program under contracts 33615-98-C-1341 and 33615-99-C-1500, managed by the Air Force Research Laboratory (AFRL). We gratefully acknowledge DARPA's Soft-ware-Enabled Control program, AFRL, and Boeing Phantom Works for their continued support. We have benefited greatly from the guidance of Helen Gill (NSF), John Bay (DARPA), and Bill Koenig (AFRL) and from interactions with Scott Clements, Cameron Craddock, Murat Guler, Eric Johnson, Suresh Kannan, Chris Lee, J.V.R. Prasad, Freeman Rufus, Sam Sander, Daniel Schrage, Balasubramanian Seshasayee, and Ilkay Yavrucuk.

References

1. G.T. Heineman and W.T. Councill, *Component-Based Software Engineering: Putting the Pieces Together*, Addison-Wesley, Reading, MA, 2001.
2. C. Szyperski, *Component Software: Beyond Object-Oriented Programming*, Addison-Wesley, Reading, MA, 1998.
3. D. Levine, S. Mungee and D. Schmidt, The design and performance of real-time object request brokers, *Comput. Commun.* **21**, 294–324, April 1998.
4. D. Schmidt and F. Kuhns, An overview of the real-time CORBA specification, *IEEE Computer* **33**, 56–63, June 2000.
5. P. Oreizy, N. Medvidovic and R.N. Taylor, Architecture-based runtime software evolution, in *Proceedings of the International Conference on Software Engineering (ICSE'98)*, Kyoto, Japan, pp. 117–186, 1998.
6. P. Oreizy, M. Gorlick, R.N. Taylor, D. Heimbigner, G. Johnson, N. Medvidovic, A. Quilici, D. Rosenblum and A. Wolf, An architecture-based approach to self-adaptive software, *IEEE Intell. Syst.* **14**, 54–62, May/June 1999.
7. J. Waldo, The Jini architecture for network-centric computing, *Commun. ACM* **42**(7), 76–82, July 1999.
8. E. Gamma, R. Helm, R. Johnson and J. Vlissides, *Design Patterns: Elements of Reusable Object-Oriented Software*, Addison-Wesley, Reading, MA, 1995.
9. G. Booch, J. Rumbaugh and I. Jacobson, *The Unified Modeling Language User Guide*, Addison-Wesley, Reading, MA, 1999.
10. D. Auslander, J.R. Ridgely and J.D. Ringgenberg, *Control Software for Mechanical Systems, Object-Oriented Design in a Real-Time World*, Prentice-Hall, Upper Saddle River, NJ, 2002.

11. M. Guler, S. Clements, L. Wills, B. Heck and G. Vachtsevanos, Generic transition management for reconfigurable hybrid control systems, in *Proceedings of 20th American Control Conference (ACC-2001)*, Arlington, VA, June 2001.
12. M. Guler, L. Wills, S. Clements, B. Heck and G. Vachtsevanos, A pattern for gradual transitioning during dynamic component replacement in extreme performance UAV hybrid control systems, in *OOPSLA 2001 Workshop Patterns and Pattern Languages Object-Oriented Distributed Real-Time and Embedded Systenbs*, Tampa, FL, October 2001 [Online]. Available: http:/www.cs.wustl.edu/~mk1/realtimepatterns/oopsla2001/submissions/muratguler.pdf
13. R. Johnson and B. Foote, Designing reusable classes, *J. Object-Oriented Programming* **1**(2), 22–35, June/July 1988.
14. W. Pree, Essential framework design patterns: Why 'hot spot' identification helps, *Object Mag.* **7**(1), 34–37, March 1997.
15. D. D'Souza and A. Wills, *Objects, Components, and Frameworks with UML: The Catalysis Approach*, Addison-Wesley, Reading, MA, 1999.
16. R. Prieto-Diaz and G. Arango, *Domain Analysis and Software Systems Modeling*, IEEE Comput. Soc. Press, Los Alamitos, CA, 1991.
17. T. Brown, A. Pesetti, W. Pree, T. Henzinger and C. Kirsch, A reusable and platform-independent framework for distributed control systems, in *Proceedings of the IEEE/AIAA 20th Digital Avionics Systems Conference (DASC'2001)*, Daytona, FL, October, Vol. 2, pp. 1–11, 2001.
18. A. Traub and R. Schraft, An object-oriented realtime framework for distributed control systems, in *Proceedings of the IEEE International Conference on Robotics and Automation*, Detroit, MI, May, pp. 3115–3121, 1999.
19. J. Fernandez and J. Gonzalez, NEXUS: A flexible, efficient and robust framework for integrating software components of a robotic system, in *Proceedings of the 1998 IEEE ICRA*, Leuven, Belgium, pp. 524–529, 1998.
20. D. Hammer, C. Becchetti, P. Narasimhan, A. Schurr and B. Selic, What are the most urgent research problems of component-based software engineering for resource-constraint systems?, in *Proceedings of the 6th International Workshop Object-Oriented Real-Time Dependable Systems (WORDS'01)*, Rome, Italy, January, pp. 97–98, 2001.
21. D. Hammer and M.R.V. Chaudron, Component-based software engineering for resource-constraint systems: What are the needs?, in Proc. 6th Int. Workshop Object-Oriented Real-Time Dependable Syst. (WORDS'01), Rome, Italy, Jan. 2001, pp. 91-96.
22. E. Lee, What's ahead for embedded software?, *IEEE Computer* **33**, 18–26, September 2000.
23. A. Tanenbaum and M. van Steen, *Distributed Systems: Principles and Paradigms*, Prentice-Hall, Upper Saddle River, NJ, 2002.
24. E. Johnson and S. Mishra, Flight simulation for the development of an experimental UAV, in *Proceedings of AIAA Modeling and Simulation Technology Conference*, Monterey, CA, No. AIAA-2002-4975, August 2002.
25. A. Tanenbaum, *Computer Networks*, Prentice-Hall, Englewood Cliffs, NJ, 1996.
26. S. Schneider, G. Pardo-Castellote and M. Hamilton, Can Ethernet be real time? [Online]. Available: http://www.rti.com/products/ndds/leads.html
27. G. Schickhuber and O. McCarthy, Distributed fieldbus and control network systems, *IEEE Comput. Control Eng. J.* **8**, 21–32, February 1997.
28. S. Kopetz, A comparison of CAN and TTP, *Annu. Rev. Control* **24**, 177–188, 2000.
29. F.-L. Lian, J. Moyne and D. Tilbury, Performance evaluation of control networks: Ethernet, ControlNet, and DeviceNet, *IEEE Contr. Syst. Mag.* **21**, 66–83, 2001.
30. C. Georgakis and W. Levine, Workshop on the research needs in modeling, dynamics, monitoring and control of complex engineering systems, Report, Anchorage, AK, May 11–12, 2002 [Online]. Available: http://www.isr.umd.edu/ISR/Complex/
31. J. Paunicka, B. Mendel and D. Corman, The OCP – An open middleware solution for embedded systems, in *Proceedings of the American Control Conference*, Arlington, VA, pp. 3345–3350, 2001.

32. J. Paunicka, D. Corman, and B. Mendel, A CORBA-based middleware solution for UAVs, in *Proceedings 4th IEEE International Symposium on Object-Oriented Real-Time Distributed Computing (ISORC 2001)*, Madgeburg, Germany, May 2001.
33. L. Wills, S. Kannan, S. Sander, M. Guler, B. Heck, J. Prasad, D. Schrage and G. Vachtsevanos, An open software infrastructure for reconfigurable control systems, *IEEE Contr. Syst. Mag.* **21**, 49–64, June 2001.
34. M. Morrison, D. Brownell and F. Broumphrey, *XML Unleashed*, Sams Publishing, Indianapolis, IN, 1999.
35. Java Community Process, Real-time specification for Java [Online]. Available: http://jcp.org/jsr/detail/1.jsp
36. L. Carnahan and M. Ruark (Eds.), *Requirements for Real-Time Extensions to the Java P atform*, NIST Pub. 500-243 [Online]. Available: http://www.itl.nist.gov/div897/ctg/real-time/rtj-final-draft.pdf
37. J. Farley, .NET from the enterprise perspective: SOAP, *O'Reilly News*, 12 September 2000 [Online]. Available: http://java.oreilly.com/news/soap_0900.html
38. J. Farley, Microsoft .NET vs. J2EE: How do they stack up?, *O'Reilly News*, August 2000 [Online]. Available: http://java.oreilly.com/news/farley_0800.html
39. G. Pardo-Castellote and S. Schneider, The network data delivery service: Real-time data connectivity for distributed control applications, in *Proceedings of 1994 IEEE International Conference on Robotics and Automation*, San Diego, CA, Vol. 4, pp. 2870–2876, 1994.
40. G. Pardo-Castellote, S. Schneider and M. Hamilton, NDDS: The real-time publish-subscribe middleware [Online]. Available: www.rti.com/prod-ucts/ndds/leads.html
41. D. Bakken, Z. Zhan, C. Jones and D. Karr, Middleware support for voting and data fusion, in *Proceedings of the International Conference on Dependable Systems and Networks*, Goteborg, Sweden, July, pp. 453–462, 2001.
42. C. Liu and J. Layland, Scheduling algorithms for multiprogramming in a hard real-time environment, *J. ACM* **20**, 46–61, January 1973.
43. A. Tanenbaum, *Modern Operating Systems*, Prentice-Hall, Upper Saddle River, NJ, 2001.
44. H. Kopetz, *Real-Time Systems Design Principles for Distributed Embedded Applications*, Kluwer, Norwell, MA, 2001.
45. C. Gill, D. Levine and D. Schmidt, The design and performance of a real-time CORBA scheduling service, *Int. J. Time-Critical Comput. Syst.* (Special Issue on Real-Time Middleware) **20**(2), 1999.
46. D. Levine, C. Gill and D. Schmidt, Dynamic scheduling strategies for avionics mission computing, in *Proc. 17th DASC. AIAA/IEEE/SAE Digital Avionics Syst. Conf.*, Vol. 1, pp. C15/1–8, 1998.
47. A. Tanenbaum, *Distributed Operating Systems*, Prentice-Hall, Englewood Cliffs, NJ, 1996.
48. D.B. Stewart and P.K. Khosla, Real-time scheduling of sensor-based control systems, in *Real-Time Programming*, W. Halang and K. Ramamritham (Eds.), Pergamon, Tarrytown, NY, 1992.
49. M. Cardei, I. Cardei, R. Jha and A. Pavan, Hierarchical feedback adaptation for real time sensor-based distributed applications, in *Proceedings of 3rd IEEE International Symposium on Object-Oriented Real-Time Distributed Computing (ISORC 2000)*, Newport Beach, CA, March, pp. 181–188, 2000.
50. B. Doerr, T. Venturella, R. Jha, C. Gill and D. Schmidt, Adaptive scheduling for real-time, embedded information systems, in *Proceedings of the 18th Digital Avionics Systems Conference*, St. Louis, MO, October, pp. 2.D.5/9, 1999.
51. C.S.R. Murthy and G. Manimaran, *Resource Management in Real-Time Systems and Networks*, MIT Press, Cambridge, MA, 2001.
52. A.M. van Tilborg and G.M. Koob, *Foundations of Real-Time Computing: Scheduling and Resource Management*, Kluwer, Norwell, MA, 1991.
53. B. Li and K. Nahrstedt, A control-based middleware framework for quality-of-service adaptation, *IEEE J. Select. Areas Commun.* **17**, 1632–1650, September 1999.
54. M. Agrawal, D. Cofer and T. Samad, Real-time adaptive resource management for multi-model control, in *Proceedings of the American Control Conference*, Arlington, VA, pp. 3451–3456, 2001.

55. S. Natarajan (Ed.), *Imprecise and Approximate Computation*, Kluwer, Norwell, MA, 1995.
56. G. Manimaran and C.S.R. Murthy, An efficient dynamic scheduling algorithm for multiprocessor real-time systems, *IEEE Trans. Parallel Distrib. Syst.* **9**, 312–319, 1998.
57. F. Petrini and W.C. Feng, Scheduling with global information in distributed systems, in *Proceedings of the International Conference on Distributed Computing Systems*, Taipei, Taiwan, April, pp. 225–232, 2000.
58. G. Walsh and H. Ye, Scheduling of networked control systems, *IEEE Contr. Syst. Mag.* **21**, 57–65, February 2001.
59. Venture Development Corp., Global markets and user needs for industrial distributed/remote I/O, section edition [Online]. Available: http://ics.pennnet.com/Articles/Article_Display.cfm?Section=OnlineArticles&SubSection=Display&PUBLICATION_ID=26&ARTICLE_ID=134229
60. Advanced Realtime Control Systems Inc. [Online]. Available: http://www.arcsinc.com/progref-manual.pdf
61. Industrial Automation Insider [Online]. Available: http://www.abpubs.demon.co.uk/scada.htm
62. X. Liu, J. Liu, J. Eker and E. Lee, Heterogeneous modeling and design of control systems, submitted for publication in *Software-Enabled Control: Information Technology for Dynamical Systems*, T. Samad and G. Balas (Eds.), IEEE Press, New York, 2003.
63. E. Lee, Computing for embedded systems, Paper presented at the IEEE Instrumentation and Measurement Technology Conference, Budapest, Hungary, May 2001.
64. M. Hiltunen and R. Schlichting, The Cactus approach to building configurable middleware services, in *Proceedings of the Workshop Dependable System Middleware and Group Communication (DSMGC 2000)*, Nuremberg, Germany, October 2000 [Online]. Available: http://www.cs.arozona.edu/cactus/public.html
65. W. Chen, M. Hiltunen and R. Schlichting, Constructing adaptive software in distributed systems, in *Proceedings of the 21st International Conference on Distributed Computing Systems (ICDCS-21)*, Mesa, AZ, April, pp. 635–643, 2001.
66. M. Hiltunen, X. Han and R. Schlichting, Real-time issues in Cactus, in *Proceedings of the Workshop Middleware Distributed Real-Time System Services*, December, pp. 214–221, 1997.
67. T.A. Henzinger, C.M. Kirsch, M.A.A. Sanvido and W. Pree, From control models to real-time code using Giotto, *IEEE Contr. Syst. Mag.* **23**, 50–64, February 2003.
68. T. Henzinger, B. Horowitz and C. Kirsch, Embedded control systems development with Giotto, in *Proceedings of the International Conference on Languages, Compilers, and Tools Embedded Systems (LCTES)*, Snowbird, UT, pp. 64–72, 2001.
69. T. Henzinger, B. Horowitz and C. Kirsch, Giotto: A time-triggered language for embedded programming, in *Proceedings of the 1st International Workshop Embedded Software (EMSOFT '01)*, Taho City, CA, pp. 166–184, 2001.
70. A. Ledeczi, M. Maroti, A. Bakay, G. Karsai, J. Garrett, C. Thomason, G. Nordstrom, J. Sprinkle and P. Volgyesi, The generic modeling environment, in *Proceedings of the IEEE International Workshop Intelligent Signal Processing (WISP'2001)*, Budapest, Hungary, May 2001 [CD-ROM].
71. P. Lutz, W. Sperling, D. Fichtner and R. Mackay, OSACA – The vendor neutral control architecture, in *Proceedings of the European Conference on Integration in Manufacturing*, Dresden, Germany, September, pp. 247–256, 1997.

Chapter 12
UGV Localization Based on Fuzzy Logic and Extended Kalman Filtering

Athanasios Tsalatsanis, Kimon P. Valavanis and Ali Yalcin

Abstract Accurate localization necessitates precise sensor information regarding the vehicle's position and orientation. In general, sensors are vulnerable to environmental changes, disturbances, faults and noise to the communication channels. To cope with the uncertainty in sensor readings, data manipulation techniques such as preprocessing and sensor fusion have been commonly utilized. The dominant sensor fusion technique in the UGV localization area is that of Kalman filtering. Kalman filters are used to estimate the vehicle's position based on the vehicle's kinematics and sensor readings, assuming zero mean Gaussian noise to both vehicle's kinematics and sensor readings. A drawback of Kalman filtering techniques is the lack of information regarding the sensor performance in different operational environments: a vision system will provide more accurate information in an experiment that is conducted under daylight than an experiment conducted in a dark room. In this work, a multi-sensor localization method for UGVs based on fuzzy logic and Extended Kalman filtering is presented. This work utilizes a set of fuzzy logic controllers to incorporate the performance of a sensor under different operational conditions into the Extended Kalman filter. The sensors considered are a GPS, an IMU, a laser range finder, a stereo vision system and the vehicle's odometer. The superiority of the proposed filter methodology is demonstrated through extensive experimentation in various environmental conditions.

Athanasios Tsalatsanis · Ali Yalcin
Department of Industrial and Management Systems Engineering, University of South Florida, Tampa, FL 33620, USA; e-mail: atsalats@mail.usf.edu

Kimon P. Valavanis
Department of Electrical and Computer Engineering, University of Denver, Denver, CO 80208, USA; e-mail: kimon.valavanis@du.edu

K.P. Valavanis (ed.), Applications of Intelligent Control to Engineering Systems, 295–328.

Nomenclature

$\mathbf{X}_k$	The state vector (system's model) at time k
$\mathbf{x}_k = [x_k \quad y_k \quad \vartheta_k]^T$	The vehicle's posture at time k
$\mathbf{u}_k = [V_k \quad \Omega_k]^T$	Velocity vector consisting of the vehicle's linear and angular velocities at time k
$\mathbf{gps}_k = [gps_{x,k} \quad gps_{y,k}]^T$	GPS readings related to vehicle's position at time k
$\mathbf{imu}_k = [ax_k \quad eax_k \quad ay_k \quad eay_k \quad angle_{z,k}]^T$	IMU readings related to the vehicle's accelerations, errors in acceleration, and steering angle at time k
$\mathbf{odo}_k = [odo_{x,k} \quad odo_{y,k} \quad odo_{\vartheta,k}]^T$	Odometer readings related to the vehicle's position at time k
$\mathbf{cam}_k^i = [dcam_{x,k}^i \quad dcam_{y,k}^i]^T$	Distance between the ith landmark and the vehicle as measured by the camera at time k
$\mathbf{las}_k^i = [dlas_{x,k}^i \quad dlas_{y,k}^i]^T$	Distance between the ith landmark and the vehicle as measured by the laser range finder at time k
$\mathbf{lan}_k^i = [lan_{x,k}^i \quad lan_{y,k}^i]^T$	Position of the ith landmark at time k
$\mathbf{d}_{j,k}^i = [dx_{j,k}^i \quad dy_{j,k}^i]^T$	Distance between the ith landmark and the vehicle as measured by the jth range sensor at time k
$\mathbf{n}_k$	Zero mean Gaussian noise with covariance $\mathbf{Q}_k$
$\mathbf{w}_k$	Zero mean Gaussian noise with covariance $\mathbf{R}_k$
$\mathbf{Z}_k$	Measurement's model at time k
$\mathbf{P}_k$	Covariance matrix at time k

12.1 Introduction

Localization is prerequisite to Unmanned Ground Vehicle's (UGV) accurate and autonomous navigation in uncertain and/or unknown environments. At any given time, the UGV should have exact knowledge, or should be able to derive, or estimate accurately its position and orientation.

Required information is basically acquired from position estimation sensors such as Global Positioning Systems (GPS), Inertial Measurement Units (IMU) and odometers. The individual use of these sensors has specific disadvantages. For example: GPS cannot provide accurate position estimation when the UGV travels small distances or it travels in indoor environments; IMU is sensitive to sadden accelerations and magnetic interferences; odometers lose track of the UGV due to wheel slippage. A method that fuses information captured from multiple sensors and provides an accurate estimation of the UGV position and orientation is required to diminish the weaknesses of individual sensors.

Commonly used sensor fusion techniques such as Kalman Filters (KF) fail to utilize information regarding the performance of a sensor under specific operational conditions. The traditional KF approach cannot distinguish when a sensor works under ideal conditions and when it is not. To the KF methodology, a sensor reading is always reliable and under a specific zero mean Gaussian distributed noise. However, this is not always the case. Consider the use of a stereo vision system. Even though a stereo vision system can provide significant amount of information regarding the vehicle's surroundings, information captured under absolute darkness is unreliable for any application.

This chapter presents a multi sensor fusion method for UGV localization based on fuzzy logic (FL) and Extended Kalman filters (EKF). The sensors considered are a Global Positioning System (GPS), an Inertial Measurement Unit (IMU), a stereo vision system, a laser range finder, and the vehicle's odometer. In this work, the stereo vision system operates as an additional range finder sensor. All sensors are mounted on top of the UGV. Position sensors such as GPS and odometer provide information related to the vehicle's posture. Inertial sensors such as the IMU compute accelerations and angular rates and range sensors such as the stereo vision system and the laser range finder compute distances between landmarks and the vehicle. The presented methodology utilizes FL to consider the performance of each sensor in various operational conditions. The FL controllers receive information such as sensor readings and environmental characteristics to compute the statistics of the error distributions in the measurements of the individual sensors. Throughout this chapter, the term error is defined as the difference between the true value and the measured value over the true value.

In contrast to the traditional EKF approach where the error in sensor readings is either preset or an additional variable in the system's model, the presented fuzzy EKF method approximates the error in sensor readings with multiple zero mean Gaussian distributions, based on the performance of a sensor in a series of experiments. In this way, the error in sensor readings is incorporated into the covariance matrix $\mathbf{R}$ of the EKF measurement's model as zero mean Gaussian distribution and not as an unknown variable in the system's model.

Experimental validation and verification of the presented method is carried out using the *ATRV-Jr* skid steering differential drive UGV, equipped with 3 GHz P IV processor and 1 GB of RAM. The sensor suit of the *ATRV-Jr* consists of a SICK laser range finder, a GPS, a Microstrain IMU, a compass unit and a stereo vision system. The vehicle runs on RedHat Linux 7.0 and Mobility interface. The experiments have been conducted indoors and outdoors, on different types of floors and multiple weather conditions.

The rest of the chapter is organized as follows: Section 12.2 discusses work related to the UGV localization problem and summarizes the contributions of presented research. Section 12.3 presents the EKF design and implementation and Section 12.4 discusses the FL extension to the EKF. Section 12.5 demonstrates the use of the fuzzy EKF in a case study and Section 12.6 concludes this work.

12.2 Related Work

The EKF has been widely used in mobile robot navigation applications, mostly to integrate data from position sensors such as GPS, INS and odometer, and/or range sensors such as laser range finder and ultrasonic sensors. The choice of the sensors to be integrated is related to the robot's usage and operational environment. For example, indoor applications may not benefit from a GPS unit due to signal scrambling and outdoor applications may not benefit from ultrasonic sensors due to their limited performance outside.

Examples of sensor integration for mobile robot localization include odometry and laser [2, 9, 12, 14]; odometry and ultrasonic sensors [7]; odometer, gyroscope and GPS [11]; INS [3]; INS and vision system [17]; GPS, INS, odometer and vision system [19]; vision system and odometry [4, 6, 13, 15, 21]; laser and vision system [1, 24], to name a few.

A drawback of the EKF is that it requires precise knowledge of the system's dynamics, and noise distributions to be Gaussians with zero mean [8]. Various techniques have been proposed to overcome these problems with some focusing on the use of FL. Research in [18] proposes an adaptive fuzzy Kalman Filter (KF) method for mobile robot localization. A weighted EKF is used to fuse measurements from a GPS and an INS, while a fuzzy logic system is used to tune weights based on the covariance and the mean value of the residuals. In this way, the EKF is prevented from diverging. In [10], two FL controllers are used to enhance the EKF. The first adjusts the measurement covariance matrix based on a covariance matching technique and the second monitors the performance of the EKF and assigns a degree of confidence on the outputs of the first fuzzy controller. In [5], a Takagi–Sugeno fuzzy model is used to linearize the EKF system model. In [16], a fuzzy KF is proposed based on the assumption that the measurement and the model uncertainty is 'possibilistic' instead of Gaussian. In [22], a Neuro-Fuzzy KF is presented to evaluate sensor measurements and measurement errors. Finally, in [18] an adaptive fuzzy logic system is discussed that modifies the Kalman gain preventing the filter from diverging.

The contribution of this research can be summarized as follows:

1. The presented localization method fuses information derived from five different sensors: GPS, IMU, odometer, stereo vision system and laser range finder, while most related work utilizes up to three sensors.

2. The FL controller design allows for real time computation of the statistics of the error in sensor readings used in the EKF without prior knowledge of the mathematical model of the sensor.
3. The presented FL controller design allows the incorporation of each sensor operational characteristics into the EKF model.
4. A comparative study between the EKF and the presented fuzzy EKF is presented based on one, two, three and five sensors.

12.3 Extended Kalman Filter

The EKF is a powerful estimation tool designed for systems described by non-linear equations. Implementation of EKF necessitates definition of the models that describe the system's dynamics and the sensor measurements. The rest of this section describes the development of the EKF model for a UGV localization application.

12.3.1 Constructing the System's Model

In localization applications, the system's model represents the belief of the vehicle's posture based on a series of mathematical equations such as the kinematics equations, landmark positions and error models. The kinematics equations for a differential drive vehicle, which provide an estimate of the vehicle's posture, are defined as:

$$\mathbf{x}_{k+1} = \begin{bmatrix} x_k + V_k \Delta t \cos \vartheta_k + \frac{1}{2} ax_k \Delta t^2 \\ y_k + V_k \Delta t \sin \vartheta_k + \frac{1}{2} ay_k \Delta t^2 \\ \vartheta_k + \Omega_k \Delta t \end{bmatrix}, \tag{12.1}$$

where $\mathbf{x}_k = [x_k \quad y_k \quad \vartheta_k]^T$ is the vehicle's position at time k with respect to a global coordinate system, $\mathbf{u}_k = [V_k \quad \Omega_k]^T$ is the control input consisting of the vehicle linear and angular velocities and ax_k, ay_k the vehicle's accelerations on axes x and y respectively (Figure 12.1). It is assumed that the vehicle travels in a 2D environment. It is also assumed that the accelerations ax_k, ay_k are constant for the Δt time interval:

$$\begin{aligned} ax_{k+1} &= ax_k, \\ ay_{k+1} &= ay_k. \end{aligned} \tag{12.2}$$

In addition to the vehicle's kinematics, range readings derived by sensors as the laser range finder, provide additional means of estimating the vehicle's position. Consider a set of landmarks in positions, $\mathbf{lan}_k^i = [lan_{x,k}^i \quad lan_{y,k}^i]^T$, $i = 1 \dots n$, with respect to a global coordinate system as depicted in Figure 12.1. These landmarks are stationary and their position in time can be expressed as:

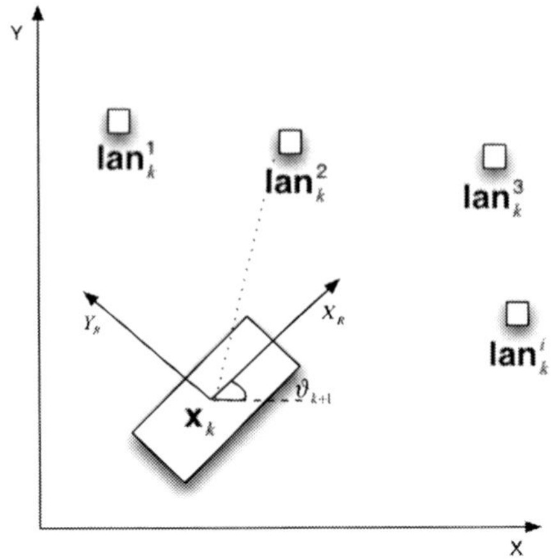

Fig. 12.1 Landmark position with respect to a global coordinate system.

$$\mathbf{lan}^i_{k+1} = \mathbf{lan}^i_k \tag{12.3}$$

Equations (12.1), (12.2) and (12.3) are used to synthesize the system's model for most differential drive UGV localization applications. The system's model is defined by

$$\mathbf{X}_{k+1} = \mathbf{f}(\mathbf{X}_k, \mathbf{u}_k) + \mathbf{n}_k, \tag{12.4}$$

where $\mathbf{n}_k$ is Gaussian noise with zero mean and covariance $\mathbf{Q}_k$ related to the uncertainty of the velocity vector $\mathbf{u}_k$. The non-linear transition function $\mathbf{f}$ describes the state updates and it is given by

$$\mathbf{f}(\mathbf{X}_k, \mathbf{u}_k) = \begin{bmatrix} \mathbf{x}_k \\ ax_k \\ ay_k \\ \mathbf{lan}^1_k \\ \vdots \\ \mathbf{lan}^n_k \end{bmatrix} = \begin{bmatrix} x_k + V_k \Delta t \cos\vartheta_k + \frac{1}{2} ax_k \Delta t^2 \\ y_k + V_k \Delta t \sin\vartheta_k + \frac{1}{2} ay_k \Delta t^2 \\ \vartheta_k + \Omega_k \Delta t \\ ax_k \\ ay_k \\ lan^1_{x,k} \\ lan^1_{y,k} \\ \vdots \\ lan^n_{x,k} \\ lan^n_{y,k} \end{bmatrix}. \tag{12.5}$$

12.3.2 Constructing the Measurement's Model

The measurement's model of the EKF shows how information derived from the vehicle's sensors ties to the system's state. In this research, five sensors are considered: a GPS, an IMU, an odometer, a laser range finder and a stereo vision system. The measurements acquired by the GPS and the odometer are related to the posture of the vehicle, namely the vector $\mathbf{x}_k$ of the system's model and can be described as

$$\begin{bmatrix} \mathbf{gps}_k \\ \mathbf{odo}_k \end{bmatrix} = \begin{bmatrix} gps_{x,k} \\ gps_{y,k} \\ odo_{x,k} \\ odo_{y,k} \\ odo_{\vartheta,k} \end{bmatrix} = \begin{bmatrix} x_k \\ y_k \\ x_k \\ y_k \\ \vartheta_k \end{bmatrix} + \begin{bmatrix} \mathbf{w}_{gps} \\ \mathbf{w}_{odo} \end{bmatrix}, \tag{12.6}$$

where **gps** and **odo** denote the readings from the GPS and the odometer respectively and $\mathbf{w}_{odo}$, $\mathbf{w}_{gps}$ are zero mean Gaussian noise associated to the odometer and GPS readings respectively.

The measurements acquired from the range sensors are the distances between various landmarks and the vehicle and can be expressed as

$$\begin{bmatrix} \mathbf{d}^1_{1,k} \\ \vdots \\ \mathbf{d}^n_{m,k} \end{bmatrix} = \begin{bmatrix} \cos\vartheta_k & \sin\vartheta_k \\ -\sin\vartheta_k & \cos\vartheta_k \\ \vdots & \vdots \\ \cos\vartheta_k & \sin\vartheta_k \\ -\sin\vartheta_k & \cos\vartheta_k \end{bmatrix} \begin{bmatrix} lan^1_{x,k} - x_k \\ lan^1_{y,k} - y_k \\ \vdots \\ lan^n_{x,k} - x_k \\ lan^n_{y,k} - y_k \end{bmatrix} + \begin{bmatrix} \mathbf{w}^1_{1,k} \\ \vdots \\ \mathbf{w}^n_{m,k} \end{bmatrix}, \tag{12.7}$$

where $\mathbf{d}^i_{j,k} = [dx^i_{j,k} \quad dy^i_{j,k}]$ is the distance between the ith landmark and the vehicle as measured by the jth range sensor, $\mathbf{w}^i_j$ is zero mean Gaussian noise associated with the measurement of the jth range sensor.

The range sensors considered in this research are a stereo vision system and a laser range finder. Thus, Equation 12.7 is modified as follows:

$$\begin{bmatrix} \mathbf{cam}^1_k \\ \vdots \\ \mathbf{cam}^n_k \\ \mathbf{las}^1_k \\ \vdots \\ \mathbf{las}^n_k \end{bmatrix} = \begin{bmatrix} \cos\vartheta_k & \sin\vartheta_k \\ -\sin\vartheta_k & \cos\vartheta_k \\ \vdots & \vdots \\ \cos\vartheta_k & \sin\vartheta_k \\ -\sin\vartheta_k & \cos\vartheta_k \end{bmatrix} \begin{bmatrix} lan^1_{x,k} - x_k \\ lan^1_{y,k} - y_k \\ \vdots \\ lan^n_{x,k} - x_k \\ lan^n_{y,k} - y_k \end{bmatrix} + \begin{bmatrix} \mathbf{w}^1_{cam,k} \\ \vdots \\ \mathbf{w}^n_{cam,k} \\ \mathbf{w}^1_{las,k} \\ \vdots \\ \mathbf{w}^n_{las,k} \end{bmatrix}, \tag{12.8}$$

where **cam** and **las** denote the range readings from the vision systems and the laser range finder respectively.

Finally, the measurements acquired from the IMU are associated to the accelerations and the steering angle of the vehicle and can be defined by

$$[\mathbf{imu}_k] = \begin{bmatrix} \cos\vartheta_k & \sin\vartheta_k & 0 \\ -\sin\vartheta_k & \cos\vartheta_k & 0 \\ 0 & 0 & 1 \end{bmatrix} \begin{bmatrix} ax_k \\ ay_k \\ \vartheta_k \end{bmatrix} + \mathbf{w}_{imu,k}. \tag{12.9}$$

Equations (12.6), (12.8) and (12.9) are used to synthesize the measurement's model of the EKF for the localization application. The measurement's model is described by

$$\mathbf{Z}_k = \mathbf{h}(\mathbf{X}_k) + \mathbf{w}_k. \tag{12.10}$$

Based on the sensor availability, the transition function **h** is defined as

$$\mathbf{h}(\mathbf{X}_k) = \begin{bmatrix} \mathbf{gps}_k \\ \mathbf{odo}_k \\ \mathbf{imu}_k \\ \mathbf{cam}_k^1 \\ \mathbf{las}_k^1 \\ \vdots \end{bmatrix}. \tag{12.11}$$

12.3.3 Linearization

The transition functions **f** and **h** are non linear functions. To linearize these functions, the first order Taylor expansion is used. The linear system's and measurement's models become

$$\mathbf{X}_{k+1} = \mathbf{F}_k\mathbf{X}_k + \mathbf{n}_k, \tag{12.12}$$

$$\mathbf{Z}_k = \mathbf{H}_k\mathbf{X}_k + \mathbf{w}_k, \tag{12.13}$$

where

$$\mathbf{F}_k = \nabla\mathbf{f}(\mathbf{X}_k, \mathbf{u}_k) \tag{12.14}$$

and

$$\mathbf{H}_k = \nabla\mathbf{h}(\mathbf{X}_k). \tag{12.15}$$

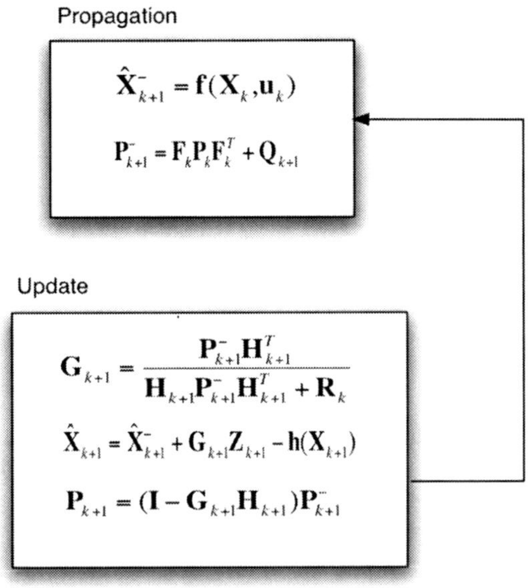

Fig. 12.2 Block diagram for the EKF implementation.

12.3.4 Implementation

The EKF is implemented in two steps. In the first step (propagation) the system's current state and the error covariance matrix are estimated based on the system's model. In the second step (update) the system's state is improved by incorporating the sensor readings. Figure 12.2 depicts the block diagram for the EKF implementation where $\mathbf{Q}_k$, $\mathbf{R}_k$ are the covariance matrices for the system's and the measurement's model respectively, and $\mathbf{P}_k$ is the error covariance.

12.4 Fuzzy Logic Controllers

The EKF estimates the state of the system, including the vehicle's position and orientation, assuming zero mean Gaussian distributions in both the control input and the sensor readings. However, EKF does not consider errors in sensor readings unless a model of the sensor is developed. In this work, instead of creating a complex mathematical model that will estimate the error in sensor readings for each sensor, FL has been employed to represent the error in sensor readings as multiple zero mean Gaussian distributions. The design of the FL controllers is based on the performance of each sensor in a series of experiments. The statistics of the error in each sensor readings are incorporated into the EKF measurement's model through

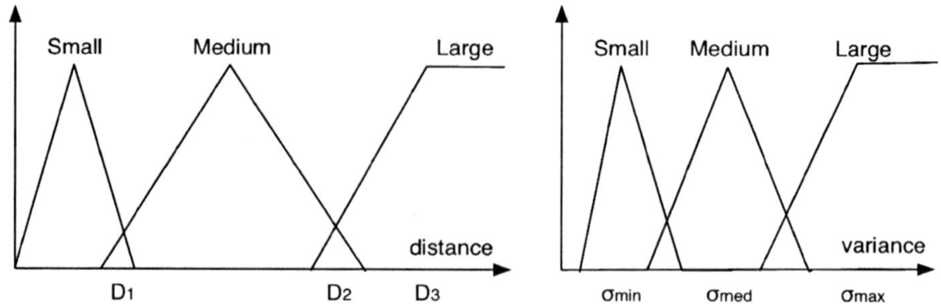

Fig. 12.3 Membership functions for the sensory readings and the variance of the error distribution.

the covariance matrix R_k. The rest of this section describes the development of a FL controller for each sensor and the implementation of the fuzzy EKF.

12.4.1 Range Sensors

It has been observed that the error in range readings is related to the distance between the target and the range sensor. For example, it has been experimentally determined that the error associated with the laser readings is described by the distribution $N(0, \sigma_{\text{max}})$ when the target's distance from the sensor is 7 to 9 m, and by the distribution $N(0, \sigma_{\text{min}})$ when the target's distance is less than 1 m. Based on this observation, a Mamdani type fuzzy controller has been developed that approximates the error in range sensor readings using multiple zero mean Gaussian distributions based on the distance between the target and the sensor. In other words, the fuzzy controller computes the variance, σ, for the error distribution for every range sensor based on the sensory readings.

Three membership functions are used to describe the range measurements and the variances of the error distributions as: *Small*, *Medium* and *Large*. The FL controller for the range sensors uses the following three rules:

1. "If the range measurement is *Small*, then the variance is *Small*"
2. "If the range measurement is *Medium*, then the variance is *Medium*"
3. "If the range measurement is *Large*, then the variance is *Large*"

Figure 12.3 depicts an example of the membership functions describing the distance between a range sensor and a target, and the variance of the zero mean Gaussian distribution. Detailed membership function figures for all the sensors are shown in Section 12.5.

12.4.2 Odometer

The error related to the odometer readings (Figure 12.4) increases as the distance the vehicle travels increases. In addition, the error in the odometer readings is closely related to the type of floor the vehicle travels on and the velocity in which the vehicle travels. For example, the error in the odometer readings is greater when the vehicle travels on tiles than when it travels on asphalt. In general, readings from the odometer cannot be incorporated into the EKF without an accurate error model that considers both the traveled distance and the type of floor the vehicle travels on. This is a significant drawback since the odometer data can provide an additional way of estimating the system's state.

In this work, the error in odometer readings is approximated by multiple zero mean Gaussian distributions using FL. A Mamdani type fuzzy controller has been developed that receives as inputs the type of floor the vehicle travels on and the distance that the vehicle has traveled to compute the error statistics. Additional inputs to the odometer FL controller may be the velocities the vehicle travels. In this work, a constant velocity is considered and thus velocity considerations are not included into the FL design.

Three membership functions have been used to describe the floor slippage, the distance traveled and the variance of the zero mean Gaussian distribution as: *Small*, *Medium* and *Large*. The FL controller designed for the odometer consists of 16 rules of the form:

1. "If the distance that the vehicle has traveled is *Small* and the floor slippage is *Small*, then the variance is *Small*".
2. "If the distance that the vehicle has traveled is *Medium* and the floor slippage is *Medium*, then the variance is *Medium*".

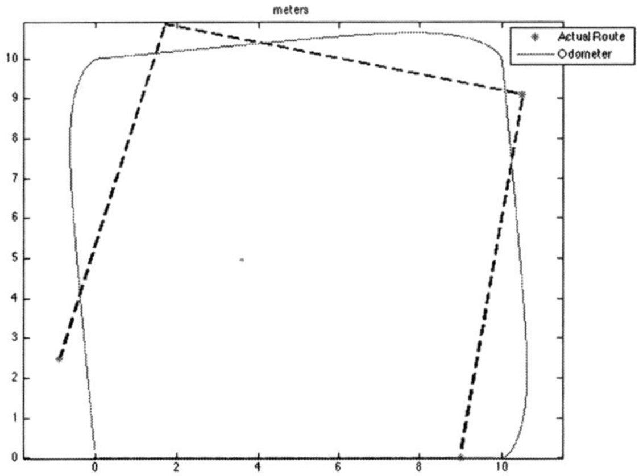

Fig. 12.4 Relation between the actual vehicle route and the route computed by the vehicle's odometer.

3. "If the distance that the vehicle has traveled is *Large* and the floor slippage is *Large*, then the variance is *Large*".

12.4.3 GPS

There are many parameters that influence the error in the GPS readings (Figure 12.5) such as the satellite coverage and the weather conditions. In this work a Mamdani type FL controller has been developed that approximates the error in GPS readings with multiple zero mean Gaussian distributions based on the satellite coverage, namely the number of satellites used for triangulation. The FL controller uses three membership functions to describe the satellite coverage and the variance of the zero mean Gaussian distributions as: *Small*, *Medium* and *Large*. There are three rules in the FL controller as follows:

1. "If the coverage is *Large*, then the variance is *Small*".
2. "If the coverage is *Medium*, then the variance is *Medium*".
3. "If the coverage is *Small*, then the variance is *Large*".

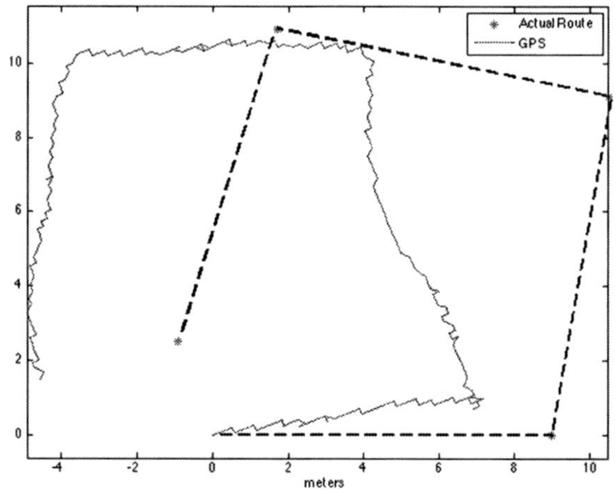

Fig. 12.5 Relation between the actual vehicle route and the route computed by the vehicle's GPS.

12.4.4 IMU

The IMU provides information on angular rate and acceleration of the vehicle. To estimate the position of the vehicle using as sole information the acceleration of the

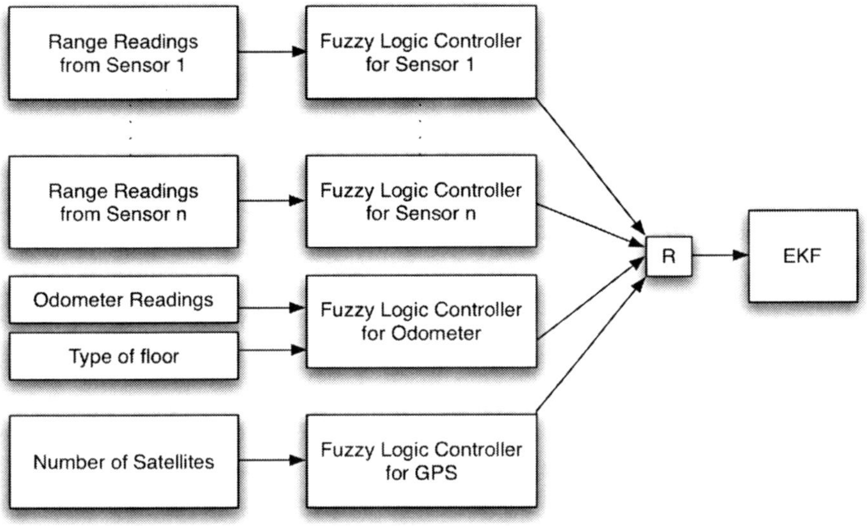

Fig. 12.6 Relation between the EKF and the fuzzy controllers.

vehicle, a double integration is required. However, the error in the IMU readings is integrated as well making the IMU readings useless in short time.

Unfortunately, there are no empirical rules that could be used to approximate the error in the IMU readings as zero mean Gaussian distributions. An error model needs to be developed in order to efficiently incorporate the IMU readings into the EKF. The developed error model is based on [3] and it is described in Section 12.5.

12.4.5 Fuzzy Controllers Implementation

The FL controllers described in the previous subsections are used to update the covariance matrix R_k of the measurement's model in the EKF. Figure 12.6 describes this procedure.

12.5 Case Study

The *ATRV-Jr* UGV has been used to test the performance of the fuzzy EKF. Experiments were run indoors and outdoors using known color landmarks. The vision system identifies the landmarks and computes the distance between the vehicle and the landmarks. A second set of range measurements to the landmarks is acquired from the laser range finder. A vision/laser registration algorithm is used to verify that both sensors provide range measurements for the same landmark. Additional

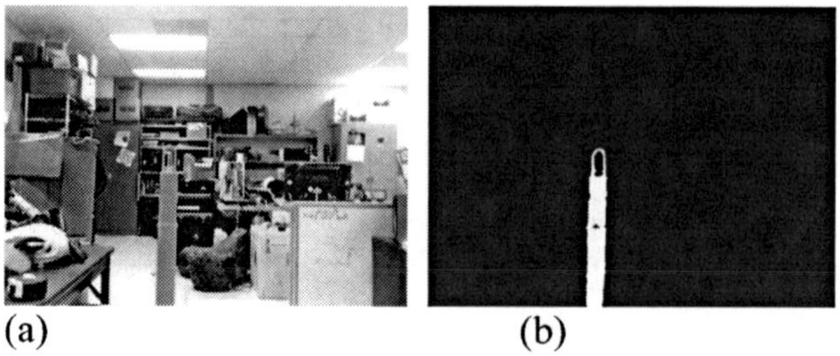

Fig. 12.7 Color threshold technique.

position readings are acquired from the vehicle's odometer and GPS and acceleration readings from the IMU.

The rest of this section describes the landmark identification technique, the stereo vision distance computation technique, the vision/laser registration algorithm, the GPS data conversion, the IMU error model and the fuzzy EKF.

12.5.1 Landmark Identification

The vision system of the ATRV-Jr consists of two pan/tilt color cameras mounted on top of the vehicle, at a distance of *35 cm* from each other. Image acquisition is achieved at a rate of 30 fps. The landmarks that have been used in these experiments have distinct color and can be separated from the image's background by a color threshold technique.

A landmark is identified by its color. Identifying color in an image instead of template matching techniques requires less computational time. The HSI color space has been used for the segmentation technique that follows. Since the landmark is a known object, the variations of the hue, saturation and intensity values of the color's representation in the HSI color space are known. A segmentation technique on H–I plane based on a region growing algorithm is used to separate the target from the image's background. The basic concept is to identify a pixel, "seed", in the image that takes the mean hue and intensity values within the area of the object's color and grow a region with similar neighboring pixels. For example, the cone shown in Figure 12.7a has hue and intensity values that vary between (10, 20) and (216, 250) respectively. Thus, the seed pixel will have a hue value 15 and an intensity value of 233. The region growing algorithm will merge neighboring to the seed pixels that have hue and intensity values in the former area. Figure 12.7b shows the result of this technique.

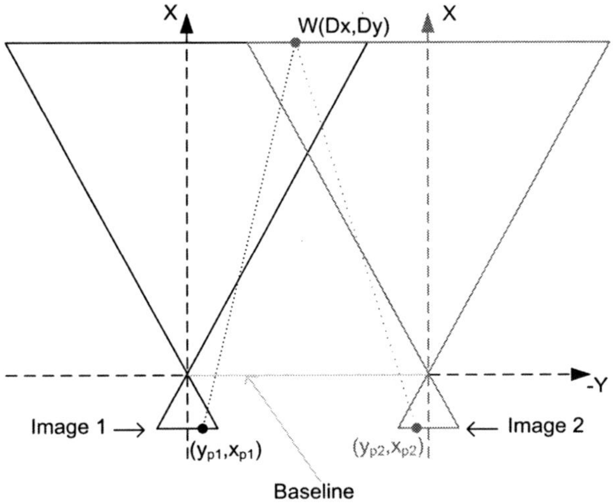

Fig. 12.8 Stereo vision system in parallel configuration.

12.5.2 Distance Computation

The method used to compute the distance between the vehicle and the landmark is presented in [23]. This method does not require any camera calibration technique that adds computational load to the system, but it is based on the image size, the field of view of the camera(s) and the relative position of the cameras (baseline). Figure 12.8 demonstrates the relationship between the fields of view of the two cameras and the perspective of a world point from each camera.

The depth measurement is derived by

$$\mathbf{cam}_k^i = \frac{320B}{2(xp_1 - xp_2)\tan(24.4)} \text{ (cm)}, \tag{12.16}$$

where $\mathbf{cam}_k^i$ is the distance between the left camera and the ith landmark, B is the relative distance between the two cameras (baseline), $(xp_1 - xp_2)$ is the pixel disparity.

The distance between the left camera and the landmark over the Y axis is derived by

$$cam_{y,k}^i = -\frac{2}{320} xp\mathbf{cam}_k^i \tan(24.4) \text{ (cm)}, \tag{12.17}$$

where xp the image location of a landmark pixel and

$$cam_{x,k}^i = \sqrt{\mathbf{cam}_k^{i2} - cam_{y,k}^{i2}}. \tag{12.18}$$

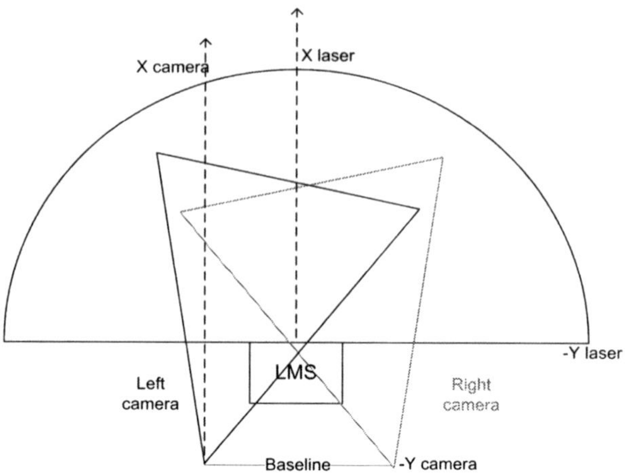

Fig. 12.9 Relation between the coordinate systems of the vision system and the laser range finder. The figure's units are cm.

12.5.3 Vision/Laser Registration

An essential issue on data fusion applications is that all sensors must provide data for the same entity. In this case study a stereo vision system and laser range finder sensors mounted on top of a vehicle are used to compute distances from a set of landmarks. Both sensors are able to recognize objects located in front of the vehicle. The purpose of the vision system/ laser registration is to verify that both sensors provide range data for the same landmark. Figure 12.9 depicts the coordinate systems of both sensors that need to be aligned.

The depth computation from the vision system is derived from (12.16). This measurement has to be transferred to the laser's coordinate system. The transformation is achieved by translations over the X and Y axes (Figure 12.9). Thus the distance of an object computed by the vision system in the laser's coordinate system will be

$$las^i_{x,k} = cam^i_{x,k} - camtr_x,$$
$$las^i_{y,k} = cam^i_{y,k} - camtr_y, \qquad (12.19)$$

where $camtr_x = 43$ cm and $camtr_y = 17$ cm.

Finally, the angle in which the ith landmark lies in the laser's coordinate system is derived by

$$\delta = \text{arch}\tan\left(\frac{las^i_{x,k}}{las^i_{y,k}}\right). \qquad (12.20)$$

The range measurement from a landmark taken by the laser sensor at angle δ corresponds to the measurement taken by the vision system for the same landmark.

Table 12.1 Values of fitting parameters.

	C_1	C_2	T
Acceleration on X	–0.037 m/s^2	0.171 m/s^2	3.787 s
Acceleration on Y	2.167 m/s^2	0.016 m/s^2	735.714 s
Acceleration on Z	–0.036 m/s^2	0.178 m/s^2	1.498 s
Angle rate around X	–0.009 rad/s	0.078 rad/s	8.329 s
Angle rate around Y	–0.002 rad/s	0.002 rad/s	1.536 s
Angle rate around Z	–0.001 rad/s	0.001 rad/s	14.141 s

12.5.4 GPS Conversions

The information derived from the GPS is converted from Latitude/Longitude to Universal Transverse Mercator (UTM) coordinates. UTM is a rectilinear mapping system in map coordinates are represented as Cartesian coordinates and distance is calculated using Euclidian distance measures [20]. The units of the UTM system are meters and the equations for the conversion could be found in [20].

12.5.5 IMU

As described in [3], the error in IMU readings is approximated by the following function:

$$\varepsilon(t) = C_1(1 - e^{-t/T}) + C_2, \tag{12.21}$$

where C_1, C_2 and T are parameters. To estimate the values of the parameters C_1, C_2 and T, IMU readings were collected while the sensor was immobilized. Using the Levenberg-Marquardt least square fit method, the IMU readings were fitted to Equation (12.21). Table 12.1 summarizes the best fitting parameter values for each of the IMU outputs.

The IMU error model will be incorporated into the system's model of the EKF. For simplicity reasons, it is assumed that the vehicle is moving in a 2D space and readings such as acceleration on Z axis, angle rates around X and Y axes are not considered. The system's model described in Equations (12.4) and (12.5) will become:

$$\mathbf{X}_{k+1} = \mathbf{f}(\mathbf{X}_k, \mathbf{u}_k) + \mathbf{n}_k, \tag{12.22}$$

$$\mathbf{f}(\mathbf{X}_k, \mathbf{u}_k) = \begin{bmatrix} \mathbf{x}_k \\ \mathbf{imu}_k \\ \mathbf{lan}_k^1 \\ \vdots \\ \mathbf{lan}_k^n \end{bmatrix} = \begin{bmatrix} x_k + V_k \Delta t \cos \vartheta_k + \frac{1}{2} a x_k \Delta t^2 \\ y_k + V_k \Delta t \sin \vartheta_k + \frac{1}{2} a y_k \Delta t^2 \\ \vartheta_k + \Omega_k \Delta t \\ a x_k + e a x_k \\ \frac{T a x_k}{T a x_k + \Delta t} \\ a y_k + e a y_k \\ \frac{T a x_y}{T a x_y + \Delta t} \\ angle_{z,k} \\ lan_{x,k}^1 \\ lan_{y,k}^1 \\ \vdots \\ lan_{x,k}^n \\ lan_{y,k}^n \end{bmatrix}, \tag{12.23}$$

where $\mathbf{imu}_k = [ax_k \quad eax_k ay_k \quad eay_k \quad angle_{z,k}]^T$ the IMU readings related to the vehicle's accelerations, errors in acceleration, and steering angle at time k, and Tax_k, Tay_k the corresponding parameters T as shown in Table 12.1.

12.5.6 Fuzzy Extended Kalman Filter

This section describes in detail the FL controllers and their involvement into the EKF. A FL logic controller has been designed for each of the sensors: GPS, odometer, stereo vision system and laser range finder. These controllers are responsible for updating the statistics of the distributions describing the error in sensor readings. The design of the FL controllers involved a number of experiments that helped identify the error in sensor readings in various conditions. To compute the error in odometer and GPS readings, experiments were run indoors and outdoors with the vehicle traveling at a constant speed of 0.5 m/s. On the other hand, the error in range measurements was computed with the vehicle immobilized while multiple readings were collected in various distances from a target.

Table 12.2 shows the statistics of the error in the range sensors readings as a function of the measured distance and Table 12.3 summarizes the performance of the odometer in different types of floors based on the traveled distance. Table 12.4 shows the performance of the error in the GPS readings based on the number of satellites available.

Figure 12.10 depicts the membership functions of the FL controllers designed for the stereo vision system (a), the laser range finder (b), the GPS (c) and the odometer (d). The rules for each FL controller follow the structure described in Section 12.4.

Table 12.2 Distributions of the error in sensor readings with respect to the range measurements.

Distance	7–9 m	3–5 m	0–2 m
Vision System	$N(0, 0.7^2)$	$N(0, 0.3^2)$	$N(0, 0.1^2)$
Laser	$N(0, 0.6^2)$	$N(0, 0.2^2)$	$N(0, 0.1^2)$

Table 12.3 Distributions of the error in odometer readings with respect to the travelled distance.

Distance	0–5 m	10–15 m	20-inf
Odometer-Tile	$N(0, 0.2^2)$	$N(0, 1.5^2)$	$N(0, 2.5^2)$
Odometer-Grass	$N(0, 0.2^2)$	$N(0, 0.7^2)$	$N(0, 1.6^2)$
Odometer-Asphalt	$N(0, 0.2^2)$	$N(0, 0.6^2)$	$N(0, 1.2^2)$

Table 12.4 Distributions of the error in GPS readings with respect to the satellite coverage.

Satellites	5	6	7
GPS	$N(0, 7^2)$	$N(0, 4^2)$	$N(0, 2^2)$

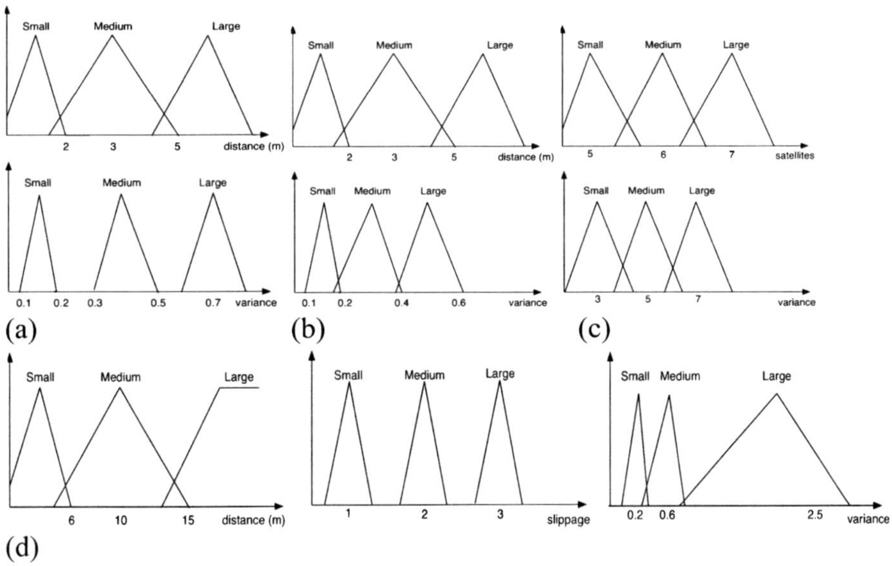

Fig. 12.10 Membership functions of the vision system (a), laser range finder (b), GPS (c) and odometer (d) FL controllers.

The fuzzy EKF is implemented by recursively computing the propagation equations, the measurements covariance matrix and the update equations. Figure 12.11 demonstrates this procedure.

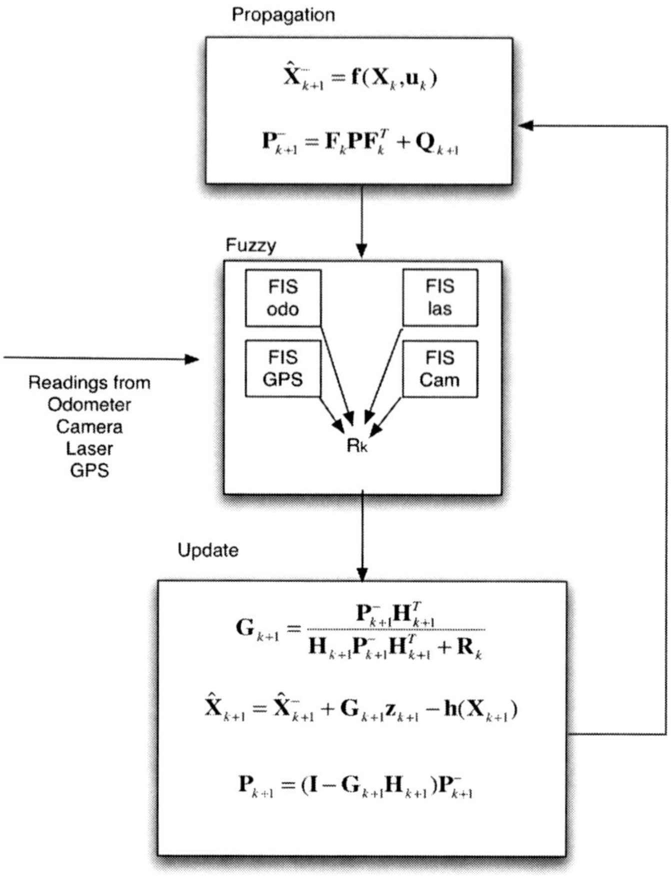

Fig. 12.11 Block diagram for the fuzzy EKF.

12.5.7 Results

To evaluate the performance of the fuzzy EKF, a set of experiments was run indoors and outdoors. The distinction between indoor and outdoor environments has to be stated because it influences considerably the sensors' performance. For example the GPS cannot establish satellite connection indoors while the IMU readings are heavily distorted due to metallic surroundings. Concerning outdoor environments, the experiments were run in different days with different weather conditions (sunny/cloudy) and satellite coverage. The performance of the presented fuzzy EKF is compared with the traditional EKF in terms of average distance from the actual vehicle's route. It is in the scope of this work to investigate whether consideration of additional sensors in the fuzzy EKF and EKF increases the performance of the filters. To that end, the filters are compared at the sensor level in the following combinations: odometry; GPS; odometry and GPS; odometry, GPS and IMU; odometry,

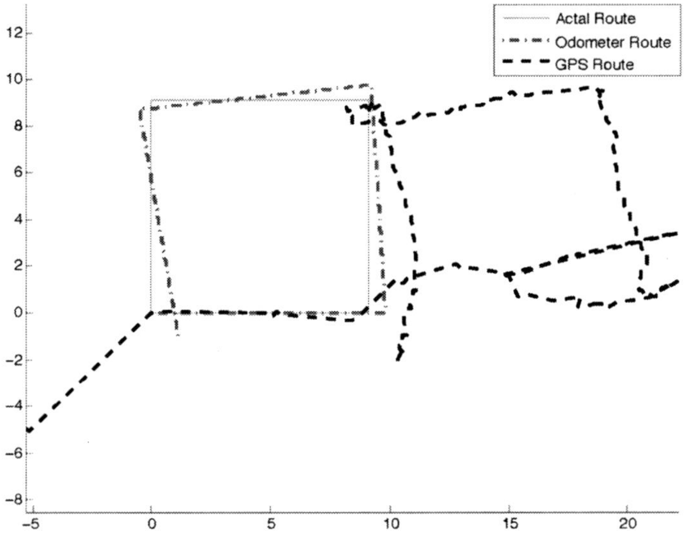

Fig. 12.12 Vehicle's route as computed using odometer and GPS readings.

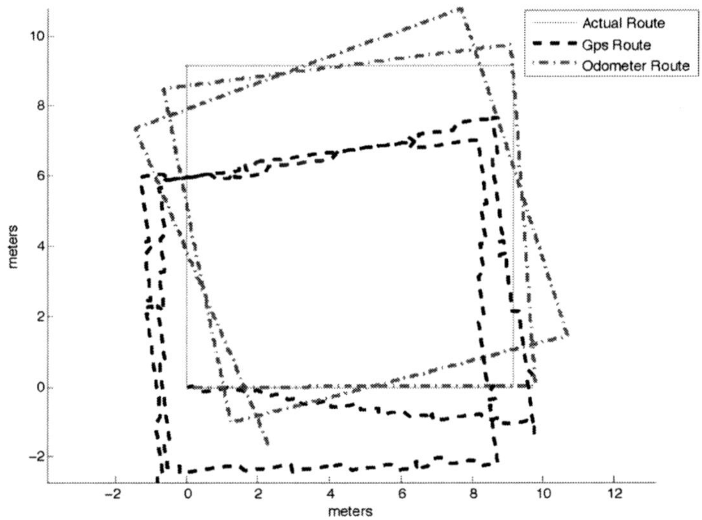

Fig. 12.13 Vehicle's route as computed using odometer and GPS readings, experiment 1.

GPS, IMU and range sensors. The rest of this section presents in detail the experiments layout and summarizes the performances of the fuzzy EKF and traditional EKF.

To demonstrate the effectiveness of the presented fuzzy EKF, a square route is assigned to the vehicle. Three different experiments are run at which the vehicle has

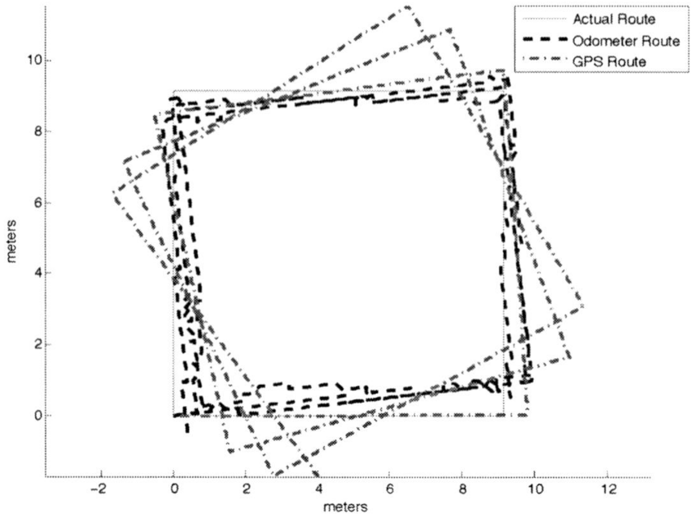

Fig. 12.14 Vehicle's route as computed by odometer readings and GPS readings, experiment 2.

Fig. 12.15 Position of the landmarks during an experiment.

to perform one square (Figure 12.12), two squares (Figure 12.13) and three squares (Figure 12.14) respectively. Each vertex of the square route is 9.15 m long. The vehicle's linear velocity was 0.5 m/s. Four known landmarks have been placed at the square edges (Figure 12.15). Figures 12.12–12.14 present the vehicle's actual route and the routes as computed by odometer readings and by GPS readings. As shown in Figures 12.12–12.14, while the route computed by the odometer is close to the vehicle's actual route, the route computed by the GPS readings varies considerably from the actual route. This is because the experiments in Figures 12.12, 12.13 and 12.14 are run with GPS readings computed by four, seven and nine satellites respectively.

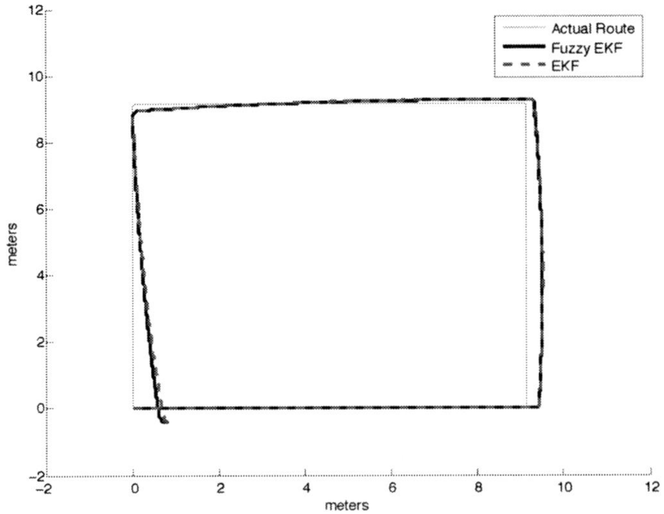

Fig. 12.16 Fuzzy EKF versus traditional EKF using odometer readings, experiment 1.

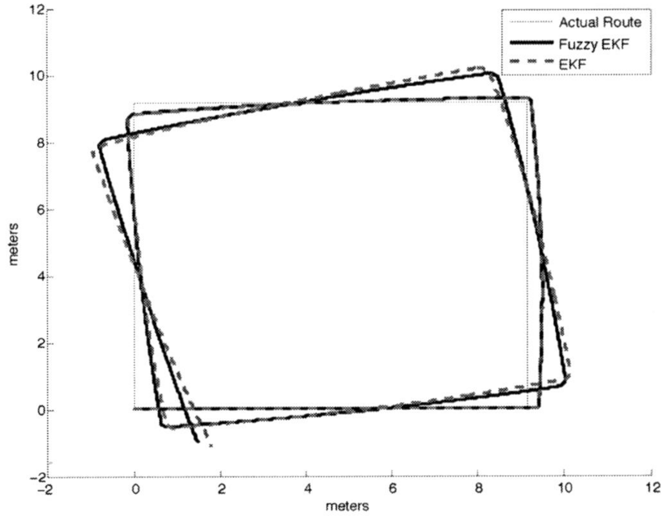

Fig. 12.17 Fuzzy EKF versus traditional EKF using odometer readings, experiment 2.

12.5.7.1 Odometer

The next figures demonstrate the performance of the fuzzy EKF and the traditional EKF in the square route considering only the odometer readings. The fuzzy EKF performs better than the traditional EKF by a few centimeters.

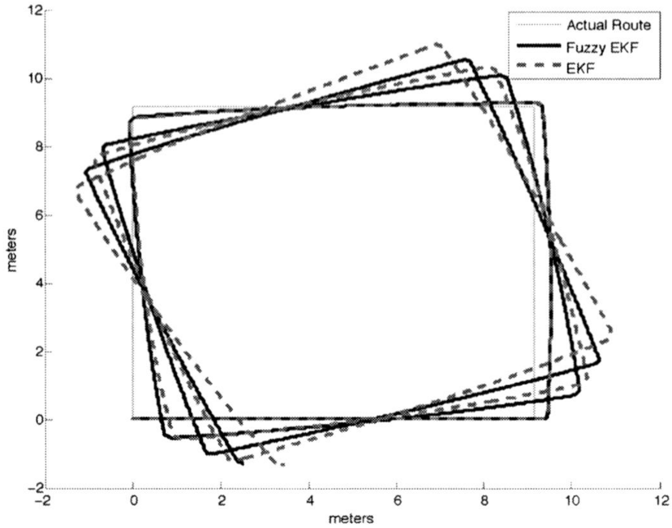

Fig. 12.18 Fuzzy EKF versus traditional EKF using odometer readings, experiment 3.

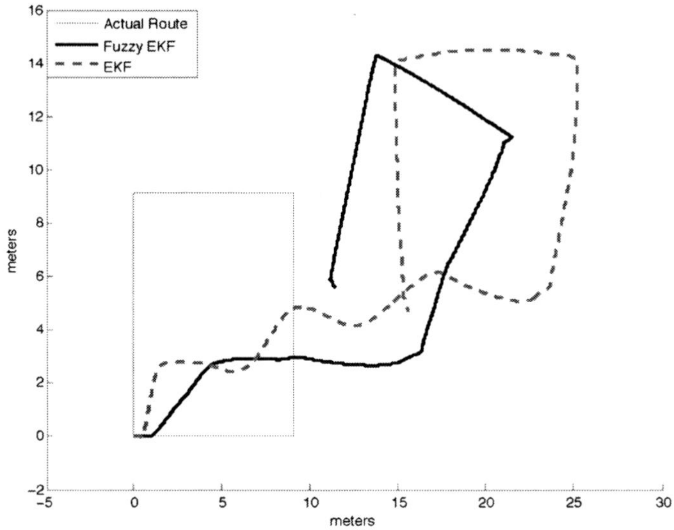

Fig. 12.19 Fuzzy EKF versus traditional EKF using GPS readings, experiment 1.

12.5.7.2 GPS

The following figures present the performance of the fuzzy EKF and the EKF considering only the GPS readings. The fuzzy EKF performs considerably better than the traditional EKF especially when the GPS readings present big errors (Fig-

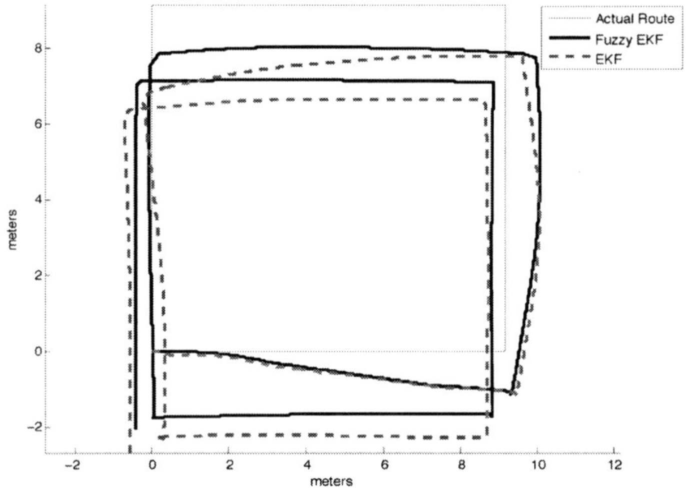

Fig. 12.20 Fuzzy EKF versus traditional EKF using GPS readings, experiment 2.

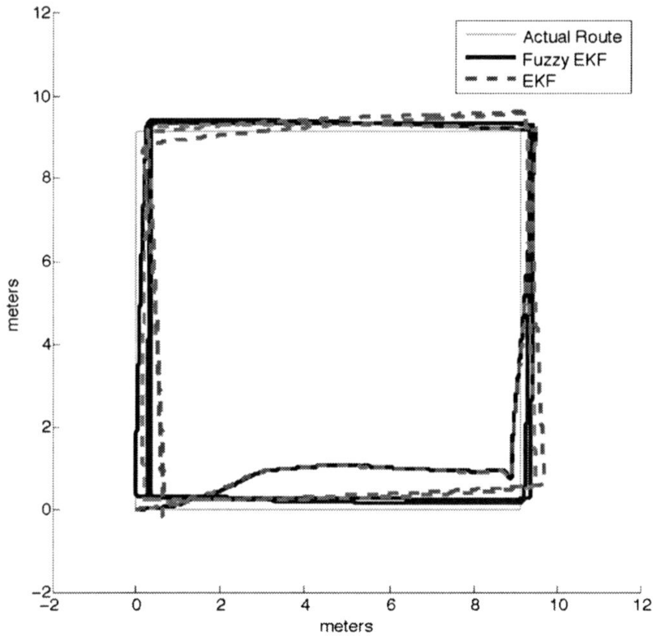

Fig. 12.21 Fuzzy EKF versus traditional EKF using GPS readings, experiment 3.

ure 12.19). However, using solely readings from a faulty sensor influences considerably the performance of both filters.

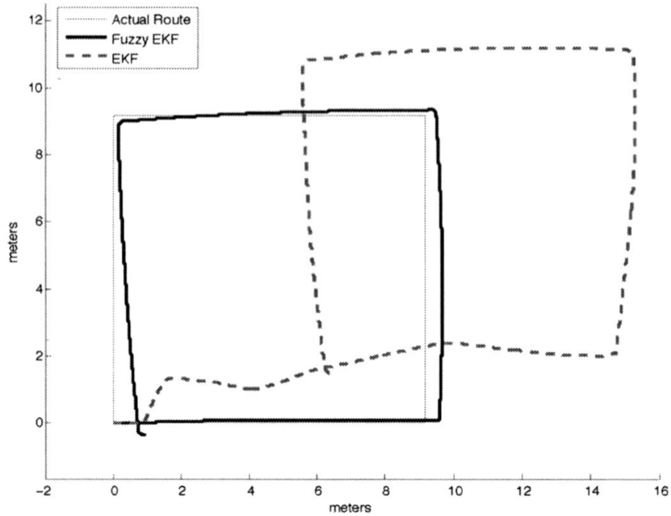

Fig. 12.22 Fuzzy EKF versus traditional EKF using odometer and GPS readings, experiment 1.

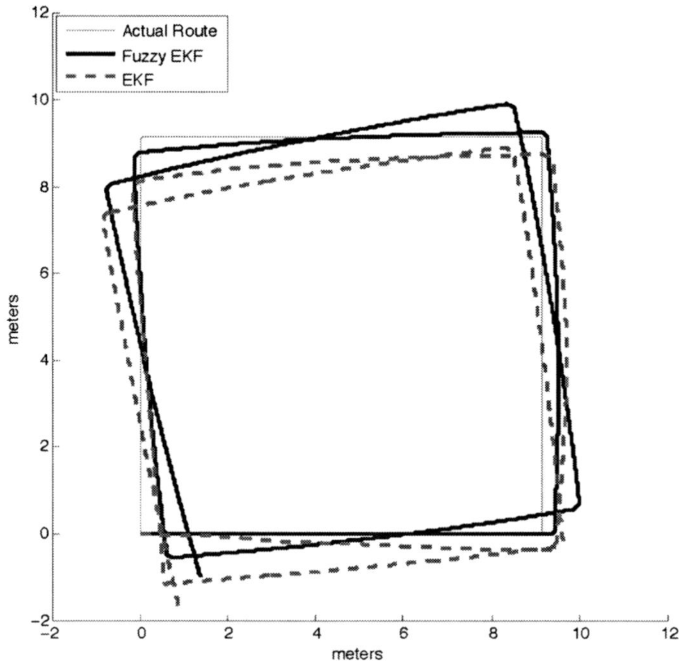

Fig. 12.23 Fuzzy EKF versus traditional EKF using odometer and GPS readings, experiment 2.

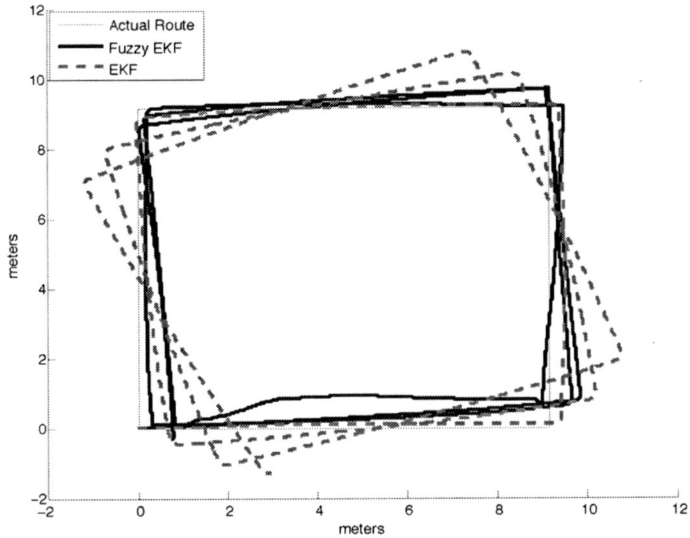

Fig. 12.24 Fuzzy EKF versus traditional EKF using odometer and GPS readings, experiment 3.

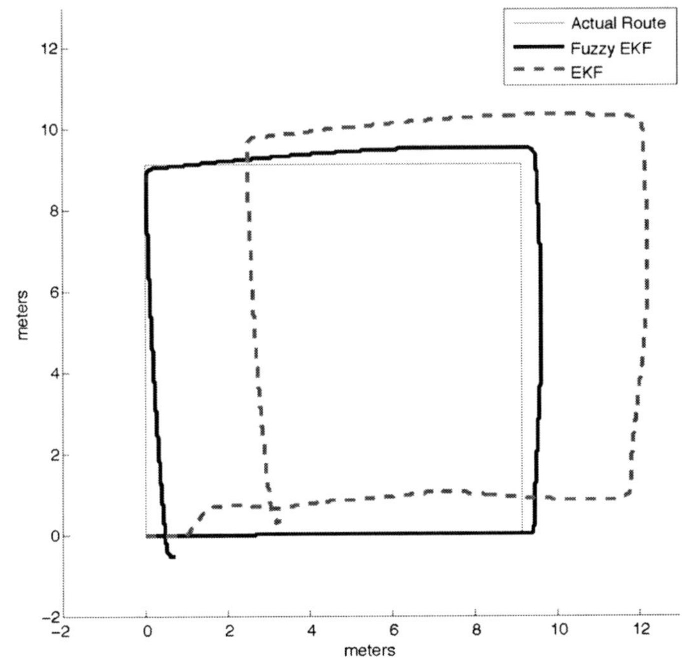

Fig. 12.25 Fuzzy EKF versus traditional EKF using odometer, GPS and IMU readings, experiment 1.

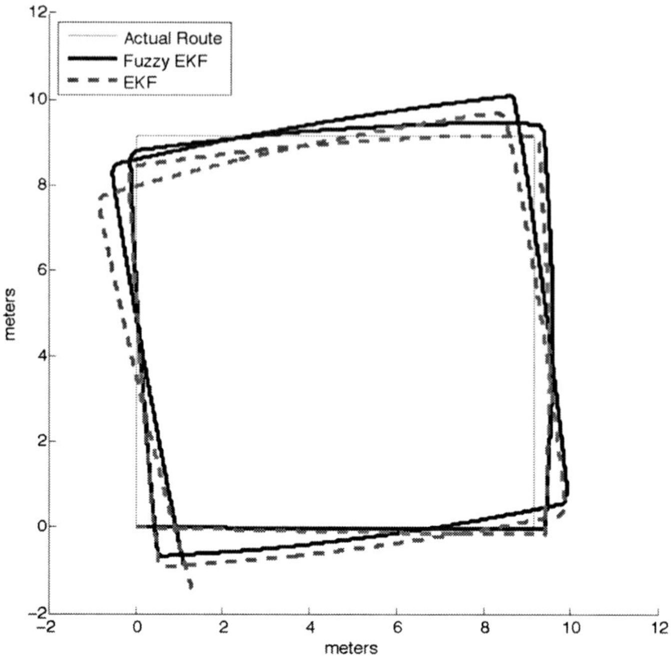

Fig. 12.26 Fuzzy EKF versus traditional EKF using odometer, GPS and IMU readings, experiment 2.

12.5.7.3 Odometer and GPS

In this and the following subsections the presented fuzzy EKF and the traditional EKF are used to fuse information from multiple sensors. As shown in Figures 12.22–12.24, the fuzzy EKF performs better than the traditional EKF.

12.5.7.4 Odometer, GPS and IMU

Figures 12.25–12.27 demonstrate the performance of the fuzzy EKF and traditional EKF using information from the odometer, the GPS and the IMU.

12.5.7.5 All Sensors

Finally, this subsection presents the performance of the fuzzy EKF and the traditional EKF using information captured by all sensors: odometer, GPS, IMU and range sensors (laser range finder and vision system).

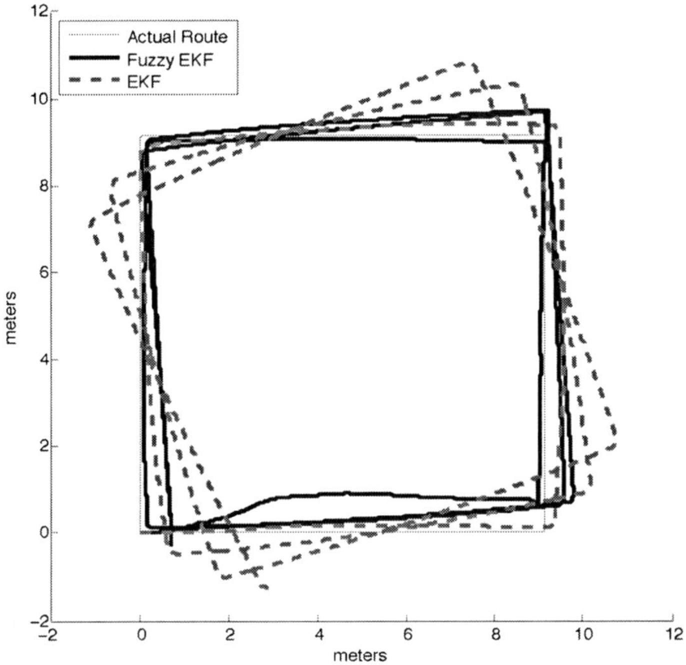

Fig. 12.27 Fuzzy EKF versus traditional EKF using odometer, GPS and IMU readings, experiment 3.

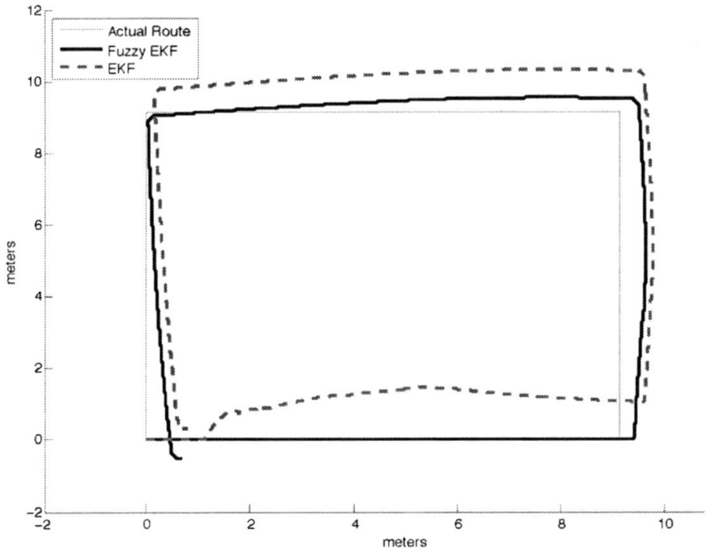

Fig. 12.28 Fuzzy EKF versus traditional EKF using readings from all sensors, experiment 1.

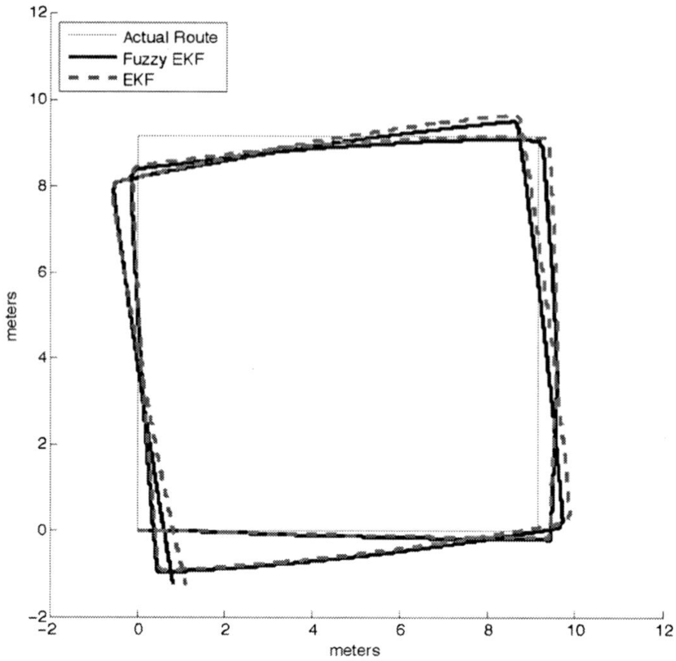

Fig. 12.29 Fuzzy EKF versus traditional EKF using readings from all sensors, experiment 2.

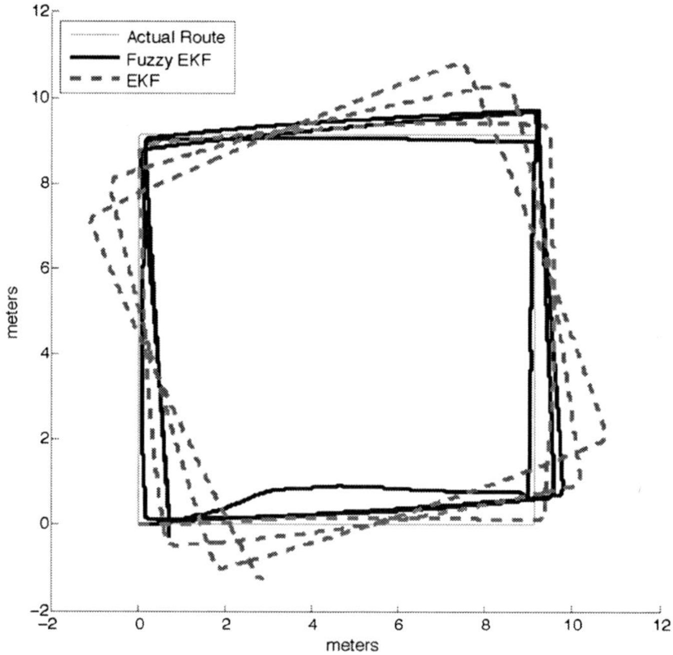

Fig. 12.30 Fuzzy EKF versus traditional EKF using readings from all sensors, experiment 3.

Table 12.5 Average distance between the actual route and the route computed by the filters in meters.

		Odo-meter	GPS	Odo & GPS	Odo, GPS & IMU	All sensors
Exp. 1	Fuzzy EKF	1.25	9.89	1.31	1.15	1.25
	EKF	1.35	15.30	5.66	2.83	1.54
Exp.2	Fuzzy EKF	0.85	1.54	0.85	0.83	0.85
	EKF	1.05	2.10	1.26	0.91	0.88
Exp. 3	Fuzzy EKF	2.00	1.20	1.51	1.40	1.40
	EKF	2.60	1.51	2.29	2.15	2.15

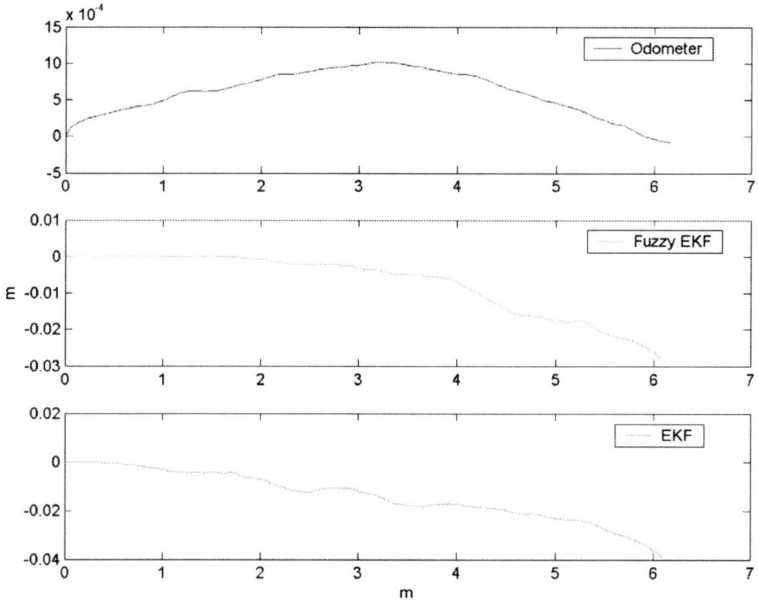

Fig. 12.31 Fuzzy EKF versus traditional EKF using readings from all sensors, experiment 4.

12.5.7.6 Summary of Results

Table 12.5 summarizes the performance of the fuzzy EKF and the traditional EKF in terms of average distance from the vehicle's actual route. As shown, the fuzzy EKF performs better than the EKF in terms of position accuracy, especially when sensors provide inaccurate readings.

Additional experiments were run in straight line (Figure 12.31) and meander (Figure 12.32) routes. Table 12.6 summarized the performance of the fuzzy EKF and the traditional EKF for these routes when all the sensor readings are utilized.

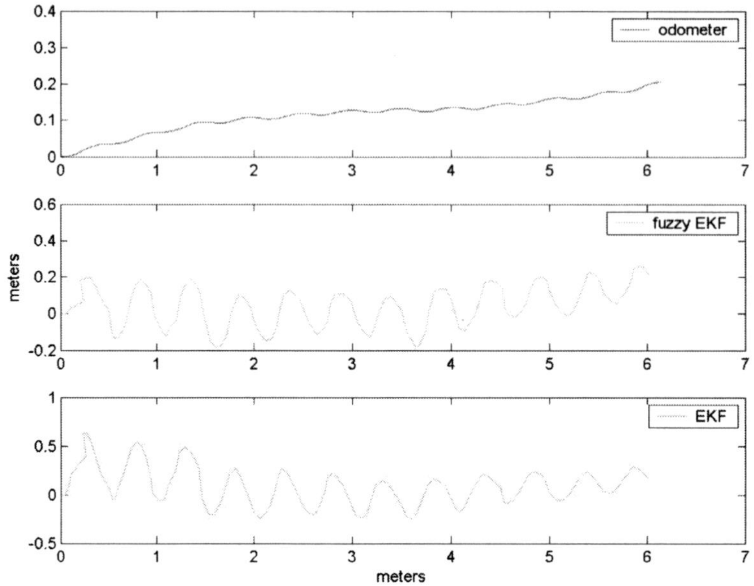

Fig. 12.32 Fuzzy EKF versus traditional EKF using readings from all sensors, experiment 5.

Table 12.6 Average distance between the actual route and the route computed by the filters in meters.

	EKF	Fuzzy EKF
Straight line route	0.16	0.12
Meander route	0.33	0.21

12.6 Conclusions

This chapter presents a method for UGVs localization using Fuzzy Logic and Kalman Filtering. Information from five different sensors is fused to provide an estimate of the vehicles position. Fuzzy Logic is used to compute and adjust the parameters of the error distribution of the sensor readings and update the covariance matrix of the measurement's model in EKF. As demonstrated the fuzzy EKF performs better than the traditional EKF in terms of position accuracy particularly when the readings from the sensors are inaccurate.

As demonstrated in the first experiment, where the GPS readings are considerably inaccurate, the fuzzy EKF performs better than the EKF since it considers the error in the sensor readings through the fuzzy logic controller. In the specific experiment the GPS readings were computed by 4 satellites and the variance of the error in sensor readings was set to *Large* by the fuzzy logic controller.

In addition, as demonstrated in Table 12.5, although consideration of additional sensors favors the performance of the EKF, it does not affect the performance of the fuzzy EKF. Since fuzzy EKF considers the errors in sensor readings, it computes the best estimate of the vehicle's posture using less sensors than the EKF.

Future work involves natural landmark selection and sensor failure detection. The vehicle should dynamically identify the most dominant objects in the surrounding environment and use their position as reference for its posture estimation. In addition, since various kinds of failures, faults and disturbances may influence the performance of the filter, a methodology that identifies and classifies uncertainties in sensor readings should be considered.

References

1. K.O. Arras, N. Tomatis, et al., Multisensor on-the-fly localization using laser and vision, in *Proceedings of International Conference on Intelligent Robots and Systems*, 2000.
2. A. Arsenio and M.I. Ribeiro, Active range sensing for mobile robot localization, in *Proceedings of International Conference on Intelligent Robots and Systems*, 1998.
3. B. Barshan and H.F. Durrant-Whyte, Inertial navigation systems for mobile robots, *IEEE Transactions on Robotics and Automation* **11**(3), 328–342, 1995.
4. E.T. Baumgartner and S.B. Skaar, An autonomous vision-based mobile robot, *IEEE Transactions on Automatic Control* **39**(3), 493–502, 1994.
5. R. Carrasco and A. Cipriano, Fuzzy logic based nonlinear Kalman filter applied to mobile robots modelling, in *Proceedings of IEEE International Conference on Fuzzy Systems*, 2004.
6. F. Chenavier and J.L. Crowley, Position estimation for a mobile robot using vision and odometry, in *Proceedings of IEEE International Conference on Robotics and Automation*, 1992.
7. T. Ching-Chih, A localization system of a mobile robot by fusing dead-reckoning and ultrasonic measurements, in *Proceedings of IEEE Instrumentation and Measurement Technology Conference*, 1998.
8. H. Choset, K.M. Lynch, et al., *Principles of Robot Motion, Theory, Algorithms and Implementation*, The MIT Press, 2005.
9. H. Chou, M. Traonmilin, et al., A simultaneous localization and mapping algorithm based on Kalman filtering, in *Proceedings of IEEE Intelligent Vehicles Symposium*, 2004.
10. P.J. Escamilla-Ambrosio and N. Mort, A hybrid Kalman filter-fuzzy logic architecture for multisensor data fusion, in *Proceedings of IEEE International Symposium on Intelligent Control*, 2001.
11. P. Goel, S.I. Roumeliotis, et al., Robust localization using relative and absolute position estimates, in *Proceedings of International Conference on Intelligent Robots and Systems*, 1999.
12. H. Huosheng and G. Dongbing, Landmark-based navigation of mobile robots in manufacturing, in *Proceedings of 7th IEEE International Conference on Emerging Technologies and Factory Automation*, 1999.
13. S. Kai-Tai and T. Wen-Hui, Environment perception for a mobile robot using double ultrasonic sensors and a CCD camera, *IEEE Transactions on Industrial Electronics* **43**(3), 372–379, 1996.
14. U. Larsson, J. Forsberg, et al., Mobile robot localization: integrating measurements from a time-of-flight laser, *IEEE Transactions on Industrial Electronics* **43**(3), 422–431, 1996.
15. M. Marron, J.C. Garcia, et al., Fusing odometric and vision data with an EKF to estimate the absolute position of an autonomous mobile robot, in *Proceedings of IEEE Conference Emerging Technologies and Factory Automation*, 2003.
16. F. Martia, A. Jimenez, et al., A novel fuzzy Kalman filter for mobile robots localization, in *Information Processing and Management of Uncertainty in Knowledge-Based Systems*, 2004.
17. S. Niwa, T. Masuda, et al., Kalman filter with time-variable gain for a multisensor fusion system, in *Proceedings of International Conference on Multisensor Fusion and Integration for Intelligent Systems*, 1999.

18. J.Z. Sasiadek and P. Hartana, Sensor data fusion using Kalman filter, in *Proceedings of Third International Conference on Information Fusion*, 2000.
19. K. Seong-Baek, L. Seung-Yong, et al., An advanced approach for navigation and image sensor integration for land vehicle navigation, in *Proceedings of IEEE 60th Vehicular Technology Conference*, 2004.
20. J.P. Snyder, *A Working Manual*, United States Government Printing Office, Washington, 1987.
21. E. Stella, G. Cicirelli, et al., Position estimation for a mobile robot using data fusion, in *Proceedings of IEEE International Symposium on Intelligent Control*, 1995.
22. A. Tiano, A. Zirilli, et al., Application of interval and fuzzy techniques to integrated navigation systems, in *Proceedings of Joint 9th IFSA World Congress and 20th NAFIPS International Conference*, 2001.
23. A. Tsalatsanis, K. Valavanis, et al., Mobile robot navigation using sonar and range measurements from uncalibrated cameras, in *Proceedings of 14th Mediterranean Conference on Control and Automation*, 2006.
24. W.S. Wijesoma, K.R.S. Kodagoda, et al., A laser and a camera for mobile robot navigation, in *Proceedings of 7th International Conference on Control, Automation, Robotics and Vision*, 2002.

Chapter 13
Adaptive Estimation of Fuzzy Cognitive Networks and Applications

T.L. Kottas, Y.S. Boutalis and M.A. Christodoulou

Abstract Fuzzy Cognitive Networks (FCN) have been introduced by the authors as an operational extension of Fuzzy Cognitive Maps (FCM), which were initially introduced by Kosko to model complex behavioral systems in various scientific areas. One important issue of their operation is the conditions under which they reach a certain equilibrium point after an initial perturbation. This is equivalent to studying the existence and uniqueness of solutions for their concept values. In this chapter, we present a study on the existence of solutions of FCMs equipped with continuous differentiable sigmoid functions having contractive or at least non-expansive properties. This is done by using an appropriately defined contraction mapping theorem and the non-expansive mapping theorem. It is proved that when the weight interconnections fulfill certain conditions the concept values will converge to a unique solution regardless the exact values of the initial concept values perturbations, or in some cases a solution exists that may not necessarily be unique. Otherwise the existence or the uniqueness of equilibrium cannot be assured. Based on these results an adaptive weight estimation algorithm is proposed which employs appropriate weight projection criteria to assure that the uniqueness of FCM solution is not compromised. Fuzzy Cognitive Networks are in the sequel invoked providing an application framework for the obtained results.

Thodoris Kottas
Democritus University of Thrace, 67100 Xanthi, Greece; e-mail: tkottas@ee.duth.gr

Yiannis Boutalis
Democritus University of Thrace, 67100 Xanthi, Greece and
Department of Electrical, Electronic and Communication Engineering, Chair of Automatic Control, University of Erlangen-Nuremberg, 91058 Erlangen, Germany; e-mail: ybout@ee.duth.gr

Manolis Christodoulou
Technical University of Crete, 73100 Chania, Crete, Greece; e-mail: manolis@ece.tuc.gr

K.P. Valavanis (ed.), Applications of Intelligent Control to Engineering Systems, 329–363.

13.1 Introduction

Fuzzy Cognitive Maps (FCM) have been introduced by Kosko [1] based on Axelrod's work on cognitive maps [2]. They are inference networks using cyclic directed graphs that represent the causal relationships between concepts. They use a symbolic representation for the description and modelling of the system. In order to illustrate different aspects in the behavior of the system, a fuzzy cognitive map consists of nodes where each one represents a system characteristic feature. The node interactions represent system dynamics. An FCM integrates the accumulated experience and knowledge on the system operation, as a result of the method by which it is constructed, i.e., by using human experts who know the operation of the system and its behavior. Different methodologies to develop FCM and extract knowledge from experts have been proposed in [3–6].

Kosko enhanced the power of cognitive maps considering fuzzy values for their nodes and fuzzy degrees of interrelationships between nodes [1,7]. He also proposed the differential Hebian rule [7] to estimate the FCM weights expressing the fuzzy interrelationships between nodes based on acquired data. After this pioneering work, fuzzy cognitive maps attracted the attention of scientists in many fields and have been used to model behavioral systems in many different scientific areas. Application examples can be found in political science [8, 9], in economic field [10, 11], in representing social scientific knowledge and describing decision making methods [12–14]. Other applications include geographical information systems [15–17], cellular automata [18], pattern recognition applications [19, 20] and numerical and linguistic prediction of time series functions [21]. Fuzzy cognitive maps have also been used to model the behavior and reactions of virtual worlds [22–26], as a generic system for decision analysis [14, 27] and as coordinator of distributed cooperative agents.

Regarding FCM weight estimation and updating, recent publications [28–32] extend the initially proposed differential Hebian rule [7] to achieve better weight estimation. Another group of methods for training FCM structure involves genetic algorithms and other exhaustive search techniques [33–37], where the training is based on a collection of particular values of input output historical examples and on the definition of appropriate fitness function to incorporate design restrictions.

Various extensions of FCMs have been proposed in the literature [38–49]. Dynamic Cognitive Networks (DCN) appear in [41], the Fuzzy Causal Networks in [42–46], while the neutrosophic cognitive maps appear in [47, 48]. Recently Fuzzy Cognitive Networks (FCN) [49] has been proposed as a complete computational and storage framework to facilitate the use of FCM in cooperation with the physical system they describe.

Fuzzy Cognitive Networks (FCNs) and their storage mechanism assume that they reach equilibrium points, each one associated with a specific operation condition of the underlying physical system. However, the conditions under which FCMs and consequently FCNs reach an equilibrium point and whether this point is unique have not been adequately studied so far. Simple FCMs have bivalent node values and trivalent edges (weights) and are equipped with binary threshold functions or sigmoid functions with very large inclination [50]. According to Kosko [50, 51], starting from an initial state, simple FCMs follow a path, which ends in a fixed point or limit cycle, while more complex ones may end in an aperiodic or "chaotic" attractor. These fixed points and attractors could represent *meta rules* of the form "If input then attractor or fixed point". A more detailed study on the performance of FCMs has been presented recently in [52], where the inference capabilities of FCMs equipped with binary, trivalent or sigmoid functions are compared. The relation of the existence of these attractors or fixed points to the weight interconnections of the FCM and FCN has not been fully investigated. This is, however, of paramount importance if one wants to use FCNs with learning capabilities in reliable adaptive system identification and control schemes.

In this chapter, we present a study on the existence of the above fixed points of FCMs equipped with continuous differentiable sigmoid functions having contractive or at least non-expansive properties. The study is based on the work presented in [53, 54] and is made by using an appropriately defined contraction mapping theorem and the non-expansive mapping theorem. It is proved that when the weight interconnections fulfill certain conditions, related to the size of the FCM and the inclination of the sigmoid functions, the concept values will converge to a unique solution regardless of their initial states, or in some cases a solution exists that may not necessarily be unique. Otherwise the existence or the uniqueness of equilibria may or may not exist, it may depend on the initial states, but it can not be assured. In case the FCM has also input nodes (that is nodes that influence but are not influenced by other nodes), the unique equilibrium does not depend solely on the weight set, as in the case of FCMs with no input nodes; it depends also on the values of the input nodes.

In the sequel, an adaptive weight estimation algorithm is proposed with guaranteed exponentially fast error convergence to zero, which uses the obtained conditions to construct appropriate weight projection rules assuring that the obtained weights do not compromise the existence of the FCM solution. In view of these results *meta rules* of the form "If weights then fixed point" are more appropriate to represent the behavior of an FCM which satisfy the above weight conditions. Fuzzy Cognitive Networks (FCN) [49, 55, 56], introduced recently as an extension of FCMs can work on the basis of such *meta rules* and provide the application framework of the obtained results. It is demonstrated that when the necessary weight conditions are fulfilled by an FCN during its updating procedure, its information storage mechanism is actually a depository of this kind of *meta rules*.

The chapter is organized as follows. Section 13.2 describes the representation and mathematical formulation of Fuzzy Cognitive Maps. Section 13.3 provides the proof of the existence solution of the concept values of a Fuzzy Cognitive Map. Section 13.4 presents the adaptive weight estimation algorithm with proven stability and parameter convergence, while Section 13.5 provides illustrative numerical examples. The FCNs as operational extension of FCMs, which rely on the obtained theoretical results are briefly invoked in Section 13.6 and some of their important aspects are presented. Finally, Section 13.7 concludes the work providing also hints for future extensions.

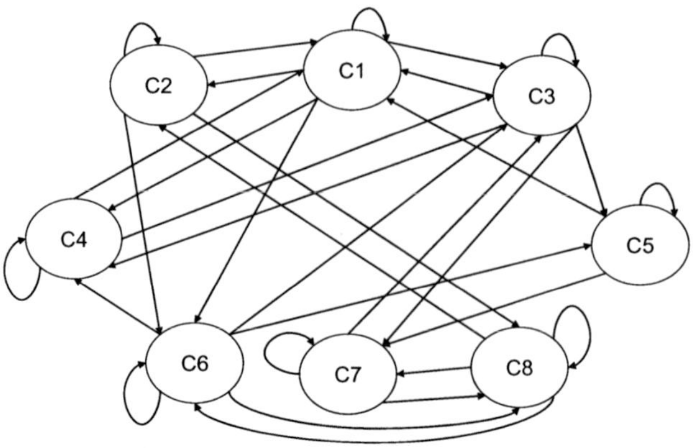

Fig. 13.1 An FCM with 8 nodes.

13.2 Fuzzy Cognitive Maps

A graphical representation of FCMs is depicted in Figure 13.1. Each concept represents a characteristic of the system; in general it represents events, actions, goals, values and trends of the system. Each concept is characterized by a number A_i that represents its value and it results from the transformation of the real value of the systems variable, represented by this concept, either in the interval [0, 1] or in the interval [−1, 1]. All concept values form vector A as follows:

$$A = \left[A_1 \; A_2 \; \ldots \; A_n \right]^T ,$$

with n being the number of the nodes (in Figure 13.1 $n = 8$). Causality between concepts allows degrees of causality and not the usual binary logic, so the weights of the interconnections can range in the interval [−1, 1].

The existing knowledge on the behavior of the system is stored in the structure of nodes and interconnections of the map. The value of w_{ij} indicates how strongly concept C_i influences concept C_j. The sign of w_{ij} indicates whether the relationship between concepts C_i and C_j is direct or inverse.

For the FCM of Figure 13.1 matrix W is equal to

$$W = \begin{bmatrix} d_{11} & w_{12} & w_{13} & w_{14} & 0 & w_{16} & 0 & 0 \\ w_{21} & d_{22} & 0 & 0 & 0 & w_{26} & 0 & w_{28} \\ w_{31} & 0 & d_{33} & w_{34} & w_{35} & 0 & w_{37} & 0 \\ w_{41} & 0 & w_{43} & d_{44} & 0 & 0 & 0 & 0 \\ w_{51} & 0 & 0 & 0 & d_{55} & 0 & w_{57} & 0 \\ 0 & 0 & w_{63} & w_{64} & w_{65} & d_{66} & 0 & w_{68} \\ 0 & 0 & w_{73} & 0 & 0 & 0 & d_{77} & w_{78} \\ 0 & w_{82} & 0 & 0 & 0 & w_{86} & w_{87} & d_{88} \end{bmatrix} .$$

The equation that calculates the values of concepts of Fuzzy Cognitive Networks, may or may not include self-feedback. In its general form it can be written as:

$$A_i(k) = f\left(\sum_{\substack{j=1 \\ j\neq i}}^{n} w_{ji}A_j(k-1) + d_{ii}A_i(k-1)\right), \tag{13.1}$$

where $A_i(k)$ is the value of concept C_i at discrete time k, $A_i(k-1)$ the value of concept C_i at discrete time $k-1$ and $A_j(k-1)$ is the value of concept C_j at discrete time $k-1$. w_{ji} is the weight of the interconnection from concept C_j to concept C_i and d_{ii} is a variable that takes on values in the interval [0, 1], depending upon the existence of "strong" or "weak" self-feedback to node i. Repetitive application of (13.1) for each node A_i will probably lead the FCN in an equilibrium point. Alternatively, it may present a limit cycle or a chaotic behavior.

Regarding the functions f used in FCMs, the following functions are usually found in the literature, allowing also different interpretation of their results:

The bivalent function [50, 52]

$$f(x) = \begin{cases} 1 & x > 0 \\ 0 & x \leq 0 \end{cases}$$

The use of this function allows the activation of each concept to either 0 or 1, leading to the development of binary FCMs, where each concept is either activated or not.

The trivalent function [50, 52]

$$f(x) = \begin{cases} 1 & x > 0 \\ 0 & x = 0 \\ -1 & x < 0 \end{cases}$$

In this case when the concept takes the value 1, it means that this concept increases, when the activation level equals -1, it means that the concept decreases and when level equals to 0, it means that the concept is remaining stable.

The hyperbolic tangent or sigmoid function [52] with saturation level -1 and 1. In a general form it can be written as

$$f(x) = \tanh(c_l x) \quad \text{or} \quad f(x) = \frac{e^{2c_l x} - 1}{e^{2c_l x} + 1},$$

where $0 < c_l \leq 1$ is a number used to adjust the inclination of the sigmoid function. This function squashes the result in the interval $[-1, 1]$.

The sigmoid function with saturation level 0 and 1 (log-sigmoid). f is a sigmoid function commonly used in the Fuzzy Cognitive Maps, which squashes the result in the interval [0, 1] and is expressed as

$$f = \frac{1}{1+e^{-c_l x}}.$$

where $0 < c_l \leq 1$ is used to adjust its inclination.

Equation (13.1) can be rewritten as:

$$A(k) = f(W^T A(k-1)). \tag{13.2}$$

In the next section we will derive conditions which determine the existence of a unique solution of (13.2), when continuous differentiable transfer functions f are used.

13.3 Existence and uniqueness of solutions in Fuzzy Cognitive Maps

In this section we check the existence of solutions in Equation (13.2), when a continuous and differentiable transfer function is used, such as sigmoid functions are. We know that the allowable values of the elements of FCM vectors A lie either in the closed interval [0, 1] or int he closed interval [−1, 1]. This is a subset of $\Re$ and is a complete metric space with the usual l_2 metric. We will define the regions where the FCM has a unique solution, which does not depend on the initial condition since it is the unique equilibrium point.

13.3.1 The Contraction Mapping Principle

We now introduce the Contraction Mapping Theorem [57].

Definition 1. *Let X be a metric space, with metric d. If φ maps X into X and there is a number $0 < c < 1$ such that*

$$d(\varphi(x), \varphi(y)) \leq cd(x, y) \tag{13.3}$$

for all $x, y \in X$, then φ is said to be a contraction of X into X.

Theorem 1 (see [57]). *If X is a complete metric space, and if φ is a contraction of X into X, then there exists one and only one $x \in X$ such that $\varphi(x) = x$.*

In other words, φ has a unique fixed point. The uniqueness follows from the fact that if $\varphi(x) = x$ and $\varphi(y) = y$, then (13.3) gives $d(x, y) \leq cd(x, y)$, which can only happen when $d(x, y) = 0$ (see [57]).

Equation (13.2) can be written as:

$$A(k) = G(A(k-1)) \tag{13.4}$$

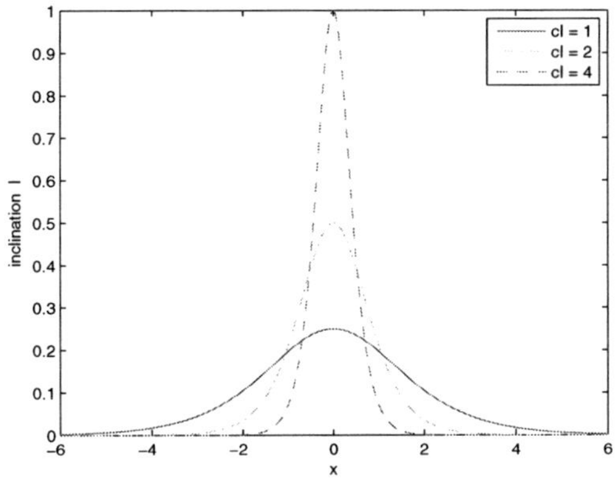

Fig. 13.2 Inclination of sigmoid function $f = 1/(1 + e^{-c_l x})$ when $c_l = 1, 2, 4$.

where $G(A(k-1))$ is equal to $f(WA(k-1))$.

In FCMs $A \in [0, 1]^n$ or $A \in [-1, 1]^n$ and it is also clear according to (13.2) that $G(A(k-1)) \in [0, 1]^n$ or $G(A(k-1)) \in [-1, 1]^n$ depending on which squashing sigmoid function is used. If the following inequality is true:

$$d(G(A), G(A')) \leq cd(A, A'),$$

where A and A' are different vectors of concept values and G is defined in (13.4), then G has a unique fixed point A such that:

$$G(A) = A.$$

Before presenting the main theorem we need to explore the role of f as a contraction function.

Theorem 2. *The scalar sigmoid function f, ($f = 1/(1 + e^{-x})$) is a contraction of the metric space X into X, were $X = [a, b], a, b$, finite, according to Definition 1, where*

$$d(f(x), f(y)) \leq cd(x, y) \tag{13.5}$$

Proof. Here f is the sigmoid function, $x,y \in X$, X is as defined above and c is a real number such that $0 < c < 1$.

The inclination l of a sigmoid function f is equal to:

$$l = \frac{\partial f}{\partial x} = \frac{c_l e^{-c_l x}}{\left(1 + e^{-c_l x}\right)^2} = \frac{c_l}{e^{c_l x}} \left(\frac{1}{1 + e^{-c_l x}}\right)^2 = \frac{c_l}{e^{c_l x}} f^2 \tag{13.6}$$

for $x \in X$. Equation (13.6) for $c_l = 1, 2, 4$ is plotted in Figure 13.2. According to Equation (13.6) one can see that the inclination l of $f(x)$ in the bounded set X depends on c_l and x. In particular, taking into account Figure 13.2 one can con-

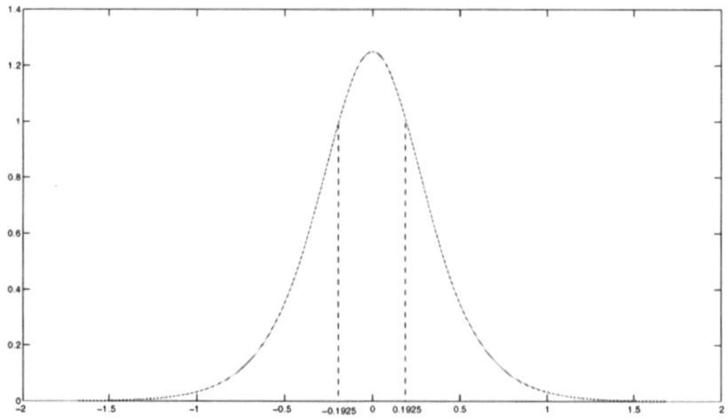

Fig. 13.3 Inclination of sigmoid function $f = 1/(1 + e^{-c_l x})$ when $c_l = 5$.

clude that when $c_l < 4$ the inclination is always smaller than 1 regardless of the value of x. Consequently, for the sigmoid with $c_l \geq 4$ the contraction mapping is not valid for every x. There is an interval, for which it is not valid. Figure 13.3 shows the inclination of the sigmoid function when $c_l = 5$. It can be seen that when $-0.1925 < x < 0.1925$ the inclination exceeds 1. In this interval, if one wants to keep the contraction property of the sigmoid used he should probably consider lowering c_l.

In deriving the results of this chapter it is assumed for simplicity that $c_l = 1$. Similar approach can be applied for other values of c_l giving analogous results. According to Figure 13.2 when $c_l = 1$ it always holds that

$$\frac{1}{4} \geq l \tag{13.7}$$

and for any x, y

$$\frac{d\,(f(x), f(y))}{d\,(x, y)} \leq 1/4. \tag{13.8}$$

From (13.7) and (13.8) we get:

$$\frac{d\,(f(x), f(y))}{d\,(x, y)} < 1. \tag{13.9}$$

Thus there is always a number c for which $0 \leq c < 1$, such that (13.9) is

$$\frac{d\,(f(x), f(y))}{d\,(x, y)} < c < 1. \tag{13.10}$$

Theorem 2 can be easily expanded for the continuous and differentiable sigmoid function $f = \tanh(c_l x)$. The inclination l of $f = \tanh(c_{li} x)$ is equal to:

$$l = c_l(1 - f^2)$$

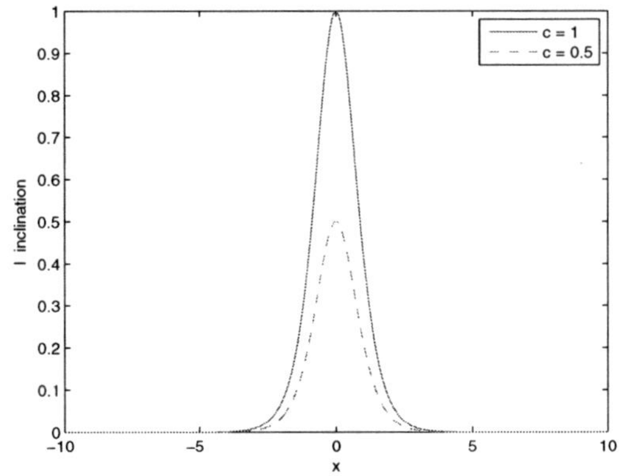

Fig. 13.4 Inclination of sigmoid function $f = \tanh(c_l x)$ when $c_l = 1$ and $c_l = 0.5$ respectively.

and its plot for $c_l = 1$ and $c_l = 0.5$ is given in Figure 13.4. According to Figure 13.4 one can see that for $c_l = 1$ the inclination l of $f(x)$ in the bounded set X is always smaller than 1. Thus for the hyperbolic tangent function we get

$$\frac{d\left(f(x), f(y)\right)}{d\left(x, y\right)} \leq 1,$$

and there is always a number c, such that

$$\frac{d\left(f(x), f(y)\right)}{d\left(x, y\right)} \leq c \leq 1.$$

□

In case where $0 < c \leq 1$, the map f becomes non-expansive and the following theorem holds:

Theorem 3. *The scalar sigmoid function* f, $(f = 1/(1 + e^{-x}))$ *is a non-expansive map of the metric space* X *into* X, *were* $X = [a, b]$, a, b *finite, where*

$$d(f(x), f(y)) \leq cd(x, y) \tag{13.11}$$

and $0 < c \leq 1$.

Proof. Use the results in the previous theorem together with the Browder–Gohde–Kirk theorem [58, p. 52] for non-expansive maps in Hilbert spaces. Here it should be noted that a solution exists but *is not unique*. □

Theorem 4. *There is one and only one solution for any concept value* A_i *of any FCM where the sigmoid function* $f = 1/(1 + e^{-x})$ *is used, if:*

$$\left(\sum_{i=1}^{n} \|w_i\|^2\right)^{1/2} < 4, \qquad (13.12)$$

where w_i *is the* i_{th} *row of matrix* W^T *and* $\|w_i\|$ *is the* l_2 *norm of* w_i.

If

$$\left(\sum_{i=1}^{n} \|w_i\|^2\right)^{1/2} = 4, \qquad (13.13)$$

there exists at least one solution for each concept value A_i *of any FCM, which is* not necessarily unique.

Proof. Let X be the complete metric space $[a,b]^n$ and $G : X \to X$ be a map such that

$$d(G(A), G(A')) \leq c d(A, A') \qquad (13.14)$$

for some $0 < c < 1$.

Vector G is equal to:

$$G = \left[\, f(w_1 \cdot A)\ f(w_2 \cdot A)\ \ldots\ f(w_n \cdot A) \,\right]^T, \qquad (13.15)$$

where n is the number of concepts of the FCM, f is the sigmoid function $f = 1/(1+e^{-x})$, w_i is the ith row of matrix W^T of the FCM, where $i = 1, 2, \ldots, n$, and by $\cdot$ we denote the inner product between two equidimensional vectors which both belong in $\Re^n$.

Assume A and A' are two different concept values for the FCM. Then we want to prove the following inequality:

$$\|G(A) - G(A')\| \leq c\|A - A'\|. \qquad (13.16)$$

But $\|G(A) - G(A')\|$ according to (13.15) is equal to

$$\|G(A) - G(A')\| = \left(\sum_{i=1}^{n} (f(w_i \cdot A) - f(w_i \cdot A'))^2\right)^{1/2}.$$

According to Theorem 2 for the scalar argument of $f(.)$, which is $w_i \cdot A$ in the bounded and closed interval $[a,b]$ with a and b being finite numbers, it is true that

$$|f(w_i \cdot A) - f(w_i \cdot A')| \leq c'_i |(w_i \cdot A) - (w_i \cdot A')|$$

for every $i = 1, 2, \ldots, n$.

Thus

$$|f(w_i \cdot A) - f(w_i \cdot A')| \leq c' |(w_i \cdot A) - (w_i \cdot A')|,$$

where $c' = \max(c'_1, c'_2, \ldots, c'_n)$.

By using the Cauchy–Schwartz inequality we get

$$c'|(w_i \cdot A) - (w_i \cdot A')| = c'|w_i \cdot (A - A')| \leq c'\|w_i\| \, \|A - A'\|.$$

Subsequently, we get:

$$\|G(A) - G(A')\| = \left(\sum_{i=1}^{n} (f(w_i \cdot A) - f(w_i \cdot A'))^2 \right)^{1/2}$$

$$\Rightarrow \|G(A) - G(A')\| \leq \left(\sum_{i=1}^{n} \left(c'\|w_i\| \, \|A - A'\| \right)^2 \right)^{1/2}.$$

Finally:

$$\|G(A) - G(A')\| \leq c'\|A - A'\| \left(\sum_{i=1}^{n} \|w_i\|^2 \right)^{1/2}.$$

A necessary condition for the above to be a contraction is

$$c' \left(\sum_{i=1}^{n} \|w_i\|^2 \right)^{1/2} < 1 \tag{13.17}$$

From Equation (13.7) we have that $c' \leq 1/4$, so that condition of Equation (13.17) now becomes

$$\left(\sum_{i=1}^{n} \|w_i\|^2 \right)^{1/2} < 4. \tag{13.18}$$

The proof of the equality according to (13.13) is similar, using the results of Theorem 3. □

Remark 1. Using the same analysis one can prove that for the hyperbolic tangent sigmoid function $f = \tanh(x)$ Equations (13.12) and (13.13) become:

$$\left(\sum_{i=1}^{n} \|w_i\|^2 \right)^{1/2} < 1 \tag{13.19}$$

and

$$\left(\sum_{i=1}^{n} \|w_i\|^2 \right)^{1/2} = 1 \tag{13.20}$$

respectively.

13.3.2 Exploring the Results

Suppose we have an FCM using log-sigmoid function with $c_l = 1$. In case the FCM has two or three nodes inequality (13.12) is *always true* because $|w_{ij}|_{i \neq j} \leq 1$ and d_{ii} can at most take the value of 1. FCMs with four or more nodes need more attention and are examined below.

13.3.2.1 An FCM with Four Concepts

Suppose that we have an FCM with four nodes. The weight matrix W_4 of this FCM is:

$$W_4 = \begin{bmatrix} d_{11} & w_{12} & w_{13} & w_{14} \\ w_{21} & d_{22} & w_{23} & w_{24} \\ w_{31} & w_{32} & d_{33} & w_{34} \\ w_{41} & w_{42} & w_{43} & d_{44} \end{bmatrix}.$$

The square root of the sum of the square l_2 norm of each row of matrix W_4^T is equal to

$$\sqrt{\sum_{i=1}^{4} \|w_i\|^2} = \sqrt{\|w_1\|^2 + \|w_2\|^2 + \|w_3\|^2 + \|w_4\|^2}. \tag{13.21}$$

The l_2 norm of each row is equal to

$$\|w_i\| = \sqrt{\sum_{j=1}^{4} w_{ij}^2},$$

where i denotes the ith row of matrix W_4^T and j denotes the column index. Equation (13.21) is now

$$\begin{aligned} \sqrt{\sum_{i=1}^{4} \|w_i\|^2} &= \sqrt{\|w_1\|^2 + \|w_2\|^2 + \|w_3\|^2 + \|w_4\|^2} \\ &= \sqrt{\sqrt{\sum_{j=1}^{4} w_{j1}^2}^{\,2} + \sqrt{\sum_{j=1}^{4} w_{j2}^2}^{\,2} + \sqrt{\sum_{j=1}^{4} w_{j3}^2}^{\,2} + \sqrt{\sum_{j=1}^{4} w_{j4}^2}^{\,2}} \\ &= \sqrt{\sum_{j=1}^{4} w_{j1}^2 + \sum_{j=1}^{4} w_{j2}^2 + \sum_{j=1}^{4} w_{j3}^2 + \sum_{j=1}^{4} w_{j4}^2} \end{aligned}$$

$$= \sqrt{\sum_{j=1}^{4}\left(\sum_{i=1}^{4} w_{ji}^{2}\right)} = \sqrt{\sum_{j=1}^{4}(d_{jj}) + \sum_{j=1}^{4}\left(\sum_{i=1,i\neq j}^{4} w_{ji}^{2}\right)}.$$

Since for the non-diagonal elements $|w_{ji}| < 1$, then $w_{ji}^2 < 1$.

Finally the above equation concludes to:

$$\sqrt{\sum_{j=1}^{4}(d_{jj}) + \sum_{j=1}^{4}\left(\sum_{i=1,i\neq j}^{4} w_{ji}^{2}\right)} \leq \sqrt{4+12} = 4.$$

Subsequently, we get:

$$\sqrt{\sum_{i=1}^{4} \|w_i\|^2} \leq 4.$$

According to Theorem 4 in order that only one solution exists for the concepts of an FCM the following inequality must be true:

$$\sqrt{\sum_{i=1}^{4} \|w_i\|^2} < 4.$$

If

$$\sqrt{\sum_{i=1}^{4} \|w_i\|^2} = 4,$$

a solution always exists which *is not necessarily unique*.

We finally obtain the next conclusion: Since generically most of the d_{ii} will be close to zero, "an FCM with four concepts has a unique solution generically".

13.3.2.2 An FCM with More Than Four Concepts

Suppose that we have an FCM with more than four nodes. The weight matrix W_n of the FCM is:

$$W_n = \begin{bmatrix} d_{11} & w_{12} & w_{13} & \dots & w_{1n} \\ w_{21} & d_{22} & w_{23} & \dots & w_{2n} \\ w_{31} & w_{32} & d_{33} & \dots & w_{3n} \\ \dots & \dots & \dots & \dots & \dots \\ w_{n1} & w_{n2} & w_{n3} & \dots & d_{nn} \end{bmatrix},$$

where $n > 4$. The square root of the sum of the square l_2 norm of each row of matrix W_n^T is given by

$$\sqrt{\sum_{i=1}^{n}\|w_i\|^2} = \sqrt{\|w_1\|^2+\|w_2\|^2+\cdots+\|w_n\|^2}$$

$$\Rightarrow \sqrt{\sum_{i=1}^{n}\|w_i\|^2} = \sqrt{\sum_{j=1}^{n}\left(\sum_{i=1}^{n} w_{ji}^2\right)}$$

$$\Rightarrow \sqrt{\sum_{i=1}^{n}\|w_i\|^2} = \sqrt{\sum_{j=1}^{n}(d_{jj})+\sum_{j=1}^{n}\left(\sum_{i=1,i\neq j}^{n} w_{ji}^2\right)}$$

$$\Rightarrow \sqrt{\sum_{i=1}^{n}\|w_i\|^2} \leq \sqrt{\sum_{j=1}^{n}(d_{jj})}+\sqrt{\sum_{j=1}^{n}\left(\sum_{i=1,i\neq j}^{n} w_{ji}^2\right)}$$

Finally we conclude that for an FCM with $n > 4$ concepts, Theorem 4 is true when:

$$\sqrt{\sum_{j=1}^{n}\left(\sum_{i=1,i\neq j}^{n} w_{ji}^2\right)} < 4-\sqrt{\sum_{j=1}^{n}(d_{jj})}. \tag{13.22}$$

When

$$\sqrt{\sum_{j=1}^{n}\left(\sum_{i=1,i\neq j}^{n} w_{ji}^2\right)} = 4-\sqrt{\sum_{j=1}^{n}(d_{jj})} \tag{13.23}$$

a *not necessarily unique solution* always exists.

Therefore, when $n > 4$ the condition for the uniqueness of solution of (13.2) depends on the number and the strength of diagonal d_{ii} elements of the FCM that are non-zero and the size of the FCM.

13.3.3 Interpreting the Results

Theorem 4 provides conditions that the weight interconnections of the FCM should fulfill in oder that it has an equilibrium condition and that condition be unique. It is clear that the above results hold only for continuous differentiable squashing functions like the log-sigmoid and the hyperbolic tangent functions. Therefore, the results are not valid for FCMs equipped with bivalent or trivalent functions; we can not conclude anything for this type of FCMs using the above analysis. Equations (13.12), (13.13) are valid for FCMs having log-sigmoid functions. Equation (13.12) provides with conditions for the existence and uniqueness of an equilibrium condition, while Equation (13.13) refers to condition that guarantees only the exist-

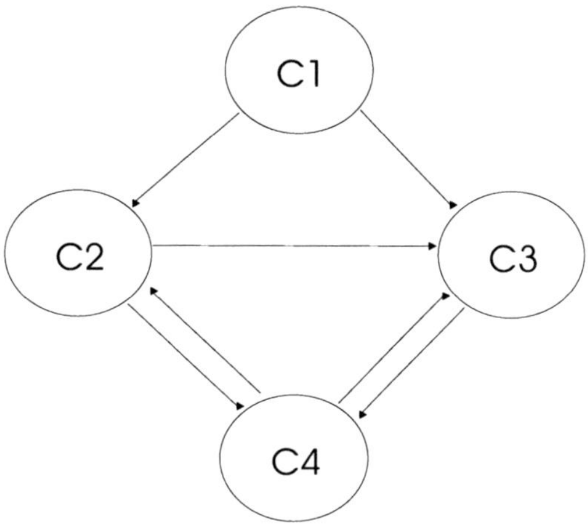

Fig. 13.5 FCM with one input node.

ence. Similar respective results arise from Equations (13.19) and (13.20) for FCMs equipped with hyperbolic tangent functions. In case none of the conditions hold, this does not imply that an equilibrium does not exist. It may or may not exist, just there is not any guarantee for its existence.

It can be observed and further verified by the exploration given in the previous section that the conditions provide with upper bounds for the FCM weights, which depend on the size of the FCM. Moreover, taking into account the observations made in the proof of Theorem 2, it can be concluded that these upper bounds depend also on the inclination of the sigmoid used. The larger the inclination becomes the smaller these bounds are, so that in the limit no weight set can be found to guarantee the uniqueness or the existence of the equilibrium. This is however expected, because sigmoids with large inclinations tend to reach the behavior of bivalent or trivalent functions and therefore no guarantee can exist based on the above analysis. FCMs that use sigmoids with large inclinations (or in the limit bivalent or trivalent functions) tend to give mainly qualitative results, while FCMs using small inclination sigmoids give both quantitative and qualitative results (see also the relevant conclusions in [52]). This is again mathematically expected since sigmoids with large inclination have very low discriminative abilities because they produce very similar outputs for inputs that may be quite dissimilar. For FCMs with sigmoids having small inclinations it is much easier to find weight sets that fulfill the existence and uniqueness conditions.

13.3.4 FCM with Input Nodes

So far we have not considered the existence of input nodes. This kind of nodes is also called "steady" nodes in [49] in the sense that they influence but are not influenced by the other nodes of the FCM. We will now show that the results obtained in the previous section are still valid. For the FCM of Figure 13.5 C_1 is such an input (or steady) node. Its weight matrix W is equal to:

$$W = \begin{bmatrix} 0 & w_{12} & w_{13} & 0 \\ 0 & d_{22} & w_{23} & w_{24} \\ 0 & 0 & d_{33} & w_{34} \\ 0 & w_{42} & w_{43} & d_{44} \end{bmatrix},$$

while vector A containing the node values is

$$A = \begin{bmatrix} U_1 & A_2 & A_3 & A_4 \end{bmatrix}^T$$

For the FCM of Figure 13.5, matrix G in Equation (13.15) now assumes the following form:

$$G = \begin{bmatrix} U_1 \; f(w_2 \cdot A) \; f(w_3 \cdot A) \; f(w_4 \cdot A) \end{bmatrix}^T,$$

where U_1 is the input to FCM nodes. In a more general form matrix G can be written as

$$G = \begin{bmatrix} U_1 \; U_2 \ldots U_m \; f(w_{m+1} \cdot A) \; f(w_{m+2} \cdot A) \ldots f(w_{m+n} \cdot A) \end{bmatrix}^T \qquad (13.24)$$

corresponding to vector $A = [\, U_1 \; U_2 \cdots U_m \; A_{m+1} \; A_{m+2} \cdots A_{m+n} \,]^T$, where m is the number of inputs and n is the number of the other concept nodes in FCM. Under this definition, Equation (13.4) assumes the same form:

$$A(k) = G(A(k-1)). \qquad (13.25)$$

The next theorem proves that for matrix G and vector A defined above, the results of Theorem 4 are still valid.

Theorem 5. *For an FCM with input nodes, with its concept values driven by (13.25), where G is described in (13.24), there is one and only one solution for any concept value A_i if Equation (13.12) is fullfiled, that is:*

$$\left(\sum_{i=1}^{n} \| w_{m+i} \|^2 \right)^{1/2} < 4,$$

where w_{m+i} is the $(m+i)$-th row of matrix W^T and $\|w_{m+i}\|$ is the l_2 norm of w_{m+i}.

If

$$\left(\sum_{i=1}^{n} \| w_{m+i} \|^2 \right)^{1/2} = 4$$

a not always unique *solution exists.*

Proof. Assume A and A' are two different concept values for the FCM having one or more inputs. Then we want to prove again inequality (13.16), that is:

$$\|G(A) - G(A')\| \le c\|A - A'\|$$

But since input node values are not influenced by the other nodes of the FCM $\|G(A) - G(A')\|$ according to (13.24) is equal to:

$$\|G(A) - G(A')\| = \left(\sum_{i=1}^{m} (U_i - U_i)^2 + \sum_{i=1}^{n} (f(w_{m+i} \cdot A) - f(w_{m+i} \cdot A'))^2\right)^{1/2},$$

where m is the number of inputs and n is the number of the other nodes in the FCM. The above equation is equivalent to the following:

$$\|G(A) - G(A')\| = \left(0 + \sum_{i=1}^{n} (f(w_{m+i} \cdot A) - f(w_{m+i} \cdot A'))^2\right)^{1/2}.$$

Therefore, $\|G(A) - G(A')\|$ assumes quite the same form with that appearing in Theorem 4 leading to the same condition, that is:

$$\left(\sum_{i=1}^{n} \|w_{m+i}\|^2\right)^{1/2} < 4.$$

Similarly, for the $=$ case the respective results of Theorem 4 still hold and the results obtained for the hyperbolic tangent function are still valid. □

The results of the above theorem should not be misinterpreted. Similar to Theorem 4, Theorem 5 states that the weights have to comply with the bounds implied by its conditions in order to assure the existence or the uniqueness of the FCM equilibrium. However, when these conditions are fulfilled the unique equilibrium does not depend solely on the weight set, as in the case of FCMs with no input nodes. It depends also on the values of the input nodes. This is a quite reasonable implication because the input node values are steady and unlike the other node values they are not changing during the repetitive application of Equation (13.1) until the FCM reaches its equilibrium. Therefore, they act as an external excitation. Different values of the excitation will drive the FCM in different equilibria. The results obtained so far are quite significant. They are valid for a class of sigmoid functions, which may present contractive or non-expansive properties and are frequently used in FCMs. Many physical systems can reach an equilibrium point after initial perturbations. If FCMs are used for the representation of such systems, it is important to know, which FCM structure may guarantee the existence and probably the uniqueness of the equilibrium. This gives rise also to the development of estimation algorithms, which can learn the structure (weights) based on real system's data. Such an algorithm will be given in the next section. Moreover, in FCMs with input nodes this unique equilibrium may be different depending on the value of the exciting input, a behavior quite similar with the behavior of stable nonlinear physical systems. Therefore FCMs could be used not only for modeling but also for the control of unknown nonlinear systems.

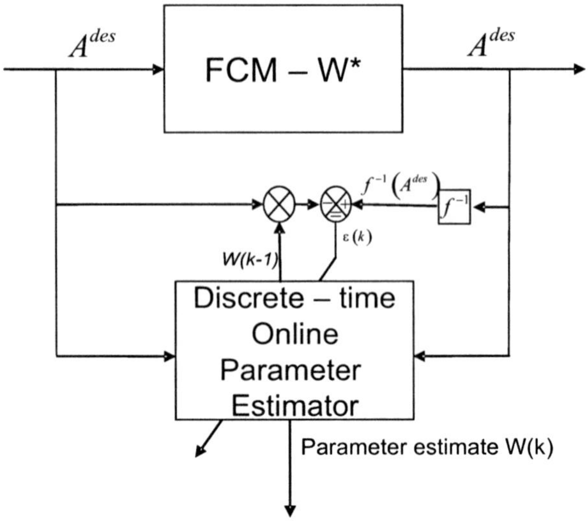

Fig. 13.6 Online parameter estimator designed for FCMs.

13.4 Online Parameter Estimation of Fuzzy Cognitive Maps

Based on the results and observations of Section 13.3 we are now proposing a method of finding appropriate weight sets related to a desired equilibrium point of the FCM. Choosing a desired state A^{des} for the FCM this is equivalent to solving the equation

$$A^{\mathrm{des}} = f(W^T A^{\mathrm{des}}) \tag{13.26}$$

in respect to W^T, where $A^{\mathrm{des}} = [A_1^{\mathrm{des}}, A_2^{\mathrm{des}}, \ldots, A_n^{\mathrm{des}}]^T$ and f is a vector valued function $f : \Re^n \rightarrow \Re$, defined as follows: $f(x) = [f_1(x_1), f_2(x_2), \cdots f_n(x_n)]^T$, where $x \in \Re^n$ and $f_i(x_i) = 1/(1 + e^{-x_i})$, for $i = 1, 2, \ldots, n$. Then,

$$f^{-1}(A^{\mathrm{des}}) = W^{*T} A^{\mathrm{des}} \tag{13.27}$$

where $f^{-1}(A^{\mathrm{des}}) = [f_1^{-1}(A_1^{\mathrm{des}}), f_2^{-1}(A_2^{\mathrm{des}}), \ldots, f_n^{-1}(A_n^{\mathrm{des}})]^T$ and $f_i^{-1}(A_i^{\mathrm{des}}) = W_i^T \cdot A^{\mathrm{des}}$, with W_i being the ith column of W^*.

In order to solve the above equation we use an adaptive estimation algorithm [59] including the parameter constraints derived in Section 13.3. The updating scheme is illustrated in Figure 13.6, demonstrating that the weight learning is performed adaptively based on the desired equilibrium point of the FCM. Taking into account that $f_i(x_i)$ is the sigmoid function weight updating laws are given as follows:

The error $\varepsilon_i(k)$ of the parametric discrete-time adaptive law is of the form:

$$\varepsilon_i(k) = \frac{f_i^{-1}(A_i^{\mathrm{des}}) - W_i^T(k-1)A^{\mathrm{des}}}{c + (A^{\mathrm{des}})^T A^{\mathrm{des}}} \tag{13.28}$$

while the updating algorithm is given by

$$W_i(k) = W_i(k-1) + \alpha\varepsilon_i(k)A^{\mathrm{des}}, \tag{13.29}$$

where $W_i(k)$ is the ith column of $W(k)$, which is the estimator of $W^*(k)$. A^{des} is constant vector and $f_i^{-1}(A_i^{\mathrm{des}})$ is also constant and scalar. $\alpha > 0$ and $c > 0$ are design parameters.

By using the above updating algorithm we are confident that the estimator converges to W^*. In Section 13.3 we proved that if inequality (13.12) is true then W^* corresponding to the designed FCM provide a unique solution and satisfies (13.26).

Proof. From (13.28), (13.29) and $\tilde{W}_i(k) = W_i(k) - W_i^*$ we obtain the error equation

$$\tilde{W}_i(k) = \left[I - \frac{aA^{\mathrm{des}}(A^{\mathrm{des}})^T}{c + (A^{\mathrm{des}})^T A^{\mathrm{des}}}\right]\tilde{W}(k-1), \tag{13.30}$$

$$\varepsilon_i(k) = -\frac{\tilde{W}_i^T(k-1)A^{\mathrm{des}}}{c + (A^{\mathrm{des}})^T A^{\mathrm{des}}}. \tag{13.31}$$

Consider the function

$$V(k) = \frac{1}{2a}\tilde{W}^T(k)\tilde{W}(k). \tag{13.32}$$

Then

$$V(k) - V(k-1) = \frac{1}{2a}\tilde{W}_i{}^T(k)\tilde{W}_i(k) - \frac{1}{2a}\tilde{W}_i{}^T(k-1)\tilde{W}_i(k-1)$$
$$= -\frac{\tilde{W}_i^T(k-1)A^{\mathrm{des}}(A^{\mathrm{des}})^T\tilde{W}_i(k-1)}{c + (A^{\mathrm{des}})^T A^{\mathrm{des}}}$$
$$+ \frac{\tilde{W}_i^T(k-1)A^{\mathrm{des}}(A^{\mathrm{des}})^T aA^{\mathrm{des}}(A^{\mathrm{des}})^T\tilde{W}_i(k-1)}{2(c + (A^{\mathrm{des}})^T A^{\mathrm{des}})^2}.$$

Using $\varepsilon_i(k)(c + (A^{\mathrm{des}})^T A^{\mathrm{des}}) = -\tilde{W}_i^T(k-1)A^{\mathrm{des}}$, we obtain

$$V(k) - V(k-1) = -\varepsilon_i^2(k)(c + (A^{\mathrm{des}})^T A^{\mathrm{des}})\left[1 - \frac{a(A^{\mathrm{des}})^T A^{\mathrm{des}}}{2(c + (A^{\mathrm{des}})^T A^{\mathrm{des}})}\right].$$

Since $a, c > 0$ we can always choose $0 < a < 2$ such that

$$\frac{a(A^{\mathrm{des}})^T A^{\mathrm{des}}}{2(c + (A^{\mathrm{des}})^T A^{\mathrm{des}})} < 2.$$

It then follows that

$$V(k) - V(k-1) \leq -c_0\varepsilon_i^2(k)(c + (A^{\mathrm{des}})^T A^{\mathrm{des}}) \leq 0 \tag{13.33}$$

for some constant $c_0 > 0$. From (13.32), (13.33) we have that $V(k)$ and therefore $W_i(k) \in \ell_\infty$ and $V(k)$ has a limit, i.e., $\lim_{k\to\infty} V(k) = V_\infty$. Consequently, using (13.33) we obtain

$$c_0 \sum_{k=1}^{\infty} (\varepsilon_i^2(k)(c + (A^{\text{des}})^T A^{\text{des}})) \leq V(0) - V_\infty < \infty,$$

which implies $\varepsilon_i(k)\sqrt{(c + (A^{\text{des}})^T A^{\text{des}})} \in \ell_2$ and $\varepsilon_i(k)\sqrt{(c + (A^{\text{des}})^T A^{\text{des}})} \to 0$ as $k \to \infty$. Since $\sqrt{(c + (A^{\text{des}})^T A^{\text{des}})} \geq c > 0$, we also have that $\varepsilon_i(k) \in \ell_2$ and $\varepsilon_i(k) \to 0$ as $k \to \infty$. We have

$$\varepsilon_i(k)A^{\text{des}} = \varepsilon_i(k)\sqrt{(c + (A^{\text{des}})^T A^{\text{des}})} \frac{A^{\text{des}}}{\sqrt{(c + (A^{\text{des}})^T A^{\text{des}})}}.$$

Since $A^{\text{des}}/\sqrt{(c + (A^{\text{des}})^T A^{\text{des}})}$ is bounded and $\varepsilon_i(k)\sqrt{(c + (A^{\text{des}})^T A^{\text{des}})} \in \ell_2$, we have that $\varepsilon_i(k)A^{\text{des}} \in \ell_2$ and $\|\varepsilon_i(k)A^{\text{des}}\| \to 0$ as $k \to \infty$. This implies (using (13.29)) that $\|W_i(k) - W_i(k-1)\| \in \ell_2$ and $\|W_i(k) - W_i(k-1)\| \to 0$ as $k \to \infty$. Now

$$\begin{aligned} W_i(k) - W_i(k-N) = {} & W_i(k) - W_i(k-1) + W_i(k-1) - W_i(k-2) + \\ & + \cdots + W_i(k-N+1) - W_i(k-N) \end{aligned}$$

for any finite N. Using the Schwartz inequality, we have

$$\begin{aligned} \|W_i(k) - W_i(k-N)\|^2 \leq {} & \|W_i(k) - W_i(k-1)\|^2 + \|W_i(k-1) - W_i(k-2)\|^2 + \\ & + \cdots + \|W_i(k-N+1) - W_i(k-N)\|^2. \end{aligned}$$

Since each term on the right-hand side of the inequality is in ℓ_2 and goes to zero with $k \to \infty$, it follows that $\|W_i(k) - W_i(k-N)\| \in \ell_2$ and $\|W_i(k) - W_i(k-N)\| \to 0$ as $k \to \infty$. □

Remark 2. The n weight updating laws given by (13.29) for each column W_i, $i = 1, \ldots, n$ of W can be written in a compact form as follows:

$$\underline{W}(k) = \underline{W}(k-1) + a\underline{\varepsilon}A^{\text{des}}, \qquad (13.34)$$

where $\underline{W} = [W_1^T, W_2^T, \ldots, W_n^T]^T$ is a $n^2 \times 1$ column vector containing all the elements of matrix W arranged in a column, row after row. Also, $\underline{\varepsilon}$ is a $n^2 \times n$ matrix having the form $\underline{\varepsilon} = [\varepsilon_1 I_n, \varepsilon_2 I_n, \ldots, \varepsilon_n I_n]^T$, where I_n is a $n \times n$ unit matrix and ε_i is given by (13.28).

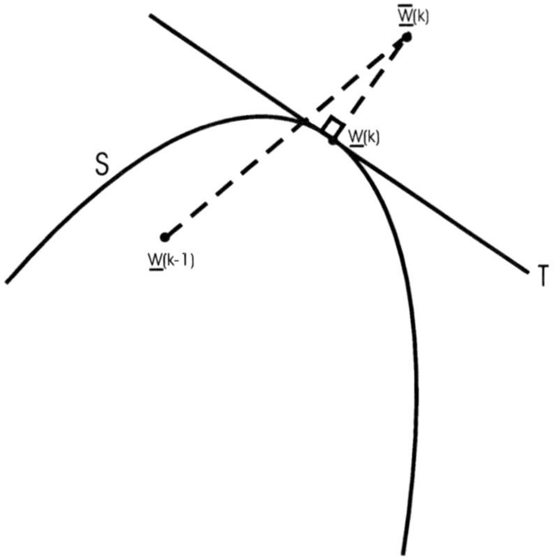

Fig. 13.7 Discrete time parameter projection.

13.4.1 Gradient Projection Method

In parameter identification problems, we have some a priori knowledge as to where the unknown parameter W^* is located in $\Re^n$. This knowledge could be the upper and the lower bounds for each element of W^*. In our parameter identification problem inequalities $-1 \le w_{ij} \le 1$ and (13.12) must be true. A convex projection method [59] is next illustrated to modify the adaptive laws of the previous section. A schematic representation of the orthogonal projection method is illustrated in Figure 13.7.

13.4.1.1 Projection Method 1

The S_1 set for projection that satisfies $-1 \le w_{ij} \le 1$ is the convex set:

$$S_1 = \{W \in \Re^n | g(w_{ij}) \le 0, \;\; g(w_{ij}) = |w_{ij}| - 1, \forall i, j \in \aleph\}.$$

The updating formula of parameters W of FCM is now given by

$$\begin{aligned} \overline{\underline{W}}(k) &= \underline{W}(k-1) + a\underline{\varepsilon} A^{\text{des}}, \\ \underline{W}(k) &= \begin{cases} \overline{\underline{W}}(k), & \text{if } \overline{\underline{W}}(k) \in S_1, \\ \Pr(\overline{\underline{W}}(k)), & \text{if } \overline{\underline{W}}(k) \notin S_1, \end{cases} \end{aligned} \tag{13.35}$$

where $\Pr(\overline{\underline{W}}(k)) = \perp$proj of $\overline{\underline{W}}(k)$ on S_1.

This formula can also be element-wise written as

$$\overline{w}_{ij}(k) = w_{ij}(k-1) + \alpha\varepsilon_i(k)\,A_i^{\text{des}}$$

$$w_{ij}(k) = \begin{cases} \overline{w}_{ij}(k), & \text{if } \left|\overline{w}_{ij}(k)\right| \le 1, \\ -1 & \text{if } \overline{w}_{ij}(k) < -1, \\ 1 & \text{if } \overline{w}_{ij}(k) > 1. \end{cases} \tag{13.36}$$

13.4.1.2 Projection Method 2

The S_2 set for projection that satisfies (13.12) is the convex set:

$$S_2 = \{\underline{W} \in \Re^{n^2} | g(\underline{W}) \le 0,\ \ g(\underline{W}) = \|\underline{W}\| - 4\}.$$

The updating equation of parameters W of FCM is now given by

$$\overline{\underline{W}}(k) = \underline{W}(k-1) + a\underline{\varepsilon}A^{\text{des}}$$

$$\underline{W}(k) = \begin{cases} \overline{\underline{W}}(k), & \text{if } \overline{\underline{W}}(k) \in S_2, \\ \Pr(\overline{\underline{W}}(k), & \text{if } \overline{\underline{W}}(k) \notin S_2, \end{cases} \tag{13.37}$$

where the orthogonal projection $\Pr(\overline{\underline{W}}(k)) = \perp\text{proj of } \overline{\underline{W}}(k)$ on S_2 is given by $4/\|\overline{\underline{W}}(k)\|\overline{\underline{W}}(k)$. Therefore, Equation (13.37) can also be written in the following compact form:

$$\overline{\underline{W}}(k) = \underline{W}(k-1) + a\underline{\varepsilon}A^{\text{des}}$$

$$\underline{W}(k) = \overline{\underline{W}}(k)\min\left(1, \frac{4}{\|\overline{\underline{W}}(k)\|}\right). \tag{13.38}$$

13.4.1.3 Concurrent Projection Method

In the previous subsections two projection methods were presented. Our objective is to combine both projection methods. The set S which satisfies both inequalities is the intersection of sets S_1 and S_2.

$$S = S_1 \bigcap S_2.$$

We replace the weight updating equation:

$$\underline{W}(k) = \underline{W}(k-1) + a\underline{\varepsilon}A^{\text{des}}$$

with

$$\overline{\underline{W}}(k) = \underline{W}(k-1) + a\underline{\varepsilon}A^{\text{des}}$$

$$\underline{W}(k) = \begin{cases} \overline{\underline{W}}(k), & \text{if } \overline{\underline{W}}(k) \in S, \\ \Pr(\overline{\underline{W}}(k), & \text{if } \overline{\underline{W}}(k) \notin S, \end{cases} \tag{13.39}$$

where $\Pr(\overline{\underline{W}}(k)) = \perp\text{proj of } \overline{\underline{W}}(k) \text{ on } S$.

13.5 Numerical Examples

For the numerical examples the FCM of Figure 13.1 is used. The initial W matrix of FCM is shown in Table 13.1. Except from the diagonal elements, which have the value 1 and the zero elements all other weights were randomly selected. This matrix fulfills Equation (13.12) because

$$\|\underline{W}\| = 3.856 < 4.$$

With this weights matrix the FCM reaches an equilibrium point given by the following vector A:

$$A = \begin{bmatrix} 0.8981\ 0.8147\ 0.9037\ 0.9387\ 0.9111\ 0.8704\ 0.8551\ 0.7666 \end{bmatrix}$$

In order to test the performance of the method in the extreme case where the self feedback weight connections are all one, we deliberately assumed that d_{ii}, $i = 1, \ldots, n$ are kept constant to $d_{ii} = 1$ and therefore they do not participate in the weight updating. The same holds for the weights with null value, because they represent non-existing connections.

13.5.1 Example 1

Suppose that for FCM of Figure 13.1 the desired state is equal to:

$$A^{\text{des}} = \begin{bmatrix} 0.76\ 0.58\ 0.69\ 0.69\ 0.58\ 0.71\ 0.63\ 0.78 \end{bmatrix}$$

Applying (13.28) and (13.29) or equivalently (13.34) the W matrix concludes to W_1 matrix of Table 13.1.

For the W_1 matrix, Equation (13.12) is true:

$$\left(\sum_{i=1}^{8} \|w_i\|^2 \right)^{1/2} = 3.2421 < 4.$$

Table 13.1 *W* matrices of numerical examples.

$$W = \begin{bmatrix} 1 & 0.4 & -0.5 & 0.7 & 0 & 0.1 & 0 & 0 \\ 0.1 & 1 & 0 & 0 & 0 & 0.5 & 0 & 0.2 \\ 0.2 & 0 & 1 & 0.9 & 0.7 & 0 & 0.7 & 0 \\ 0.5 & 0 & 0.8 & 1 & 0 & 0 & 0 & 0 \\ 0.6 & 0 & 0 & 0 & 1 & 0 & 0.4 & 0 \\ 0 & 0 & 0.5 & 0.4 & 0.9 & 1 & 0 & 0.2 \\ 0 & 0 & 0.7 & 0 & 0 & 0 & 1 & 0.1 \\ 0 & 0.4 & 0 & 0 & 0 & 0.7 & -0.1 & 1 \end{bmatrix}$$

$$W_1 = \begin{bmatrix} 1.0000 & -0.1596 & -0.8335 & 0.0527 & 0 & -0.2627 & 0 & 0 \\ -0.0771 & 1.0000 & 0 & 0 & 0 & 0.2232 & 0 & 0.2772 \\ -0.0107 & 0 & 1.0000 & 0.3123 & -0.2709 & 0 & 0.3432 & 0 \\ 0.2893 & 0 & 0.4972 & 1.0000 & 0 & 0 & 0 & 0 \\ 0.4229 & 0 & 0 & 0 & 1.0000 & 0 & 0.1001 & 0 \\ 0 & 0 & 0.1884 & -0.2048 & -0.0990 & 1.0000 & 0 & 0.2945 \\ 0 & 0 & 0.4235 & 0 & 0 & 0 & 1.0000 & 0.1838 \\ 0 & -0.1743 & 0 & 0 & 0 & 0.3277 & -0.5034 & 1.0000 \end{bmatrix}$$

$$W_2 = \begin{bmatrix} 1.0000 & 0.0141 & 0.4543 & -0.1199 & 0 & -0.2079 & 0 & 0 \\ 0.4438 & 1.0000 & 0 & 0 & 0 & 0.2819 & 0 & 0.3227 \\ 0.6904 & 0 & 1.0000 & 0.0716 & 0.1397 & 0 & 0.4161 & 0 \\ 0.7983 & 0 & 1.0000 & 1.0000 & 0 & 0 & 0 & 0 \\ 0.9943 & 0 & 0 & 0 & 1.0000 & 0 & 0.1717 & 0 \\ 0 & 0 & 1.0000 & -0.2405 & 0.4668 & 1.0000 & 0 & 0.3353 \\ 0 & 0 & 1.0000 & 0 & 0 & 0 & 1.0000 & 0.2317 \\ 0 & 0.0744 & 0 & 0 & 0 & 0.4402 & -0.3370 & 1.0000 \end{bmatrix}$$

$$W_3 = \begin{bmatrix} 1.0000 & 0.0873 & -0.2034 & 0.1830 & 0 & -0.1342 & 0 & 0 \\ 0.4217 & 1.0000 & 0 & 0 & 0 & 0.3286 & 0 & 0.5129 \\ 0.6304 & 0 & 1.0000 & 0.3937 & 0.9123 & 0 & 0.6765 & 0 \\ 0.8579 & 0 & 1.0000 & 1.0000 & 0 & 0 & 0 & 0 \\ 1.0000 & 0 & 0 & 0 & 1.0000 & 0 & 0.3770 & 0 \\ 0 & 0 & 0.7416 & -0.0210 & 1.0000 & 1.0000 & 0 & 0.5482 \\ 0 & 0 & 0.9599 & 0 & 0 & 0 & 1.0000 & 0.4747 \\ 0 & 0.1131 & 0 & 0 & 0 & 0.4851 & -0.1220 & 1.0000 \end{bmatrix}$$

$$W_4 = \begin{bmatrix} 1.0000 & 0.0904 & -0.0440 & 0.1882 & 0 & -0.0654 & 0 & 0 \\ 0.4561 & 1.0000 & 0 & 0 & 0 & 0.2987 & 0 & 0.5031 \\ 0.6641 & 0 & 1.0000 & 0.3584 & 0.9228 & 0 & 0.6110 & 0 \\ 0.8232 & 0 & 0.9248 & 1.0000 & 0 & 0 & 0 & 0 \\ 0.9689 & 0 & 0 & 0 & 1.0000 & 0 & 0.3665 & 0 \\ 0 & 0 & 0.7009 & 0.0151 & 0.9874 & 1.0000 & 0 & 0.5415 \\ 0 & 0 & 0.8857 & 0 & 0 & 0 & 1.0000 & 0.4891 \\ 0 & 0.1098 & 0 & 0 & 0 & 0.4340 & -0.0413 & 1.0000 \end{bmatrix}$$

13.5.2 Example 2

Suppose that for FCM of Figure 13.1 the desired state is:

$$A^{\text{des}} = \left[0.96\ 0.68\ 0.97\ 0.59\ 0.78\ 0.75\ 0.73\ 0.81\right]$$

Applying (13.28) and (13.36) the W matrix concludes to W_2 matrix of Table 13.1. During the updating procedure the value of weight w_{34} was moving outside the set S_1. By using the projection method given by (13.36) the weights were projected onto the set S_1 and now w_{34} has the value one, which is desirable.

For the W_2 matrix, Equation (13.12) is true:

$$\left(\sum_{i=1}^{8} \|w_i\|^2\right)^{1/2} = 3.8379 < 4.$$

13.5.3 Example 3

Suppose that for FCM of Figure 13.1 the desired is:

$$A^{\text{des}} = \left[0.97\ 0.71\ 0.95\ 0.79\ 0.93\ 0.79\ 0.85\ 0.89\right].$$

Applying (13.28) and (13.29) or equivalently (13.34) the W matrix concludes to W_3 matrix of Table 13.1.

For the W_3 matrix, Equation (13.12) is not true. Therefore, $g(\underline{W}) = \|\underline{W}\| - 4 > 0$, which means that $\underline{W} \notin S_2$.

Applying (13.28) and (13.39) we project the $\underline{W}$ matrix to set S, which is the intersection of S_1 and S_2. Matrix W finally concludes to W_4 matrix of Table 13.1.

For the W_4 matrix, Equation (13.12) is now true:

$$\|\underline{W}\| = 3.9982 < 4.$$

13.6 The Fuzzy Cognitive Network Approach

As shown in Section 13.4 the concepts values of the FCM with a specified matrix W have a unique solution as far as (13.12) and consequently (13.23) is fulfilled. The perspective of transforming FCMs into a modeling and control alternative requires, first to update its weight matrix W so that the FCM can capture different mappings of the real system and second to store these different kind of mappings. The Fuzzy Cognitive Network (FCN) [49] has been proposed as an operational extension framework of FCM, which updates its weights and reaches new equilibrium points based on the continuous interaction with the system it describes. Moreover, for each equilibrium point a fuzzy rule based storage mechanism of the form "If weights then fixed point" is provided, which facilitates and speeds-up its operation. FCNs constitute the proper framework exploiting the theoretical results obtained so far. The components of FCN are briefly presented below.

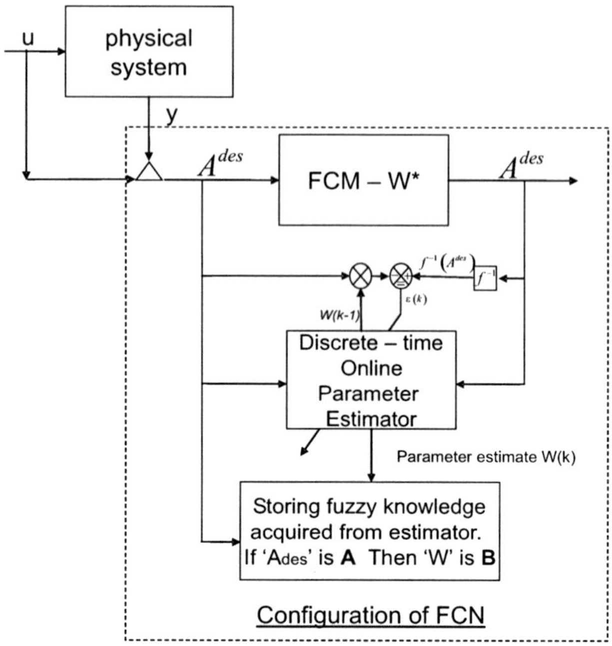

Fig. 13.8 Interactive operation of the FCN with the physical system.

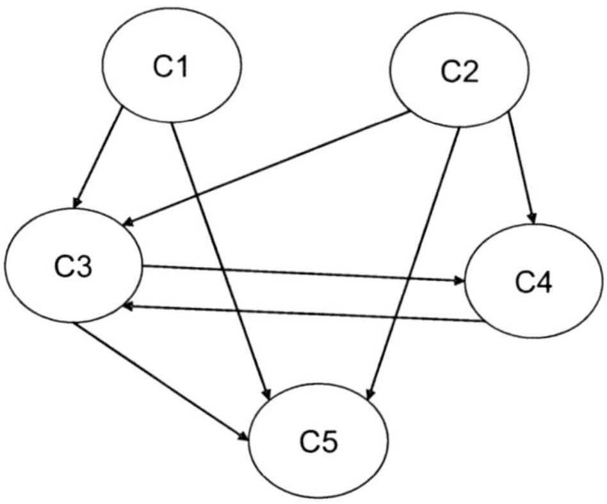

Fig. 13.9 An FCN with two input nodes.

13.6.1 Training Using Close Interaction with the Real System

The operation of the FCN in close cooperation with the real system it describes, might require continuous changes in the weight interconnections, depending on the convex input received from the real system. Figure 13.8 presents the interactive operation of the FCN with the physical system it describes. The weight updating procedure is described below.

13.6.2 Weight Updating Procedure

The updating method takes into account input u and output y from the real system. This u and y values combine to give the A^{des} vector which is equal to:

$$A^{\text{des}} = \begin{bmatrix} U_1 & \cdots & U_m & A_{m+1} & \cdots & A_{m+n} \end{bmatrix}^T$$

Using the updating algorithm described in Section 13.4 one estimates W^* matrix for which (13.26) is true.

13.6.3 Storing Knowledge from Previous Operating Conditions

The procedure described in the previous subsection modifies FCN's knowledge about the system by continuously modifying the weight interconnections and consequently the node values. During the repetitive updating operation the procedure uses input from the system variables, producing a new weight matrix for each new equilibrium state. It is desirable to device a storage mechanism for keeping these weights for probable future use in a decision making or control application. It would be also preferable this storage to allow weight retrieval even in situations, where the equilibrium conditions were not been exactly met during the training phase. To this end we propose storing the previous acquired operational situations in a fuzzy if-then rule database, which associates in a fuzzy manner the various weights with the corresponding equilibrium node values. The procedure is explained as follows.

Consider, for example, the FCN of Figure 13.9. All nodes of this FCN assume self feedback with strength 1, which however is not shown in the figure. Two nodes are input (steady) nodes. Let it also be that this FCN has a unique equilibrium point

$$A = \begin{bmatrix} 0.7 \; 0.8 \; 0.6 \; 0.67 \; 0.43 \end{bmatrix}^T,$$

which is associated with the weight matrix W:

$$W = \begin{bmatrix} 1.0000 & 0 & -0.0585 & 0 & 0.0307 \\ 0 & 1.0000 & 0.0188 & -0.2455 & -0.2507 \\ 0 & 0 & 1.0000 & 0.3909 & -0.8880 \\ 0 & 0 & -0.2517 & 1.0000 & 0 \\ 0 & 0 & 0 & 0 & 1.0000 \end{bmatrix}$$

in order that A is a unique solution of (13.26) weight matrix W has to be such that inequality (13.12) is fulfilled. For weight matrix W inequality (13.12) takes the form:

$$\left(\sum_{i=1}^{n} \|w_i\|^2 \right)^{1/2} = 2.4784 < 4,$$

where $n = 5$ is the number of concepts of the FCN.

Suppose also that the FCM in another operation point A' is related to weight matrix W' both given below.

$$A' = \left[0.6\ 0.7\ 0.6\ 0.67\ 0.53\right]^T,$$

$$W' = \begin{bmatrix} 1.0000 & 0 & -0.0513 & 0 & 0.1984 \\ 0 & 1.0000 & 0.0235 & -0.2368 & -0.0685 \\ 0 & 0 & 1.0000 & 0.3399 & -0.8016 \\ 0 & 0 & -0.2689 & 1.0000 & 0 \\ 0 & 0 & 0 & 0 & 1.0000 \end{bmatrix},$$

matrix W' also fulfills (13.12):

$$\left(\sum_{i=1}^{5} \|w_i'\|^2\right)^{1/2} = 2.4359 < 4.$$

The fuzzy rule database, which is obtained using the information from the two previous equilibrium points, is resolved as follows. There are two rules for the description of the above two different equilibrium situations:

Rule 1
if node C1 is mf1 *and* node C2 is mf1
then w_{13} is mf1 *and* w_{15} is mf1 *and* w_{23} is mf1 *and* w_{24} is mf1 *and* w_{25} is mf1 *and* w_{34} is mf1 *and* w_{35} is mf1 *and* w_{43} is mf1

Rule 2
if node C1 is mf2 *and* node C2 is mf2
then w'_{13} is mf2 *and* w'_{15} is mf2 *and* w'_{23} is mf2 *and* w'_{24} is mf2 *and* w'_{25} is mf2 *and* w'_{34} is mf2 *and* w'_{35} is mf2 *and* w'_{43} is mf2 where each membership function is derived from the two equilibrium points aquired from matrices A, W, A' and W' and presented below:

$$A = \left[0.7_{mf1}\ 0.8_{mf1}\ 0.6\ 0.67\ 0.43\right]^T$$

$$W = \begin{bmatrix} 1.0000 & 0 & -0.0585_{mf1} & 0 & 0.0307_{mf1} \\ 0 & 1.0000 & 0.0188_{mf1} & -0.2455_{mf1} & -0.2507_{mf1} \\ 0 & 0 & 1.0000 & 0.3909_{mf1} & -0.8880_{mf1} \\ 0 & 0 & -0.2517_{mf1} & 1.0000 & 0 \\ 0 & 0 & 0 & 0 & 1.0000 \end{bmatrix}$$

$$A' = \left[0.6_{mf2}\ 0.7_{mf2}\ 0.6\ 0.67\ 0.53\right]^T$$

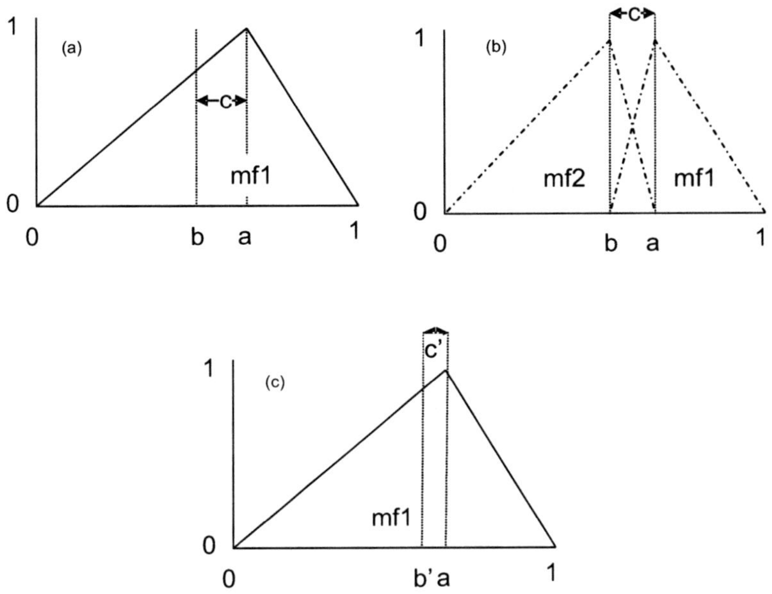

Fig. 13.10 Conditions for creating new triangular membership function. (a) Present triangular with pick corresponding to the value a of the variable, (b) $c >$ specified threshold (creation of new triangular) and (c) $c <$ specified threshold (non-creation of new triangular).

$$W' = \begin{bmatrix} 1.0000 & 0 & -0.0513_{mf2} & 0 & 0.1984_{mf2} \\ 0 & 1.0000 & 0.0235_{mf2} & -0.2368_{mf2} & -0.0685_{mf2} \\ 0 & 0 & 1.0000 & 0.3399_{mf2} & -0.8016_{mf2} \\ 0 & 0 & -0.2689_{mf2} & 1.0000 & 0 \\ 0 & 0 & 0 & 0 & 1.0000 \end{bmatrix}.$$

The number and shape of the fuzzy membership functions of the variables of both sides of the rules are gradually modified as new desired equilibrium points appear to the system during its training. To add a new triangular membership function in the fuzzy description of a variable, the new value of the variable must differ from one already encountered value more than a specified threshold. Figure 13.10 depicts this procedure. The initial value, a, of the variable determines the pick of the triangular membership function ($mf1$) for the fuzzy description of the variable. When a new value, b, of the variable appears in a new equilibrium condition, it is compared with the previously encountered value a. If $|a - b|$ exceeds a specified threshold, c, then a new triangular membership function ($mf2$) is created and the initial triangular function ($mf1$) is modified as shown in Figure 13.10b. If $|a - b|$ does not exceed the threshold the initial fuzzy partition of the variable remains unchanged (Figure 13.10c). The threshold comes usually as a compromise between the maximum number of allowable rules and the detail in fuzzy representation of each variable.

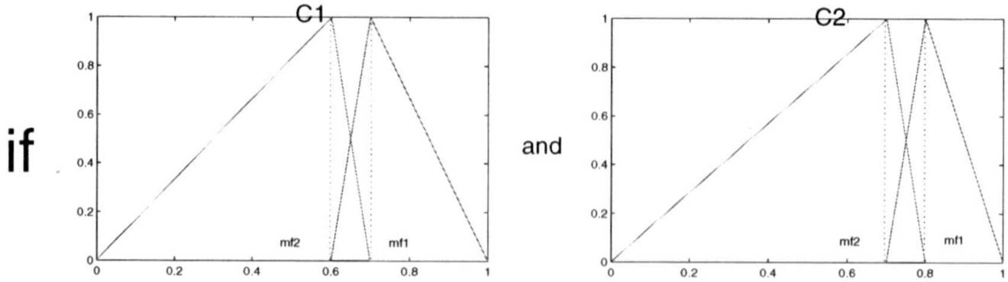

Fig. 13.11 Left-hand side (if-part).

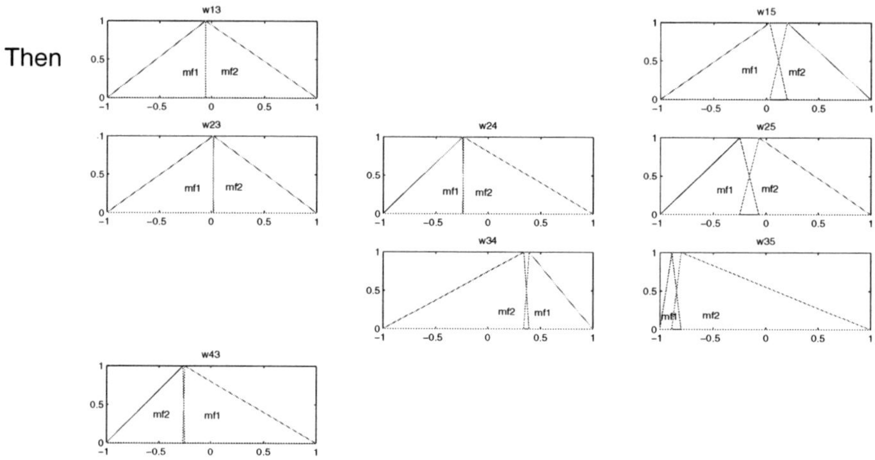

Fig. 13.12 Right-hand side (then-part).

Instead of triangular membership functions one can use gaussian or trapezoidal memberhip functions to create the fuzzy rule database.

The created fuzzy memerbship functions of the fuzzy rule databases *Rule1* and *Rule2* are depicted in Figures 13.11 and 13.12.

13.6.4 Using the Acquired Knowledge

After the FCN has been trained and the knowledge acquired has been stored in the Fuzzy Rule Database, it can be used in place of the unknown system to predict its behavior in changing condition determined by u. This kind of usage is illustrated in Figure 13.13, where the FCN estimates the behavior of the system after it recalls the relative weight values using the Fuzzy Rule database and conventional Mamdani-type inference procedure. Such a kind of operation has been employed in [60], where

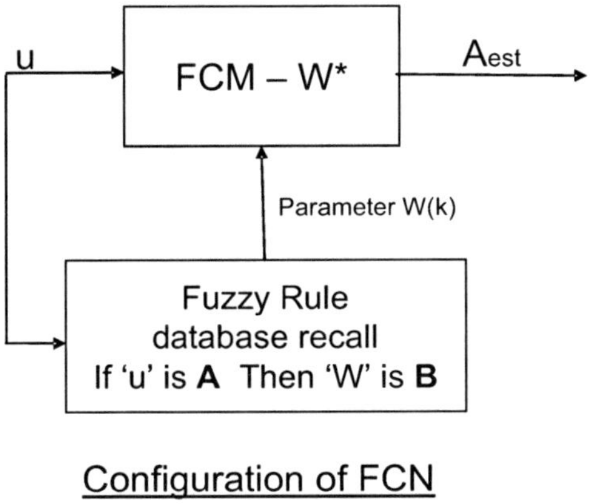

Fig. 13.13 An estimator of the real system using the trained FCN.

however the weight estimation was not performed using the proposed in this chapter algorithm. FCNs equipped with the proposed storage mechanism can also be used as a part of an adaptive control scheme of unknown nonlinear system predicting the required system inputs which will drive the system to a desired performance. This way of operation has been demonstrated in [61] where the method was used to control a pilot anaerobic wastewater treatment unit. However, the weight updating in [61] is using a conventional delta rule, which does not guarantee that the weights reach always values fullfiling the convergence conditions.

Suppose now that the desired state for the FCN of Figure 13.9 is:

$$A^{\mathrm{des}} = \left[\,0.65\ 0.75\ 0.60\ 0.67\ 0.48\,\right]^T .$$

Applying (13.28) and (13.36), the W matrix concludes to:

$$W_{\mathrm{actual}} = \begin{bmatrix} 1.0000 & 0 & -0.0553 & 0 & 0.1119 \\ 0 & 1.0000 & 0.0208 & -0.2421 & -0.1632 \\ 0 & 0 & 1.0000 & 0.3663 & -0.8506 \\ 0 & 0 & -0.2601 & 1.0000 & 0 \\ 0 & 0 & 0 & 0 & 1.0000 \end{bmatrix},$$

where the self feedback weights were excluded from the updating.

For weight matrix W_{actual} inequality (13.12) is true:

$$\left(\sum_{i=1}^{5} \|w_i\|^2\right)^{1/2} = 2.4549 < 4.$$

Suppose now that for the FCN of Figure 13.9 the input nodes $C1 = 0.65$ and $C2 = 0.75$ are given as input to the Fuzzy Rule database created from *Rule1* and *Rule2*. By using the Mamdani deffuzification method the output parameters w concludes to

$$W_{\text{output}} = \begin{bmatrix} 1.0000 & 0 & -0.0549 & 0 & 0.11455 \\ 0 & 1.0000 & 0.02115 & -0.24115 & -0.1596 \\ 0 & 0 & 1.0000 & 0.3654 & -0.8448 \\ 0 & 0 & -0.2603 & 1.0000 & 0 \\ 0 & 0 & 0 & 0 & 1.0000 \end{bmatrix}.$$

For weight matrix W_{output} inequality (13.12) is also true:

$$\left(\sum_{i=1}^{5} \|w_i\|^2 \right)^{1/2} = 2.4546 < 4.$$

Using matrix W_{output} the FCN concludes to the equilibrium point given below:

$$A_{\text{output}} = \left[0.65\ 0.75\ 0.6001\ 0.6701\ 0.4815 \right]^T,$$

which is close to the actual one.

13.7 Conclusions

In this chapter the existence and uniqueness of the equilibrium values of the concepts of FCMs were studied. This study concerns FCMs equipped with continuous differentiable sigmoid functions having contractive or at least non-expansive properties and is performed using an appropriately defined contraction mapping theorem and the non-expansive theorem. It was proved that when the weight interconnections fulfill certain conditions, related to the size of the FCM and the inclination of the sigmoids used, the concept values will converge to a unique solution regardless their initial values. The condition is further resolved by exploring its application in FCMs of various representative sizes. When the conditions are met, the convergence point of the concepts depends exclusively on the values of weight matrix W of the FCM giving rise to an FCM representation using *meta rules* of the form "If weights then equilibrium point". For the estimation of the weights corresponding to a new equilibrium point an adaptive estimation algorithm is developed and proofs for its stability and convergence properties are given. The algorithm takes into account the convergence conditions and incorporates them in parameter projection schemes. In view of the obtained results a new alternative scheme called Fuzzy Cognitive Network (FCN) was also presented, which exploits these theoretical results. It is designed to work in close interaction with the physical system it describes and stores information related to different operational points using fuzzy *meta rules* of this kind. Applications of FCNs were also referenced. Future work will include the development of reliable control schemes which will use the concept of FCN and employ the theoretical results presented here.

References

1. B. Kosko, Fuzzy cognitive maps, *International Journal of Man-Machine Studies* **24**(1), 65–75, 1986.
2. R. Axelrod, *Structure of Decision, The Cognitive Maps of Political Elites*, Princeton University Press, New Jersey, 1976.
3. C. Stylios and P. Groumpos, Fuzzy cognitive maps in modelling supervisory control systems, *Journal of Intelligent and Fuzzy Systems* **8**, 83–98, 2000.
4. C. Stylios and P. Groumpos, A soft computing approach for modelling the supervisor of manufacturing systems, *Journal of Intelligent and Robotics Systems* **26**(34), 389–403, 1999.
5. C. Stylios, P. Groumpos and V. Georgopoulos, A fuzzy cognitive maps approach to process control systems, *Journal of Intelligent and Robotics Systems* **26**(34), 389–403, 1999.
6. M. Schneider, E. Shnaider, A. Kandel and G. Chew, Constructing fuzzy cognitive maps, in *International Joint Conference of the Fourth IEEE International Conference on Fuzzy Systems and The Second International Fuzzy Engineering Symposium*, Vol. 4, No. 1, pp. 2281–2288, 1995.
7. B. Kosko, Differential Hebbian learning, in *Proceedings American Institute of Physics, Neural Networks for Computing*, pp. 277–282, 1986.
8. P. Craiger and M.D. Coovert, Modeling dynamic social and psychological processes with fuzzy cognitive maps, in *IEEE World Congress on Computational Intelligence and Fuzzy Systems*, Vol. 3, pp. 1873–1877, 1994.
9. A. Tsadiras and I. Kouskouvelis, *Using Fuzzy Cognitive Maps as a Decision Support System for Political Decisions: The Case of Turkey's Integration into the European Union*, Lecture Notes in Computer Science, Vol. 3746, Springer, Berlin, pp. 371–381, 2005.
10. D.E. Koulouriotis, I.E. Diakoulakis and D.M. Emiris, A fuzzy cognitive map-based stock market model: Synthesis, analysis and experimental results, in *Proceedings of 10th IEEE International Conference on Fuzzy Systems*, pp. 465–468, 2001.
11. J.P. Carvalho and J.A.B. Tome, Qualitative modelling of an economic system using rule-based fuzzy cognitive maps, in *Proceedings of IEEE International Conference on Fuzzy Systems*, Vol. 2, pp. 659–664, 2004.
12. T. Kottas, Y. Boutalis, G. Devedzic and B. Mertzios, A new method for reaching equilibrium points in fuzzy cognitive maps, in *Proceedings of 2nd International IEEE Conference of Intelligent Systems*, Varna, Bulgaria, pp. 53–60, 2004.
13. V. Georgopoulos, G. Malandraki and C. Stylios, A fuzzy cognitive map approach to differential diagnosis of specific language impairment, *Artificial Intelligence in Medicine* **29**(3), 261–278, 2003.
14. W. Zhang, S. Chen and J. Bezdek, Pool2: A generic system for cognitive map development and decision analysis, *IEEE Transactions on Systems, Man, and Cybernetics* **19**(1), 31–39, 1989.
15. R. Satur and Zhi-Qiang Liu, A contextual fuzzy cognitive map framework for geographic information systems, *IEEE Transactions on Fuzzy Systems* **7**(5), 481–494, 1999.
16. Zhi-Qiang Liu and R. Satur, Contextual fuzzy cognitive map for decision support in geographic information systems, *IEEE Transactions on Fuzzy Systems* **7**(5), 495–507, 1999.
17. R. Satur and Zhi-Qiang Liu, Contextual fuzzy cognitive maps for geographic information systems, in *IEEE International Conference on Fuzzy Systems*, Vol. 2, pp. 1165–1169, 1999.
18. J.P. Carvalho, M. Carola and J.A.B. Tome, Using rule-based fuzzy cognitive maps to model dynamic cell behavior in Voronoi based cellular automata, in *IEEE International Conference on Fuzzy Systems*, pp. 1687–1694, 2006.

19. G. Papakostas, Y. Boutalis, D. Koulouriotis and B. Mertzios, Fuzzy cognitive maps for pattern recognition applications, *International Journal at Pattern Recognition and Artificial Intelligence*, 2008, in press.
20. G. Papakostas, Y. Boutalis, D. Koulouriotis and B. Mertzios, A first study of pattern classification using fuzzy cognitive maps, in *International Conference on Systems, Signals and Image Processing – INSSIP'06*, pp. 369–374, 2006.
21. W. Stach, L.A. Kurgan and W. Pedrycz, Numerical and linguistic prediction of time series with the use of fuzzy cognitive maps, *IEEE Transactions on Fuzzy Systems* **16**(1), 61–72, 2008.
22. P.C. Silva, Fuzzy cognitive maps over possible worlds, in *Proceedings of International Joint Conference of the Fourth IEEE International Conference on Fuzzy Systems and The Second International Fuzzy Engineering Symposium*, Vol. 2, pp. 555–560, 1995.
23. J.A. Dickerson and B. Kosko, Virtual worlds as fuzzy cognitive maps, in *Proceedings of Virtual Reality Annual International Symposium*, pp. 471–477, 1993.
24. D.E. Koulouriotis, I.E. Diakoulakis and D.M. Emiris, Anamorphosis of fuzzy cognitive maps for operation in ambiguous and multi-stimulus real world environments, in *Proceedings of 10th IEEE International Conference on Fuzzy Systems*, pp. 1156–1159, 2001.
25. M. Parenthoen, P. Reignier and J. Tisseau, Put fuzzy cognitive maps to work in virtual worlds, in *Proceedings of 10th IEEE International Conference on Fuzzy Systems*, Vol. 1, pp. 252–255, 2001.
26. Jianqiang Xin, J.E. Dickerson and J.A. Dickerson, Fuzzy feature extraction and visualization for intrusion detection, in *Proceedings of 12th IEEE International Conference on Fuzzy Systems*, pp. 1249–1254, 2003.
27. W. Zhang, S. Chen, W. Wang and R. King, A cognitive map based approach to the coordination of distributed cooperative agents, *IEEE Transactions on Systems, Man, and Cybernetics*, **22**(1), 103–114, 1992.
28. A. Huerga, A balanced differential learning algorithm in fuzzy cognitive maps, in *Proceedings of the Sixteenth International Workshop on Qualitative Reasoning*, poster, 2002.
29. E. Papageorgiou and P. Groumpos, A weight adaptation method for fuzzy cognitive maps to a process control problem, in *Proceedings of an International Conference on Computational Science, ICCS 2004*, Krakow, Poland, 6-9 June, M. Budak et al. (Eds.), Lecture Notes in Computer Science, Vol. 3037 (II), Springer Verlag, pp. 515–522, 2004.
30. E. Papageorgiou, C. Stylios and P. Groumpos, Active Hebbian learning algorithm to train fuzzy cognitive maps, *International Journal of Approximate Reasoning* **37**(3), 219–247, 2004.
31. J. Aguilar, Adaptive random fuzzy cognitive maps, in *IBERAMIA 2002*, F.J. Garijio, J.C. Riquelme and M. Toro (Eds.), Lecture Notes in Artificial Intelligence, Vol. 2527, Springer-Verlag, Berlin/Heidelberg, pp. 402-410, 2002.
32. W. Stach, L. A. Kurgan and W. Pedrycz, Data-driven nonlinear Hebbian learning method for fuzzy cognitive maps, in *Proceedings of 2008 World Congress on Computational Intelligence (WCCI'08)*, 2008.
33. D.Koulouriotis, I. Diakoulakis and D. Emiris, Learning fuzzy cognitive maps using evolution strategies: A novel schema for modeling a simulating high-level behavior, in *Proceedings of IEEE Congress on Evolutionary Computation*, Vol. 1, pp. 364–371, 2001.
34. E. Papageorgiou, K. Parsopoulos, C. Stylios, P. Groumpos and M. Vrahatis, Fuzzy cognitive maps learning using particle swarm optimization, *International Journal of Intelligent Information Systems* **25**(1), 95–121, 2005.
35. M. Khan, S. Khor and A. Chong, Fuzzy cognitive maps with genetic algorithm for goal-oriented decision support, *Int. J. Uncertainty, Fuzziness and Knowledge-based Systems* **12**, 31–42, 2004.
36. W. Stach, L. Kurgan, W. Pedrycz and M. Reformat, Evolutionary development of fuzzy cognitive maps, in *Proceedings of 14th IEEE International Conference on Fuzzy Systems*, pp. 619–624, 2005.
37. W. Stach, L. Kurgan, W. Pedrycz and M. Reformat, Genetic learning of fuzzy cognitive maps, *Fuzzy Sets and Systems* **153**(3), 371–401, 2005.
38. M. Hagiwara, Extended fuzzy cognitive maps, in *Proceedings of IEEE International Conference on Fuzzy Systems*, pp. 795–801, 1992.

39. Jian Ying Zhang, Zhi-Qiang Liu and Sanming Zhou, Quotient FCMs – A decomposition theory for fuzzy cognitive maps, *IEEE Transactions on Fuzzy Systems* **11**(5), 593–604, 2003.
40. Ban Ying Zhang and Zhi-Qiang Liu, Quotient fuzzy cognitive maps, in *Proceedings 10th IEEE International Conference on Fuzzy Systems*, Vol. 1, pp. 180–183, 2001.
41. Y. Miao, Z. Liu, C. Siew and C. Miao, Dynamical cognitive network – An extension of fuzzy cognitive map, *IEEE Transactions on Fuzzy Systems* **9**(5), 760–770, 2001.
42. J. Zhang, Zhi-Qiang Liu and S. Zhou, Dynamic domination in fuzzy causal networks, *IEEE Transactions on Fuzzy Systems* **14**(1), 42–57, 2006.
43. Yuan Miao and Zhi-Qiang Liu, On causal inference in fuzzy cognitive maps, *IEEE Transactions on Fuzzy Systems* **8**(1), 107–119, 2000.
44. Jian Ying Zhang and Zhi-Qiang Liu, Dynamic domination for fuzzy cognitive maps, in *IEEE International Conference on Fuzzy Systems*, Vol. 1, pp. 145–149, 2002.
45. Zhi-Qiang Liu and Yuan Miao, Fuzzy cognitive map and its causal inferences, in *Proceedings of IEEE International Conference on Fuzzy Systems*, Vol. 3, pp. 1540–1545, 1999.
46. Sanming Zhou, Zhi-Qiang Liu and Jian Ying Zhang, Fuzzy causal networks: General model, inference, and convergence, *IEEE Transactions on Fuzzy Systems* **14**(3), 412–420, 2006.
47. F. Smarandache, An introduction to neutrosophy, neutrosophic logic, neutrosophic set, and neutrosophic probability and statistics, in *Proceedings of the First International Conference on Neutrosophy, Neutrosophic Logic, Neutrosophic Set, Neutrosophic Probability and Statistics*, University of New Mexico, Gallup, No. 1–3, pp. 5–22, 2001.
48. V. Kandasamy and F. Smarandache, Fuzzy cognitive maps and neutrosophic cognitive maps, *ProQuest Information & Learning* (University of Microfilm International), 2003.
49. T.L. Kottas, Y.S. Boutalis and M.A. Christodoulou, Fuzzy cognitive networks: A general framework, *Inteligent Desicion Technologies* **1**(4), 183–196, 2007.
50. J. Dickerson and B. Kosko, Virtual worlds as fuzzy cognitive maps, *Presence* **3**(2), 173–189, 2006.
51. B. Kosko, *Fuzzy Engineering*, Prentice Hall, 1997.
52. A. Tsadiras, Comparing the inference capabilities of binary, trivalent and sigmoid fuzzy cognitive maps, *Information Science* **178**, 3880–3894, 2008.
53. Y. Boutalis, T. Kottas and M. Christodoulou, On the existence and uniqueness of solutions for the concept values in fuzzy cognitive maps, in *Proceedings of 47th IEEE Conference on Decision and Control – CDC'08*, Cancun, Mexico, December 9–11, pp. 98–104, 2008.
54. Y. Boutalis, T. Kottas and M. Christodoulou, Adaptive estimation of fuzzy cognitive maps with proven stability and parameter convergence, *IEEE Transactions on Fuzzy Systems*, DOI 10.1109TFUZZ.2009.2017519, 2009.
55. T. Kottas, Y. Boutalis and M. Christodoulou, A new method for weight updating in Fuzzy cognitive Maps using system feedback, in *Proceedings of 2nd International Conference on Informatics in Control, Automation and Robotics (ICINCO05)*, Barcelona, Spain, pp. 202–209, 2005.
56. T.L. Kottas, Y.S. Boutalis and A.D. Karlis, A new maximum power point tracker for PV arrays using fuzzy controller in close cooperation with fuzzy cognitive networks, *IEEE Transactions on Energy Conversion* **21**(3), 793–803, 2006.
57. W. Rudin, *Principles of Mathematical Analysis*, McGraw-Hill, pp. 220–221, 1964.
58. A. Kranas and J.Dugundji, *Fixed Point Theory*, Springer-Verlag, New York, 2003.
59. P. Ioannou and B. Fidan, *Adaptive Control Tutorial*, Society for Industrial and Applied Mathematics (SIAM), Philadelphia, 2006.
60. A.D. Karlis, T.L. Kottas and Y.S. Boutalis, A novel maximum power point tracking method for PV systems using fuzzy cognitive networks (FCN), *Electric Power System Research* **77**(3–4), 315–327, 2007.
61. T. Kottas, Y. Boutalis, V. Diamantis, O. Kosmidou and A. Aivasidis, A fuzzy cognitive network based control scheme for an anaerobic digestion process, in *Proceedings of 14th Mediterranean Conference on Control and Applications*, Session TMS Process Control 1, Ancona, Italy, 2006.

Chapter 14
An Improved Method in Receding Horizon Control with Updating of Terminal Cost Function

Hongwei Zhang, Jie Huang and Frank L. Lewis

Abstract In this chapter, we propose a modified receding horizon control (RHC) scheme for discrete linear time-invariant systems, called updated terminal cost RHC (UTC-RHC). The standard receding horizon control, also known as model predictive control (MPC) or moving horizon control (MHC), is a suboptimal control scheme that solves a finite horizon open-loop optimal control problem in an infinite horizon context and yields a state feedback control law. The ability to deal with constraints on controls and states makes RHC popular in industry, especially in the process control industry. A lot of successful applications have been reported in the past three decades. The stability issue has been an important topic in the literature and a lot of approaches have been proposed. Most of these approaches impose constraints on either the terminal state, or the terminal cost, or the horizon size, or their different combinations.

Unlike the standard RHC, UTC-RHC algorithm updates the terminal cost function of the corresponding finite horizon optimal control problem at every stage. It ensures the closed-loop system to be uniformly exponentially stable without imposing any constraint on the terminal state, the horizon size, or the terminal cost, thus makes the design of a stabilizing controller more flexible. Moreover, the control law generated by UTC-RHC algorithm converges to the optimal control law of the infinite horizon optimal control problem.

Hongwei Zhang · Jie Huang
Department of Mechanical and Automation Engineering, The Chinese University of Hong Kong, Hong Kong; e-mail: {hwzhang, jhuang}@mae.cuhk.edu.hk

Frank L. Lewis
Automation and Robotics Research Institute, The University of Texas at Arlington, Fort Worth, TX 76118-7115, USA; e-mail: lewis@uta.edu

K.P. Valavanis (ed.), Applications of Intelligent Control to Engineering Systems, 365–393.

14.1 Introduction

Receding horizon control (RHC), also known as model predictive control (MPC) or moving horizon control, is a suboptimal control scheme that solves a finite horizon optimal control problem repeatedly in an infinite horizon context. This is an on-line method that effectively provides a measured state feedback control law.

The ability to deal with constraints on controls and states makes RHC popular in industry, especially in the process control industry, where the economical operating point often lies as near as possible to the constraints without violating them [28]. A lot of successful applications have been reported in the past three decades, see [25] and [28] for excellent surveys of industrial MPC technology. Another reason for its wide applications in industry is probably the open-loop stability property of most of the plants in process industry. Advantages of RHC are summarized in [17].

Extensive attentions have been paid to the stability issue of RHC (see [23] for a survey), leading to various stability conditions involving constraints on either the terminal state, or the terminal cost, or the horizon size, or their different combinations. Some approaches require the terminal state to be zero or to enter a neighborhood of the origin [10, 16, 18, 19, 24, 31]; some approaches impose constraints on the terminal cost [5, 9, 11, 15, 21]; some researchers also show that by taking a sufficiently long horizon, the stability of RHC algorithm is guaranteed without imposing constraints on the terminal cost (see e.g. [5, 12, 14]).

Inspired by the idea that a long enough horizon ensures the closed-loop stability of RHC, we propose a modified RHC algorithm called updated terminal cost RHC (UTC-RHC). This algorithm was first proposed in our pervious work [38] under the name adaptive terminal cost RHC (ATC-RHC). In our algorithm, instead of using the same terminal cost function all the time, we *update* it at each stage. It is shown that by updating the terminal cost function, the horizon size is effectively lengthened. The advantage of UTC-RHC is that it ensures the closed-loop stability while imposing constraints on neither terminal state, nor terminal cost function, nor the horizon size. Also, by updating the terminal cost, the resulting feedback control law approaches the optimal value corresponding to the infinite horizon optimal control problem.

Some preliminary results in this chapter have been reported in [38]. The results in this chapter are more detailed and more comprehensive. Additionally, one more example is added to illustrate the stability and convergence of UTC-RHC.

This chapter is organized as follows. First we present as a background the algorithm of receding horizon control, together with some relevant stability results and derive some performance bounds of RHC; Section 14.3 concerns the algorithm of UTC-RHC and its stability and convergence, and this section consists of the main part of this chapter; Two examples comparing the performance of RHC and UTC-RHC are shown in Section 14.4; Finally we end this chapter with a conclusion in Section 14.5.

14.2 Receding Horizon Control

In this chapter, we consider the discrete linear time-invariant (LTI) system

$$x_{k+1} = Ax_k + Bu_k, \tag{14.1}$$

where $x_k \in \mathbb{R}^n$ is the state vector with the initial state x_0, $u_k \in \mathbb{R}^m$ is the control input, $A \in \mathbb{R}^{n\times n}$ and $B \in \mathbb{R}^{n\times m}$ are constant matrices.

Without further indication, we assume the following assumptions hold throughout the chapter.

Assumption 1.

(A1) $Q \in \mathbb{R}^{n\times n}$, $R \in \mathbb{R}^{m\times m}$, $Q = Q^T \geq 0$, $R = R^T > 0$.
(A2) (A, B) *is stabilizable.*
(A3) $(A, \sqrt{Q})$ *is detectable.*

To be self contained, before stepping into RHC, we give a review of linear quadratic (LQ) optimal control or linear quadratic regulator (LQR) design [22], to which RHC is closely related. The LQ optimal control problem is often put into either infinite horizon or finite horizon framework.

14.2.1 Infinite Horizon Linear Quadratic Optimal Control

The infinite horizon LQ optimal control problem can be formulated as

$$\mathcal{P}(x_k, k) : V^{IH}(x_k, k) = \min_{\bar{u}} \left\{ \sum_{i=k}^{\infty} (x_i^T Q x_i + u_i^T R u_i) \right\}, \tag{14.2}$$

where $\bar{u} = \{u_k, u_{k+1}, \ldots\}$ is the resulting optimal control sequence, and the arguments mean the current state x_k and current stage k. Note that the bar as in $\bar{u}$ is defined similarly throughout the chapter.

By the well known LQR theory [5, 22], the optimal infinite horizon state feedback control law is given by

$$u^{IH}(x_k) = -(B^T P_\infty B + R)^{-1} B^T P_\infty A x_k = L_\infty x_k, \tag{14.3}$$

where P_∞ is the unique maximal positive semidefinite solution to the associated algebraic Riccati equation (ARE)

$$P = A^T P A + Q - A^T P B (B^T P B + R)^{-1} B^T P A. \tag{14.4}$$

The optimal cost is

$$V^{IH}(x_k, k) = x_k^T P_\infty x_k. \tag{14.5}$$

Under control law (3), the closed-loop system is

$$x_{x+1} = (A - B(B^T P_\infty B + R)^{-1} B^T P_\infty A)x_k. \tag{14.6}$$

Lemma 1 ([5, theorem 4.1] or [22, theorem 2.4–2]). *Suppose assumptions A1–A3 hold. Then*

(a)There exists a unique positive semidefinite solution P_∞ to the algebraic Riccati equation (14.4).
(b)The closed-loop system (14.6) is asymptotically stable, i.e.,

$$A - B(B^T P_\infty B + R)^{-1} B^T P_\infty A$$

has all its eigenvalues strictly within the unit circle.

This property is often referred to as the stability of ARE.

14.2.2 Finite Horizon Linear Quadratic Optimal Control

The finite horizon LQ optimal control problem can be formulated as

$$\begin{aligned} \mathcal{P}(x_k, k, N, P_0) &: V^{FH}(x_k, k, N, P_0) \\ &= \min_{\bar{u}} \left\{ \sum_{i=k}^{k+N-1} (x_i^T Q x_i + u_i^T R u_i) + x_{k+N}^T P_0 x_{k+N} \right\}, \end{aligned} \tag{14.7}$$

where x_k is the current state, k is the current stage, N is the horizon length and $P_0 \geq 0$ is the terminal cost weighting matrix.

Denote the optimal control sequence as

$$\begin{aligned} \bar{u}^{FH}(\cdot; x_k, k, N, P_0) = \{&u^{FH}(k; x_k, k, N, P_0), u^{FH}(k+1; x_k, k, N, P_0), \ldots, \\ &u^{FH}(k+N-1; x_k, k, N, P_0)\}. \end{aligned} \tag{14.8}$$

By the LQR theory [22],

$$\begin{aligned} u^{FH}(k+j; x_k, k, N, P_0) &= -(B^T P_{N-j-1} B + R)^{-1} B^T P_{N-j-1} A x_{k+j} \\ &= L_{N-j-1} x_{k+j}, \quad j = 0, 1, \ldots, N-1, \end{aligned} \tag{14.9}$$

where L_{N-j-1} denotes the optimal control gain at stage $k+j$, and the optimal cost is

$$V^{FH}(x_k, k, N, P_0) = x_k^T P_N x_k, \tag{14.10}$$

where P_t is the t-th term in the solution sequence of the Riccati difference equation (RDE) with initial condition P_0.

$$P_{t+1} = A^T P_t A + Q - A^T P_t B(B^T P_t B + R)^{-1} B^T P_t A. \tag{14.11}$$

Lemma 2 ([5, theorem 4.2). *Suppose assumptions A1–A3 hold and* $P_0 \geq 0$. *Consider the RDE (14.11). Then* $P_t \to P_\infty$ *as* $t \to \infty$, *where* P_∞ *is the unique maximal positive semidefinite solution to the ARE (14.4).*

This is the well known convergence property of RDE.

Lemma 3 ([5, theorems 4.4 and 4.5]). *If the positive semidefinite solution* P_t *of the RDE (14.11) is monotonically non-increasing (non-decreasing) at one time, then* P_t *is monotonically non-increasing (non-decreasing) for all subsequent times.*

This is the well known monotonicity property of RDE.

14.2.3 Receding Horizon Control

The finite horizon optimal control can deal with the input and state constraints by using mathematical programming, such as quadratic programming and semidefinite programming. However, the industry processes, especially the chemical reaction processes, are often slow and run for hours, days or even weeks. In these cases, infinite horizon optimal control is more practical than finite horizon optimal control. Although the infinite horizon optimal control algorithm ensures stability for any linear system under mild conditions (stability and detectability), it cannot deal with constraints, model uncertainty, etc., which are common in industry [28]. This leads to a mixed-mode optimal control problem between finite horizon and infinite horizon, i.e., receding horizon control.

Receding horizon control is a suboptimal control strategy for an infinite horizon optimal control problem. It applies the finite horizon optimal control in an infinite time context. At each stage, an open-loop finite, say N, horizon LQ optimal control problem is solved, yielding an optimal control sequence. Only the first control in the resulting control sequence is applied at the current stage. At the next stage, the same procedure is repeated. The detailed scheme is stated as follows.

At the current stage $k (k \geq 0)$, a finite horizon LQ optimal control problem $\mathcal{P}(x_k, k, N, P_0)$ is solved, yielding the optimal control sequence (14.8). Then the receding horizon optimal control law at current stage k is

$$\begin{aligned} u^{RH}(x_k, k, N, P_0) &= u^{FH}(k; x_k, k, N, P_0) \\ &= -(B^T P_{N-1} B + R)^{-1} B^T P_{N-1} A x_k = L_N^{RH} x_k, \end{aligned} \tag{14.12}$$

where L_N^{RH} is the receding horizon control gain at stage k. The controller (14.12) drives the system (14.1) to the next stage $k + 1$.

At the next stage $k + 1$, we measure the state x_{k+1} and solve the new N-horizon LQ optimal control problem $\mathcal{P}(x_{k+1}, k + 1, N, P_0)$, where the *measured* new state x_{k+1} is the initial state, to get a new optimal control sequence. Again the first control in the control sequence is applied to the stage $k + 1$. Thus,

$$\begin{aligned} u^{RH}(x_{k+1}, k+1, N, P_0) &= u^{FH}(k+1; x_{k+1}, k+1, N, P_0) \\ &= -(B^T P_{N-1} B + R)^{-1} B^T P_{N-1} A x_{k+1} = L_N^{RH} x_{k+1}. \end{aligned} \tag{14.13}$$

For a fixed horizon N and the same P_0, the receding horizon control gain L_N^{RH} is constant.

The RHC closed-loop system takes the following form

$$x_{k+1} = (A - B(B^T P_{N-1}\ B + R)^{-1} B^T P_{N-1}\ A)x_k. \tag{14.14}$$

By Lemma 2, $P_{N-1} \to P_\infty$ as $N \to \infty$, thus $L_N^{RH} \to L_\infty$ as $N \to \infty$. This also means the receding horizon control is not optimal unless $N \to \infty$.

Denote for later use $V^{RH}(x_k, k, N, P_0)$ as the cost-to-go by applying the RHC control law from time k onwards, i.e.,

$$V^{RH}(x_k, k, N, P_0) = \sum_{i=k}^{\infty} \{x_i^T Q x_i + (L_N^{RH} x_i)^T R (L_N^{RH} x_i)\}. \tag{14.15}$$

14.2.4 Stability of RHC

Closed-loop stability has been an important issue in the RHC literature. Almost all the stability results put requirements on either the terminal state x_{k+N}, or the terminal cost weighting matrix P_0, or the horizon length N, or their different combinations. Readers may refer to [23] for an excellent survey about the RHC stability issue. Here we only list some of the stability approaches that are related to our algorithm and put them into the following three categories.

14.2.4.1 Constraint on the Terminal State

This approach falls into two subcategories: one requires the terminal state to vanish at the origin, i.e., $x_{k+N} = 0$, which is also known as the terminal equality constraint (see, for example, [16, 18]); the other requires the terminal state to enter a neighborhood of the origin, and in this neighborhood, a local stabilizing controller is applied [24].

14.2.4.2 Constraint on the Terminal Cost Function

The choice of terminal cost weighting matrix P_0 is important for the stability of RHC. Ideally, one should choose $P_0 = P_\infty$. In that case the finite horizon LQ optimal control problem (14.7) would be the same as infinite horizon LQ optimal control problem (14.2), therefore the closed-loop stability can be automatically guaranteed.

When P_∞ is hard to get due to nonlinearities or constraints, researchers show that closed-loop stability can still be obtained by choosing the terminal cost properly [5, 6, 15]. For discrete linear time-invariant systems, Bitmead et al. [5] obtained the following result:

Lemma 4 ([5, theorem 4.7]). *Consider the RDE (14.11) and suppose that assumptions A1–A3 hold. If* $P_{j+1} \leq P_j$ *for some* $j \geq 0$*, then the closed-loop system (14.14) is stable for all* $N \geq j+1$.

A special case is when the terminal cost weighting matrix satisfies $P_0 \geq P_1$, then the RHC closed-loop system (14.14) is asymptotically stable for all $N \geq 1$. Theorem 1 in [21] implies that if $Q > 0$ and P_0 satisfies the condition

$$P_0 \geq (A+BH)^T P_0 (A+BH) + Q + H^T R H \tag{14.16}$$

for some $H \in \mathbb{R}^{m\times n}$, the stabilizing RHC controller can also be obtained for any $N \geq 1$. This is a more relaxed condition than $P_0 \geq P_1$, where $H = -(B^T P_0 B + R)^{-1} B^T P_0 A$. Condition (14.16) says the terminal cost $V(x_k) = x_k^T P_0 x_k$ is a control Lyapunov function (CLF) with the associated stabilizing feedback control gain H, which provides an incremental upper bound on the cost-to-go in the sense that $\Delta V(x_k) \leq -(x_k^T Q x_k + u_k^T R u_k)$.

14.2.4.3 Constraint on the Horizon Length

When there is no terminal cost (i.e., $P_0 = 0$) or for a general terminal cost (i.e., $P_0 \geq 0$), the relationship between stability and the horizon size is tricky. Since $\{P_j\}$ will eventually approach P_∞, which leads to a stabilizing closed-loop system, one may conjecture that if for some N, P_N leads to a stable closed-loop system, then for any $n \geq N$, P_n will also lead to a stable closed-loop system. However this is not true. Readers may refer to [7, 27] for some examples.

But research also shows that when the horizon size N is chosen long enough, the stability of RHC scheme can be guaranteed. Primbs and Nevistic [26] showed that, for constrained discrete LTI systems, picking the terminal cost as $P_0 = Q$, there exists a finite horizon length N^* such that for all $N \geq N^*$ the RHC closed-loop system is asymptotically stable. They also indicate that the same result holds even for arbitrary $Q \leq P_0$; Jadbabaie et al. [14] showed that, for input constrained nonlinear systems, exponential stability can be obtained for RHC scheme with general positive semidefinite terminal cost if a sufficiently long horizon is adopted; Teel's group [12] showed that, for both unconstrained discrete time nonlinear system and input-constraint linear system, a long enough horizon can ensure the exponential stability of the MPC scheme without particular requirements on the terminal cost.

For unconstrained discrete LTI systems, Bitmead et al. [6] obtained the following result:

Lemma 5 ([6, theorem 10.17]). *Suppose that assumptions A1–A3 hold and* $P_0 \geq 0$. *Then there exists an finite positive integer* N^* *such that the RHC closed-loop system (14.14) is asymptotically stable for all* $N \geq N^*$.

This result only guarantees the eventual stability of RHC scheme for arbitrary $P_0 \geq 0$, yet does not indicate how large N^* should be. Recently, Quan et al. [29] gave an explicit expression of stability guaranteed RHC horizon size (SgHS) for discrete LTI systems, and the result holds for arbitrary $P_0 \geq 0$.

14.2.5 Performance Bounds of RHC

Approximate dynamic programming (ADP) [1, 2, 4, 13, 34, 35, 36] is a technique to solve dynamic programming problems forward-in-time. It has attracted a lot of attentions recently. Interested readers may refer to [32] and the references therein for recent status of ADP. One goal of this chapter is to relate RHC more closely to methods of ADP and reinforcement learning (RL) [33]. Therefore in this section, we present the next result, inspired by proposition 3.1 in [3], which shows a cost improvement property for rollout algorithms [4].

For this purpose, let us first point out that under assumptions A1–A3, for any stabilizing feedback control law $u_i = Lx_i$, there exists a $P_0 \geq 0$ such that for all k and x_k,

$$x_k^T P_0 x_k = \sum_{i=k}^{\infty} (x_i^T Q x_i + u_i^T R u_i). \tag{14.17}$$

This fact can be established as follows.

Consider the Lyapunov equation

$$P_0 = (A + BL)^T P_0 (A + BL) + (Q + L^T RL). \tag{14.18}$$

Since L is a stabilizing control gain, i.e., $A + BL$ has all its eigenvalues strictly inside the unit disc, by Lyapunov theory, $P_0 \geq 0$ exists and is unique. Furthermore, the solution P_0 can be expressed as

$$P_0 = \sum_{j=0}^{\infty} ((A + BL)^T)^j \ (Q + L^T RL))(A + BL)^j. \tag{14.19}$$

Then $\forall x_k$, noting $x_{j+k} = (A + BL)^j x_k$ for $j \geq 0$, we have

$$\begin{aligned} x_k^T P_0 x_k &= \sum_{j=0}^{\infty} x_k^T ((A + BL)^T)^j (Q + L^T RL)(A + BL)^j x_k \\ &= \sum_{i=k}^{\infty} x_i^T (Q + L^T RL) x_i = \sum_{i=k}^{\infty} (x_i^T Q x_i + u_i^T Ru). \end{aligned}$$

Lemma 6. *Consider system (14.1). Suppose assumptions A1–A3 hold. Let* $u_i = Lx_i$ *be a stabilizing feedback control law and* P_0 *satisfies (14.17). Then the cost-to-go* $V^{RH}(x_k, k, N, P_0)$ *of the RHC control law satisfies*

$$V^{RH}(x_k, k, N, P_0) \leq x_k^T P_N x_k \leq x_k^T P_{N-1}\, x_k \leq \ldots \leq x_k^T P_0 x_k, \quad \forall N \geq 1, \tag{14.20}$$

where $P_j\ (j = 0, 1, \ldots, N)$ *is the* j*-th term in the solution sequence of RDE (14.11) with initial value* P_0.

Proof. Since (14.17) can be written in the recursive form

$$x_k^T P_0 x_k = x_k^T Q x_k + u_k^T R u_k + x_{k+1}^T P_0 x_{k+1}, \tag{14.21}$$

then for all k and x_k,

$$\min_u \{x_k^T Q x_k + u^T R u + x_{k+1}^T P_0 x_{k+1}\} \leq x_k^T P_0 x_k. \tag{14.22}$$

By (14.7) and (14.10) with $N = 1$ and P_0 being replaced by P_j, we have

$$\min_u \{x_k^T Q x_k + u^T R u + x_{k+1}^T P_j x_{k+1}\} = x_k^T P_{j+1} x_k, \quad j = 0, 1, 2, \ldots. \tag{14.23}$$

Equations (14.22) and (14.23) imply $P_1 \leq P_0$, and it follows that $P_{j+1} \leq P_j$ for $j = 0, 1, 2, \ldots$, by the monotonicity property of RDE (e.g., theorem 4.4 in [5]). Thus $x_k^T P_{j+1} x_k \leq x_k^T P_j x_k$ for all k and x_k, i.e.,

$$\min_u \{x_k^T Q x_k + U^T R u + x_{k+1}^T P_j x_{k+1}\} \leq x_k^T P_j x_k, \quad j = 0, 1, 2, \ldots. \tag{14.24}$$

Denote the one-step lookahead policy

$$u_k^j = \arg\min_u \{x_k^T Q x_k + u^T R u + x_{k+1}^T P_j x_{k+1}\}, \quad j = 0, 1, 2 \ldots. \tag{14.25}$$

then by proposition 3.1 in [3], the cost-to-go corresponding to u_k^j satisfies for all x_k and k,

$$\sum_{i=k}^{\infty} \{x_i^T Q x_i + (u_i^j)^T R u_i^j\} \leq x_k^T P_{j+1} x_k \leq x_k^T P_j x_k, \quad \forall j = 0, 1, 2, \ldots. \tag{14.26}$$

Observing $u_k^j = u^{RH}(x_k, k, j+1, P_0)$, we have

$$\sum_{i=k}^{\infty} \{x_i^T Q x_i + (u_i^j)^T R u_i^j\} = V^{RH}(x_k, k, j+1, P_0).$$

Thus (14.16) implies

$$V^{RH}(x_k, k, j+1, P_0) \leq x_k^T P_{j+1} x_k \leq x_k^T P_j x_k, \quad \forall j = 0, 1, 2, \ldots. \tag{14.27}$$

With non-increasing monotonicity of the sequence $\{P_j\}$, we can draw the conclusion that

$$V^{RH}(x_k, k, N, P_0) \le x_k^T P_N x_k \le x_k^T P_{N-1} x_k \le \cdots \le x_k^T P_0 x_k, \quad \forall N \ge 1.$$

□

Remark 1. The importance of this result is that it shows RHC can always improve on, or at least is as good as the performance of any initial stabilizing feedback control law for any $u_i = Lx_i$ for any $N \ge 1$, in the sense that $V^{RH}(x_k, k, N, P_0) \le x_k^T P_0 x_k$.

Remark 2. One may conjecture from (14.20) that under the condition $P_1 \le P_0$, RHC may perform better in the sense that $V^{RH}(x_k, k, N, P_0)$ gets smaller, as N gets larger. However, this is not the case. In fact (14.20) only means that $V^{RH}(x_k, k, N, P_0)$ is upper bounded by $x_k^T P_N x_k$ and says nothing about the monotonicity of $V^{RH}(x_k, k, N, P_0)$, although $\{P_j\}$ is monotonically non-increasing, as shown by the following counterexample.

Counterexample 1. Consider the open-loop unstable system [29]

$$x_{k+1} = \begin{bmatrix} 0 & 3 \\ -4 & 5 \end{bmatrix} x_k + \begin{bmatrix} 2 \\ 2 \end{bmatrix} u_k. \tag{14.28}$$

Let $Q = \left[\begin{smallmatrix} 1.3 & 0 \\ 0 & 1.4 \end{smallmatrix}\right] > 0$, $R = 1$. Obviously, A, B, R and Q satisfy assumptions A1–A3. The solution of the ARE (14.4) is $P_\infty = \left[\begin{smallmatrix} 114.6619 & -56.4269 \\ -56.4269 & 29.7112 \end{smallmatrix}\right] > 0$. Take $P_0 = 2P_\infty$. Computing RDE (14.11) gives $P_1 = \left[\begin{smallmatrix} 227.7368 & -112.9642 \\ -112.9642 & 57.9799 \end{smallmatrix}\right]$. It is easy to check that $P_1 < P_0$. Taking an initial state $x_0 = [-10 \;\; 18]^T$, by applying the RHC control law from the beginning $k = 0$, one obtains $V^{RH}(x_0, 0, 1, P_0) = 41407$, $V^{RH}(x_0, 0, 2, P_0) = 42369$ and $V^{RH}(x_0, 0, 3, P_0) = 41413$ for $N = 1$, $N = 2$ and $N = 3$, respectively. Obviously, $V^{RH}(x_0, 0, 2, P_0) > V^{RH}(x_0, 0, 3, P_0)$, so the performance does not necessarily improve as the horizon gets longer.

14.3 Updated Terminal Cost Receding Horizon Control

14.3.1 Algorithm Description

Different from that of the *standard* RHC (here by *standard* we mean both the terminal cost weighting matrix and the horizon size N are fixed), the corresponding finite horizon LQ optimal control problem of UTC-RHC scheme has its terminal cost *updated at each stage*. The detailed algorithm is stated as follows. Owing to the shift invariant property of LTI system (14.1), the finite horizon LQ optimal control problem is time-invariant in the sense that $V^{FH}(x, k, N, P) = V^{FH}(x, 0, N, P)$. So without loss of generality, we can assume the initial stage to be $k = 0$.

Consider the system (14.1). At the initial stage $k = 0$, the corresponding N-horizon LQ optimal control problem solved by UTC-RHC is

$$\mathcal{P}(x_0, 0, N, P_0) : V^{FH}(x_0, 0, N, P_0)$$
$$= \min_{\bar{u}} \left\{ \sum_{i=0}^{N-1} (x_i^T Q x_i + u_i^T R u_i) + x_N^T P_0 x_N \right\} \tag{14.29}$$

Solving problem $\mathcal{P}(x_0, 0, N, P_0)$ yields an optimal control sequence

$$\bar{u}^{FH}(\cdot; x_0, 0, N, P_0) = \{u^{FH}(0; x_0, 0, N, P_0), u^{FH}(1; x_0, 0, N, P_0), \ldots,$$
$$u^{FH}(N-1; x_0, 0, N, P_0)\}. \tag{14.30}$$

Then the UTC-RHC control law at stage 0 is given by

$$u^{RH}(x_0, 0, N, P_0) = u^{FH}(0; x_0, 0, N, P_0)$$
$$= -(B^T P_{N-1} B + R)^{-1} B^T P_{N-1} A x_0. \tag{14.31}$$

At stage $k = 1$, we modify the terminal cost weighting matrix of the corresponding N-horizon LQ optimal control problem to P_N, which is obtained at stage $k = 0$ as the N-th term in the solution sequence of RDE (14.11) with initial condition P_0. Thus,

$$\mathcal{P}(x_1, 1, N, P_N) : V^{FH}(x_1, 1, N, P_N)$$
$$= \min_{\bar{u}} \left\{ \sum_{i=1}^{N} (x_i^T Q x_i + u_i^T R u_i) + x_{N+1}^T P_N x_{N+1} \right\}. \tag{14.32}$$

Solving the problem $\mathcal{P}(x_1, 1, N, P_N)$ yields an optimal control sequence

$$\bar{u}^{FH}(\cdot; x_1, 1, N, P_N) = \{u^{FH}(1; x_1, 1, N, P_N), u^{FH}(2; x_1, 1, N, P_N), \ldots,$$
$$u^{FH}(N; x_1, 1, N, P_N)\}. \tag{14.33}$$

Then the UTC-RHC control law at stage 1 is given by

$$u^{RH}(x_1, 1, N, P_N) = u^{FH}(1; x_1, 1, N, P_N). \tag{14.34}$$

Similarly, at stage $k = j$ ($j \geq 1$), the corresponding N-horizon LQ optimal control problem solved by UTC-RHC is the following problem with *modified terminal cost*

$$\mathcal{P}(x_j, j, N, P_{jN}) : V^{FH}(x_j, j, N, P_{jN})$$
$$= \min_{\bar{u}} \left\{ \sum_{i=1}^{j+N-1} (x_i^T Q x_i + u_i^T R u_i) + x_{j+N}^T P_{jN} x_{j+N} \right\}. \tag{14.35}$$

Note that the terminal cost weighting matrix is *updated at each stage* j to P_{jN} ($j \geq 1$), which is obtained at stage $k = j - 1$ as the N-th term in the solution sequence of RDE (14.11) with initial condition $P_{(j-1)N}$. Solving problem $\mathcal{P}(x_j, j, N, P_{jN})$ yields the following optimal control sequence:

$$\bar{u}^{FH}(\cdot; x_j, j, N, P_{jN}) = \{u^{FH}(j; x_j, j, N, P_{jN}), u^{FH}(j+1; x_j, j, N, P_{jN}), \ldots, u^{FH}(j+N-1; x_j, j, N, P_{jN}). \tag{14.36}$$

The UTC-RHC control law at stage j ($j \geq 1$) is

$$u^{RH}(x_j, j, N, P_{jN}) = u^{FH}(j; x_j, j, N, P_{jN}). \tag{14.37}$$

The following derivation shows that updating the terminal cost weighting matrix is equivalent to lengthening the horizon. By knowledge of LQR, it is straightforward to have the result that $V^{FH}(x_j, j, N, P_t) = x_j^T P_{t+N} x_j$ for all $j \geq 0$ and all $t \geq 0$. Define $\bar{u}_i$ and $\bar{u}_t$ in the same way as is defined in Section 14.2.1. Then by principle of optimality [22], one obtains

$$\begin{aligned}
& V^{FH}(x_j, j, N, P_{jN}) \\
&= \min_{\bar{u}_i} \left\{ \sum_{i=j}^{j+N-1} (x_i^T Q x_i + u_i^T R u_i) + x_{j+N}^T P_{jN} x_{j+N} \right\} \\
&= \min_{\bar{u}_i} \left\{ \sum_{i=j}^{j+N-1} (x_i^T Q x_i + u_i^T R u_i) + V^{FH}(x_{j+N}, j+N, N, P_{(j-1)N}) \right\} \\
&= \min_{\bar{u}_i} \left\{ \sum_{i=j}^{j+N-1} (x_i^T Q x_i + u_i^T R u_i) \right. \\
&\qquad \left. + \min_{\bar{u}_t} \left\{ \sum_{t=j+N}^{j+2N-1} (x_t^T Q x_t + u_t^T R u_t) + x_{j+2N}^T P_{(j-1)N} x_{j+2N} \right\} \right\} \\
&= \min_{\bar{u}_i} \left\{ \sum_{i=j}^{j+2N-1} (x_i^T Q x_i + u_i^T R u_i) + x_{j+2N}^T P_{(j-1)N} x_{j+2N} \right\} \\
&= V^{FH}(x_j, j, 2N, P_{(j-1)N}).
\end{aligned} \tag{14.38}$$

By induction,

$$\begin{aligned}V^{FH}(x_j, j, N, P_{jN}) &= V^{FH}(x_j, j, 2N, P_{(j-1)N})\\ &= V^{FH}(x_j, j, 3N, P_{(j-2)N})\\ &= \cdots = V^{FH}(x_j, j, (j+1)N, P_0),\end{aligned} \tag{14.39}$$

i.e.,

$$\begin{aligned}V^{FH}(x_j, j, N, P_{jN}) &= V^{FH}(x_j, j, (j+1)N, P_0)\\ &= \min_{\bar{u}_i}\left\{\sum_{i=j}^{(j+1)N-1}(x_i^T Q x_i + u_i^T R u_i) + x_{(j+1)N}^T P_0 x_{(j+1)N}\right\}.\end{aligned} \tag{14.40}$$

Noting Equation (14.40), the UTC-RHC control law at stage j ($j \geq 1$) can be written as

$$\begin{aligned}u^{RH}(x_j, j, N, P_{jN}) &= u^{FH}(j; x_j, j, (j+1)N, P_0)\\ &= -(B^T(P_{(j+1)N-1}B + R)^{-1}B^T P_{(j+1)N-1}Ax_j\end{aligned} \tag{14.41}$$

Clearly, when $j = 0$, equality (14.41) still holds.

Therefore, at stage j ($\forall j \geq 0$), the UTC-RHC control law is

$$\begin{aligned}u^{RH}(x_j, j, N, P_{jN}) &= -(B^T P_{(j+1)N-1}B + R)^{-1}B^T P_{(j+1)N-1}Ax_j\\ &= L_{(j+1)N}^{RH} x_j,\end{aligned} \tag{14.42}$$

and the optimal cost of the corresponding finite horizon LQ optimal control problem is

$$V^{FH}(x_j, j, N, P_{jN}) = x_j^T P_{(j+1)N} x_j, \tag{14.43}$$

and the UTC-RHC closed-loop system is

$$x_{j+1} = (A - B(B^T P_{(j+1)N-1}B + R)^{-1}B^T P_{(j+1)N-1}A)x_j, \quad j = 0, 1, 2, \ldots, \tag{14.44}$$

where P_t is the t-th term in the solution sequence of the RDE (14.11) with the initial condition P_0.

Remark 3. Equation (14.35) shows that the terminal cost weighting matrix is updated at each stage j ($j \geq 1$) to the N-horizon LQ optimal cost weighting matrix computed at the previous stage. Equation (14.40) shows that this is in fact equivalent to effectively making the horizon longer at each stage. As will be seen in Theorem 2, for arbitrary initial terminal cost weighting matrix $P_0 \geq 0$, our algorithm may generate a trajectory that converges to the origin exponentially.

14.3.2 Stability and Convergence

The next result shows that UTC-RHC guarantees stability under a standard assumption made in RHC. This result is established using a proof technique adapted from [23] which shows the closed-loop stability under RHC scheme. The objective is to tie UTC-RHC firmly to the mode of thinking prevalent in standard RHC.

Theorem 1. *Consider the UTC-RHC closed-loop system (14.44) with the initial state x_0. Suppose that assumptions A1–A3 hold and $Q > 0$. Let P_0 satisfy the following inequality:*

$$P_0 \geq A^T P_0 A + Q - A^T P_0 B (B^T P_0 B + R)^{-1} B^T P_0 A. \tag{14.45}$$

Then $\forall N \geq 1$, the closed-loop system (14.44) is uniformly exponentially stable.

Proof. Taking

$$\mathcal{J}(x_j, j) = V^{FH}(x_j, j, N, P_{jN}) = V^{FH}(x_j, j, (j+1)N, P_0) = x_j^T P_{(j+1)N} x_j$$

as a Lyapunov function candidate, where $P_{(j+1)N}$ is the $((j+1)N)$-th term in the solution sequence of RDE (14.11) with initial condition P_0. By Lemma 3, condition (14.45), i.e., $P_0 \geq P_1$, implies the non-increasing monotonicity of sequence $\{P_j\}$. Thus $\mathcal{J}(x_j, j) \leq x_j^T P_0 x_j, \forall x_j \in \mathbb{R}^n$.

To simplify the notations, let $u(x_j) = u^{RH}(x_j, j, N, P_{jN})$ and $l(x, u) = x^T Q x + u^T R u$. It is trivial to show that

$$\mathcal{J}(x_j, j) \geq l(x_j, u(x_j)) \geq x_j^T Q x_j, \quad \forall x_j \in \mathbb{R}^n.$$

Hence, $\mathcal{J}(x_j, j)$ is positive definite and decrescent.

Now we show $\Delta \mathcal{J}(x_j, j) = \mathcal{J}(x_{j+1}, j+1) - \mathcal{J}(x_j, j)$ is negative definite. By principle of optimality [22],

$$\mathcal{J}(x_j, j) = l(x_j, u(x_j)) + V^{FH}(x_{j+1}, j+1, (j+1)N - 1, P_0). \tag{14.46}$$

Then

$$\begin{aligned}
&\mathcal{J}(x_{j+1}, j+1) - \mathcal{J}(x_j, j) + l(x_j, u(x_j)) \\
&\quad = \mathcal{J}(x_{j+1}, j+1) - [l(x_j, u(x_j)) + V^{FH}(x_{j+1}, j+1, (j+1)N - 1, P_0)] \\
&\qquad + l(x_j, u(x_j)) \\
&\quad = \mathcal{J}(x_{j+1}, j+1) - V^{FH}(x_{j+1}, j+1, (j+1)N - 1, P_0) \\
&\quad = x_{j+1}^T P_{(j+2)N} x_{j+1} - x_{j+1}^T P_{(j+1)N-1} x_{j+1} \\
&\quad = x_{j+1}^T (P_{(j+2)N} - P_{(j+1)N-1}) x_{j+1} \leq 0,
\end{aligned} \tag{14.47}$$

since the sequence $\{P_j\}$ is monotonically non-increasing. Therefore,

$$\Delta \mathcal{J}(x_j, j) = \mathcal{J}(x_{j+1}, j+1) - \mathcal{J}(x_j, j) \le -l(x_j, u(x_j)) \le -x_j^T Q x_j, \quad (14.48)$$

i.e., $\Delta \mathcal{J}(x_j, j)$ is negative definite.

By the Lyapunov stability criteria [30, theorem 23.3], the closed-loop system (14.44) is uniformly exponentially stable. □

Remark 4. Theorem 1 shows the stability of our algorithm under condition $P_0 \ge P_1$, which is a standard assumption in RHC [15, 23]. One should note that the condition $P_0 \ge P_1$ may not be replaced by $P_0 \ge P_\infty$. From the monotonicity and convergence properties of RDE, one may naturally conjecture that $P_0 \ge P_\infty$ implies $P_j \ge P_{j+1}$ for all $j \ge 0$. However, this is not the case. Readers may refer to [7] for some interesting counterexamples. Since the non-increasing monotonicity of the sequence $\{P_j\}$ is required in the stability proof, the condition $P_0 \ge P_1$ cannot be replaced by $P_0 \ge P_\infty$.

Remark 5. If there exists *a priori* a stabilizing control gain L, then letting the terminal cost to be $x_k^T P_0 x_k$ defined by Equation (14.17), the condition (14.45) is automatically satisfied.

The next theorem shows that UTC-RHC guarantees the uniform exponential stability even under more relaxed conditions.

Theorem 2. *Consider the UTC-RHC closed-loop system (14.44) with initial state x_0. Suppose that assumptions A1–A3 hold. Then $\forall N \ge 1$ and $\forall P_0 \ge 0$, the closed-loop system (14.44) is uniformly exponentially stable in the sense that, for any initial time k_0 and any initial value x_0, there exists an integer $K \ge k_0$, such that, for all $k \ge K$, x_k will approach exponentially.*

A detailed proof of Theorem 2 is given in [38]. Here we only show the main idea of the proof. We decompose the closed-loop system matrix into two parts (i.e., a constant stable part and a perturbation part), chose a time-invariant Lyapunov function candidate, and show that the difference of this function along the trajectory of the closed-loop system is ultimately negative, hence guarantees the uniform exponential stability of the closed-loop system. The success of our proof is rooted in the convergence property of RDE and the stability property of ARE. Note that we do not use the finite horizon optimal cost $\mathcal{J}(x_j, j) = V^{FH}(x_j, j, N, P_{jN})$ as a Lyapunov function, which is prevalently adopted in the literature (e.g. [21, 23]). In fact, $\mathcal{J}(x_j, j)$ does not qualify as a Lyapunov function for Theorem 2, because the arbitrary $P_0 \ge 0$ cannot ensure the non-increasing monotonicity of the sequence $\{P_j\}$, which is a must for $\Delta J(x_j, j)$ to be negative definite.

Remark 6. UTC-RHC can be treated as a kind of RHC with varying terminal cost weighting matrix. We note there is a generalized stability result for RHC with varying terminal cost weighting matrix by Lee et al. [21, theorem 1], which implies that if the terminal cost weighting matrix satisfies the following inequality for some $H_j \in \mathbb{R}^{m \times n}$,

$$P_{jN} \ge (A+BH_j)^T P_{(j+1)N}(A+BH_j)+Q+H_j^T R H_j, \quad \forall j \ge 0, N \ge 1, \quad (14.49)$$

where $Q > 0$, then UTC-RHC control law exponentially stabilizes the system (14.1). However, the exponential stability conditions for UTC-RHC scheme is more relaxed than the stability conditions required in [21] in the following aspects:

(a) Matrix Q is required to be positive definite in [21] to ensure the positive definiteness of the Lyapunov function they choose. While we choose a time-invariant Lyapunov function, whose positive definiteness is guaranteed by Assumption 1, in which $Q \geq 0$ and $(A, \sqrt{Q})$ is detectable, see [38] for detail.
(b) Theorem 1 in [21] requires *all* the terminal cost weighting matrices P_{jN} ($\forall j \geq 0$, $N \geq 1$) to be positive definite, more exactly, $P_{jN} \geq Q$ for condition (14.49) to hold. This is because they force the Lyapunov function to decrease monotonically from the very beginning. While our result holds for arbitrary $P_0 \geq 0$, even for $P_0 = 0$, and this benefits from the convergence property of RDE and the specific Lyapunov function we choose.
(c) Even when $Q > 0$ and $P_{jN} \geq Q$, to investigate the stability of UTC-RHC in [21], one has to check the condition (14.49) for every $j \geq 0$ and $N \geq 1$. This is not a trivial job, since for UTC-RHC both $P_0 \geq 0$ and $N \geq 1$ can be arbitrary.

Remark 7. Compared with the standard RHC, the advantage of our algorithm lies in the fact that the closed-loop stability property holds without requiring the terminal state constraint, or constraint on the initial terminal cost weighting matrix P_0, or constraint on the horizon size N. Therefore UTC-RHC makes design of a stabilizing controller more flexible.

To our best knowledge, traditional RHC schemes pay more attention to stability than optimality of the closed-loop system. Many RHC schemes have been proposed only to guarantee the close-loop stability, yet not to obtain the optimality in the sense of minimizing the infinite horizon cost. The following proposition shows our algorithm eventually achieves the optimal performance.

Proposition 1. *Assume that A1–A3 hold and $P_0 \geq 0$. Then the UTC-RHC state feedback control gain converges to the optimal control gain of the infinite horizon optimal control problem (14.2), i.e., $L^{RH}_{(j+1)N} \to L_\infty$ as $j \to \infty$.*

Proof. By Lemma 1, under assumptions A1–A3, the ARE (14.4) has a unique positive semidefinite solution P_∞. Lemma 2 implies $P_{(j+1)N-1} \to P_\infty$ as $j \to \infty$. Therefore, $L^{RH}_{(j+1)N} \to L_\infty$ as $j \to \infty$. □

14.4 Simulation Results

In this section, two examples are presented to compare the stability and optimality performances of the standard RHC and UTC-RHC algorithm. The first example is a single input open-loop unstable LTI system which is adopted directly from [29]. In the second example, we apply our algorithm to a discretized two-mass translational mechanical system. This is a two-input two-output system.

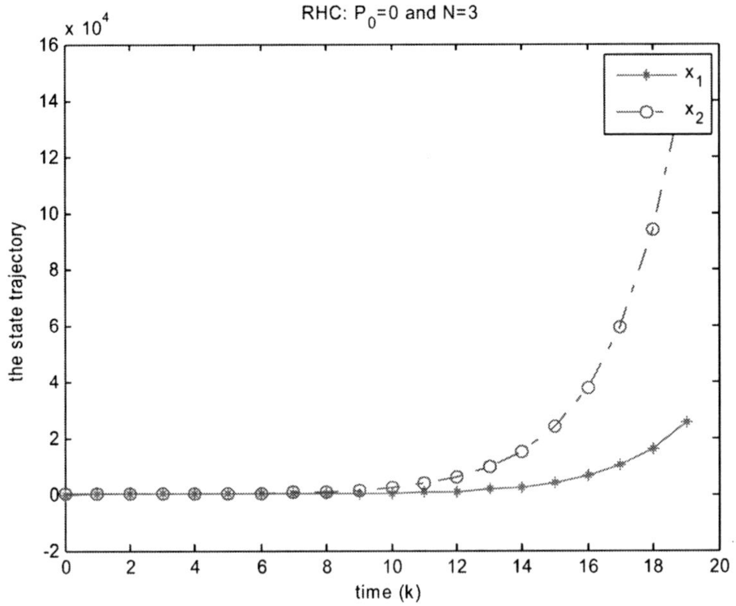

Fig. 14.1 State trajectory of RHC closed-loop system for $N = 3$.

14.4.1 Example 1

Consider the single input open-loop unstable system (14.28). Q and R take the same values as those in Counterexample 1. P_∞ solves the ARE (14.4) and corresponds to the optimal infinite horizon cost, and L_∞ is the optimal control gain given by Equation (14.3).

$$P_\infty = \begin{bmatrix} 114.6619 & -56.4269 \\ -56.4269 & 29.7112 \end{bmatrix},$$

$$L_\infty = [-1.6938 \quad -0.6519].$$

For RHC, Quan et al. [29] show that the stability guaranteed horizon size takes its maximal value when the terminal cost weighting matrix $P_0 = 0$. So in this example, we choose $P_0 = 0$ and the initial state $x_0 = [-6 \ \ 4]^T$, while noting that x_0 can be chosen arbitrarily.

14.4.1.1 Stability of RHC and UTC-RHC

For this example, by standard RHC algorithm, Quan et al. [29] obtain the smallest stability guaranteed horizon size $N^* \geq 5$ and it also points out that the proposed horizon size may not be conservative. In fact, this system is also stable for $N \geq 4$

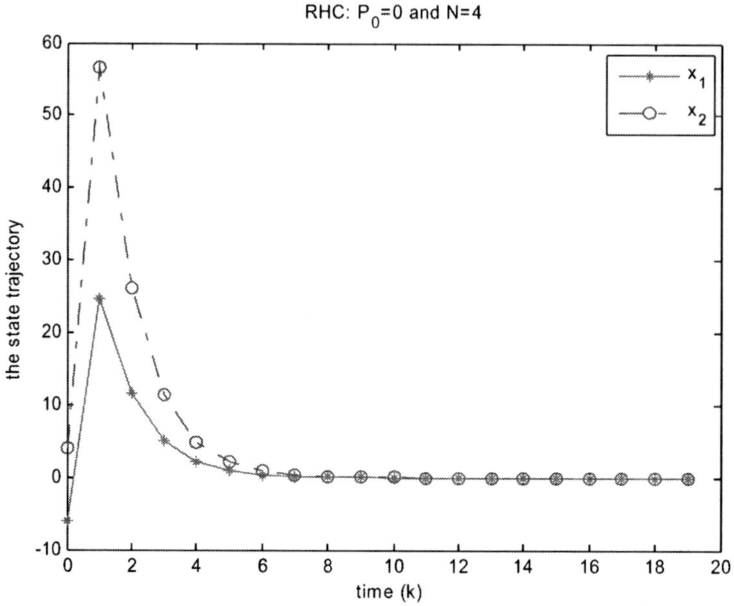

Fig. 14.2 State trajectory of RHC closed-loop system for $N = 4$.

but unstable for $N \leq 3$, as shown in Figures 14.1 and 14.2. While by applying the UTC-RHC algorithm, the closed-loop system is always stable for arbitrary $N \geq 1$, as shown in Figures 14.3 and 14.4.

14.4.1.2 Convergence of RHC and UTC-RHC

When $N = 3$, the RHC control gain is given by Equation (14.12) as $L = [-0.3988\ -1.2938]$; when $N = 4$, the RHC control gain is $L = [-1.5397\ \ -0.7283]$. For a fixed horizon N, the RHC control gain is constant and not optimal unless $N \to \infty$. On the other hand, the UTC-RHC control gain eventually approaches its optimal value L_∞ for all $N \geq 1$, which is illustrated in Figures 14.5 and 14.6.

14.4.2 Example 2

In this example, we consider a two-mass translational mechanical system [37] shown in Figure 14.7, where m_1 and m_2 are two masses; k_1 and k_2 are spring coefficients; c_1 and c_2 are damping coefficients; u_1 and u_2 are force inputs; y_1 and y_2 are displacement outputs of the two masses.

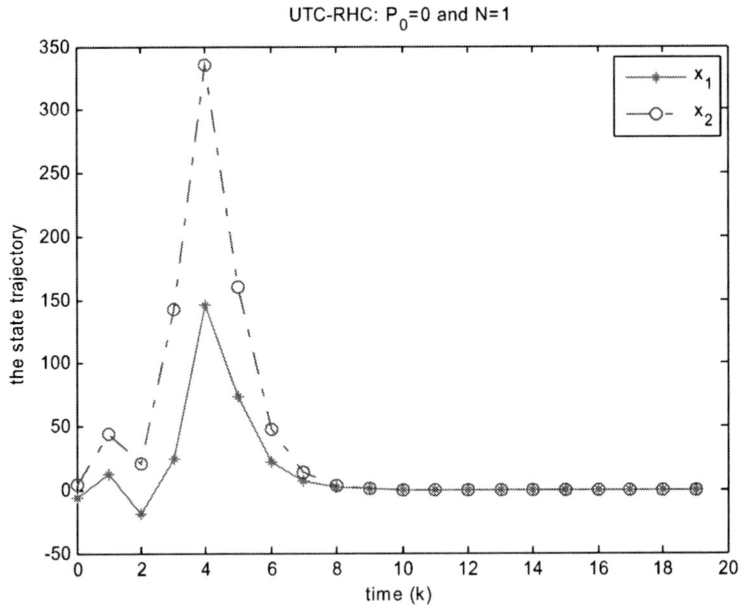

Fig. 14.3 State trajectory of UTC-RHC closed-loop system for $N = 1$.

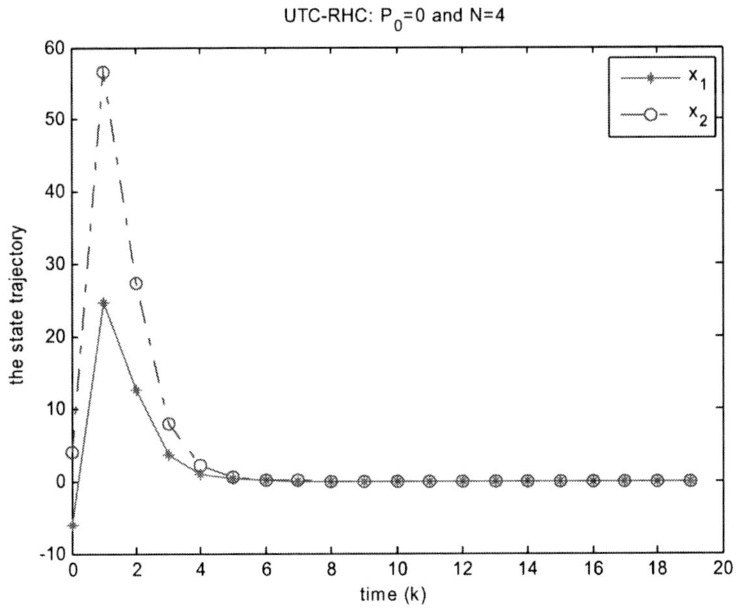

Fig. 14.4 State trajectory of UTC-RHC closed-loop system for $N = 4$.

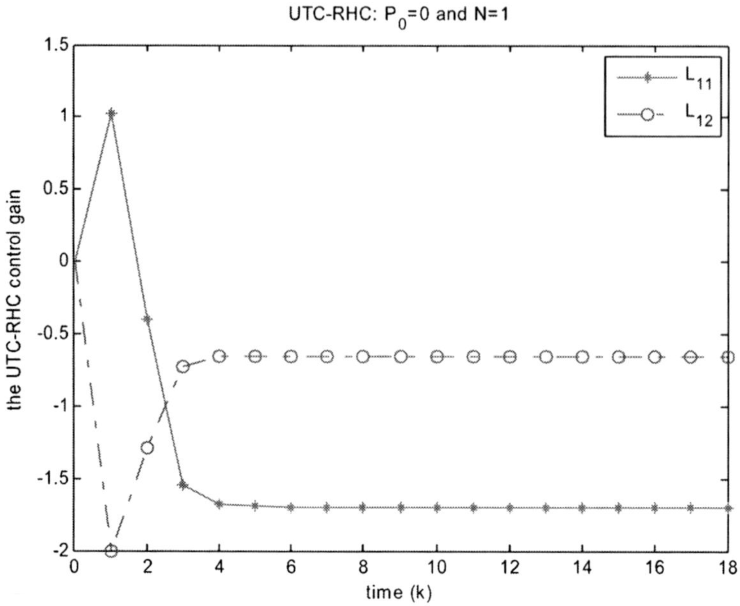

Fig. 14.5 Control gain of UTC-RHC algorithm for $N = 1$.

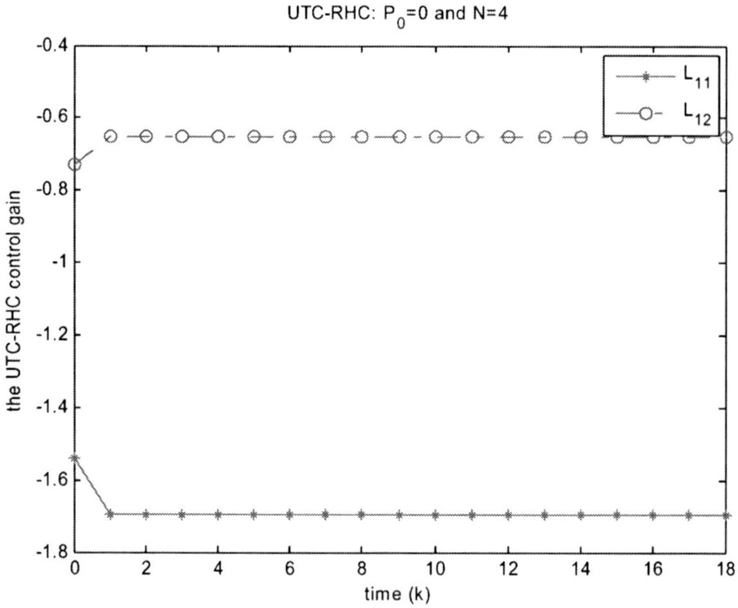

Fig. 14.6 Control gain of UTC-RHC algorithm for $N = 4$.

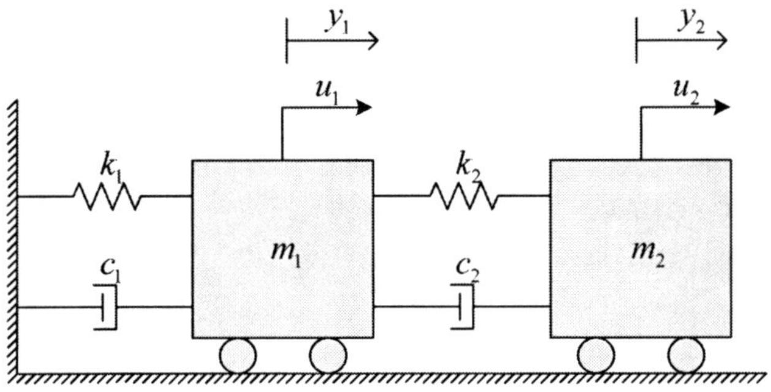

Fig. 14.7 The two-mass translational mechanical system.

Define the states as $x_c = [x_{c1}, x_{c2}, x_{c3}, x_{c4}]^T$, where $x_{c1} = y_1$, $x_{c2} = \dot{y}_1 = \dot{x}_{c1}$, $x_{c3} = y_2$ and $x_{c4} = \dot{y}_2 = \dot{x}_{c3}$. Define the output as $y = [y_1, y_2]^T$. The state space representation of this mechanical system is

$$\begin{aligned} \dot{x}_c &= A_c x_c + B_c u_c, \\ y &= C_c x_c + D_c u_c, \end{aligned} \tag{14.50}$$

where

$$A_c = \begin{bmatrix} 0 & 1 & 0 & 0 \\ \frac{-(k_1+k_2)}{m_1} & \frac{-(c_1+c_2)}{m_1} & \frac{k_2}{m_1} & \frac{c_2}{m_1} \\ 0 & 0 & 0 & 1 \\ \frac{k_2}{m_2} & \frac{c_2}{m_2} & -\frac{k_2}{m_2} & -\frac{c_2}{m_2} \end{bmatrix}, \quad B_c = \begin{bmatrix} 0 & 0 \\ \frac{1}{m_1} & 0 \\ 0 & 0 \\ 0 & \frac{1}{m_2} \end{bmatrix},$$

$$C_c = \begin{bmatrix} 1 & 0 & 0 & 0 \\ 0 & 0 & 1 & 0 \end{bmatrix} \quad \text{and} \quad D_c = \begin{bmatrix} 0 & 0 \\ 0 & 0 \end{bmatrix}.$$

This is a fourth-order two-input two-output LTI system. For the convenience of illustration, we let the system parameters be $c_1 = c_2 = 0$, $m_1 = 1$ kg, $m_2 = 1$ kg, $k_1 = 1$ N/m and $k_2 = 1$ N/m, then

$$A_c = \begin{bmatrix} 0 & 1 & 0 & 0 \\ -2 & 0 & 1 & 0 \\ 0 & 0 & 0 & 1 \\ 1 & 0 & -1 & 0 \end{bmatrix} \quad \text{and} \quad B_c = \begin{bmatrix} 0 & 0 \\ 1 & 0 \\ 0 & 0 \\ 0 & 1 \end{bmatrix}.$$

The open-loop system is marginally stable, for all the eigenvalues of A_c are on the imaginary axis.

Choosing a sampling time $T_S = 0.1$ s, the discrete time system is obtained by using the zero-order-hold technique as follows:

$$A = \begin{bmatrix} 0.9900 & 0.0997 & 0.0050 & 0.0002 \\ -0.1992 & 0.9900 & 0.0095 & 0.0050 \\ 0.0050 & 0.0002 & 0.9950 & 0.0998 \\ 0.0995 & 0.0050 & -0.0997 & 0.9950 \end{bmatrix}, \quad B = \begin{bmatrix} 0.0005 & 0 \\ 0.0997 & 0.0002 \\ 0 & 0.0050 \\ 0.002 & 0.0998 \end{bmatrix},$$

$$C = \begin{bmatrix} 1 & 0 & 0 & 0 \\ 0 & 0 & 1 & 0 \end{bmatrix} \quad \text{and} \quad D = \begin{bmatrix} 0 & 0 \\ 0 & 0 \end{bmatrix}. \tag{14.51}$$

We consider the discrete-time system (14.51). Choose Q and R to be the identity matrices with appropriate dimensions. P_∞ and L_∞ are obtained similarly as in Example 1,

$$P_\infty = \begin{bmatrix} 28.8229 & 3.2635 & -7.2967 & 2.2451 \\ 3.2635 & 13.2579 & 2.2451 & 1.6630 \\ -7.2967 & 2.2451 & 21.5262 & 5.5086 \\ 2.2451 & 1.6630 & 5.5086 & 14.9209 \end{bmatrix},$$

$$L_\infty = \begin{bmatrix} -0.1868 & -1.2057 & -0.2610 & -0.1568 \\ -0.2610 & -0.1568 & -0.4478 & -1.3625 \end{bmatrix}.$$

14.4.2.1 Open-Loop System

Obviously, the discrete-time open-loop system is marginally stable, for the eigenvalues of the system matrix are on the unit circle. Given the initial state $x_0 = [-2 \;\; 2 \;\; 1 \;\; -1]^T$, the state trajectory of the open-loop system is shown in Figure 14.8. This is a periodic oscillation movement.

14.4.2.2 Stability of RHC and UTC-RHC

Let the terminal cost weighting matrix $P_0 = 0$. We plot the state trajectory of the RHC closed-loop system and UTC-RHC closed-loop system for $N = 1$ and $N = 2$, respectively, as shown in Figures 14.9–14.12.

As illustrated by these figures, the RHC closed-loop system is not stable when $N = 1$, while the UTC-RHC closed-loop system is stable. In fact, when $N = 1$, the RHC control gain is 0, which will be shown in Figure 14.13, and in this case the closed-loop system is the same as the open-loop system, see Figures 14.8 and 14.9. When $N = 2$, both the RHC closed-loop system and UTC-RHC closed-loop system are stable, but the state approaches the origin much faster for UTC-RHC closed-loop system than for RHC closed-loop system.

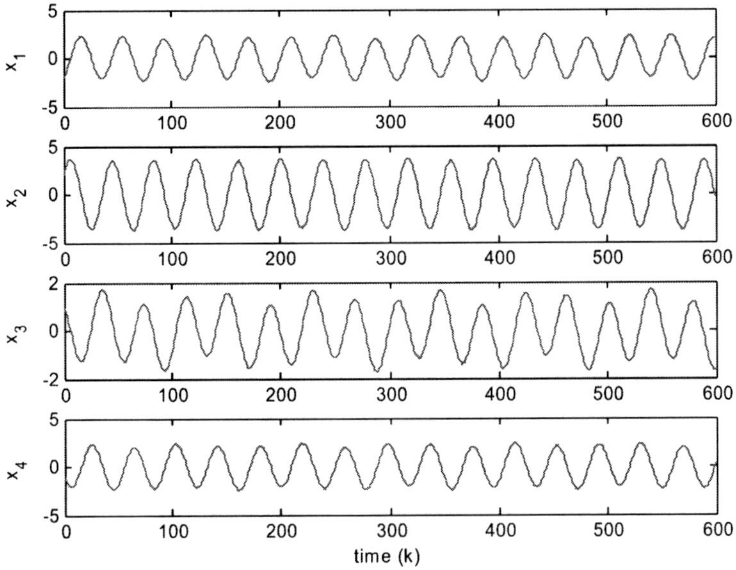

Fig. 14.8 State trajectory of the open-loop system.

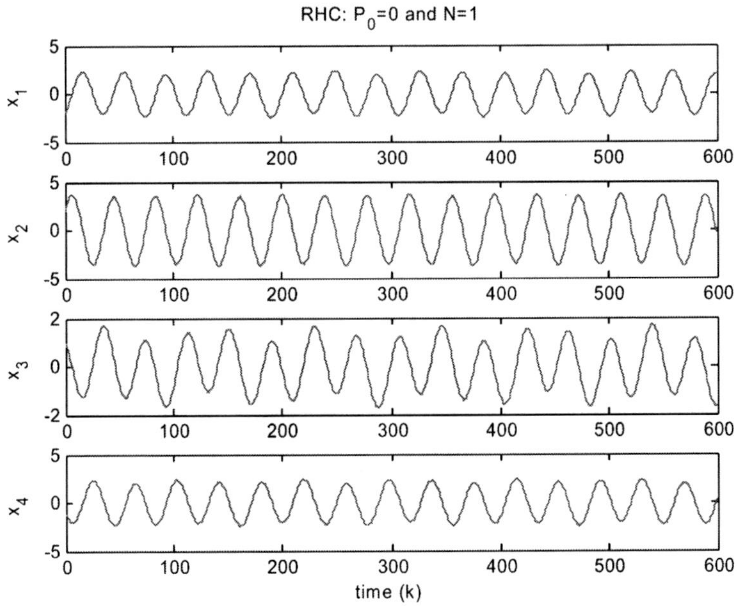

Fig. 14.9 State trajectory of the RHC closed-loop system for $N = 1$.

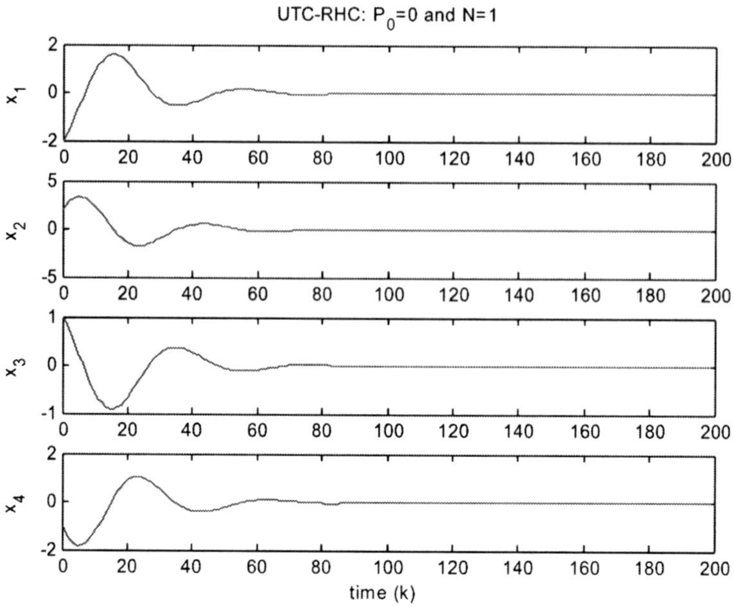

Fig. 14.10 State trajectory of the UTC-RHC closed-loop system for $N = 1$.

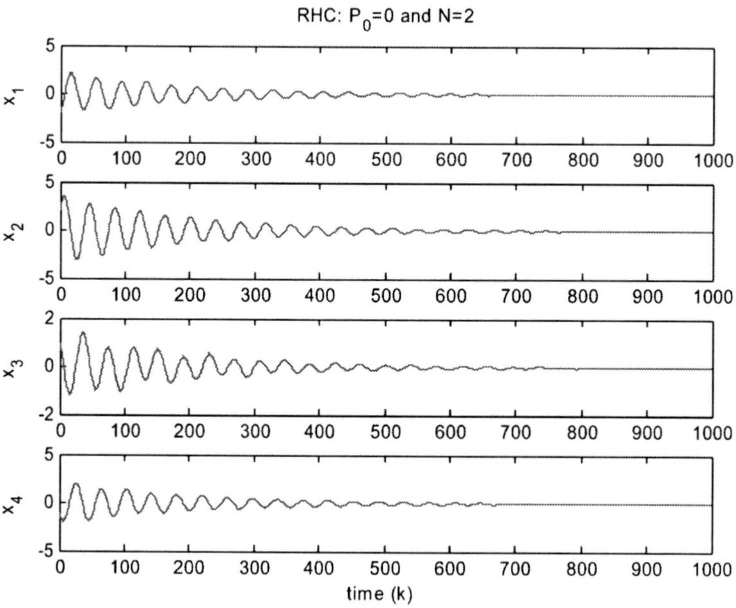

Fig. 14.11 State trajectory of the RHC closed-loop system for $N = 2$.

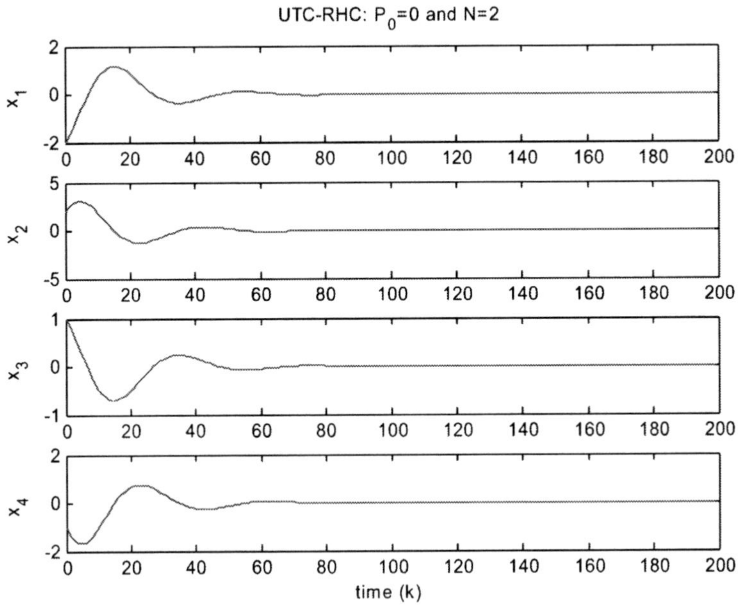

Fig. 14.12 State trajectory of the UTC-RHC closed-loop system for $N = 2$.

14.4.2.3 Convergence of RHC and UTC-RHC

As shown in Figures 14.13 and 14.14, the control gain of the RHC algorithm is constant and not optimal for $N = 1$ and $N = 2$. But for any positive integer N, the control gain of the UTC-RHC algorithm is time-varying and converges to the optimal value L_∞ as $k \to \infty$, which is illustrated by Figures 14.15 and 14.16 for the case $N = 1$ and $N = 2$.

14.5 Conclusion

In this chapter, we propose an updated terminal cost RHC (UTC-RHC) algorithm. The closed-loop system under the UTC-RHC scheme is proved to be uniformly exponentially stable without imposing any constraints on terminal state, or terminal cost, or the horizon size. Thus the design of a stable RHC closed-loop system becomes more flexible. Moreover, the control gain generated by UTC-RHC algorithm converges to the optimal control gain associated with the infinite horizon optimal control problem. Two examples illustrate the performance of UTC-RHC.

This chapter only focuses on discrete linear time-invariant systems. Extending this algorithm to discrete-time nonlinear systems and to continuous-time systems is under investigation.

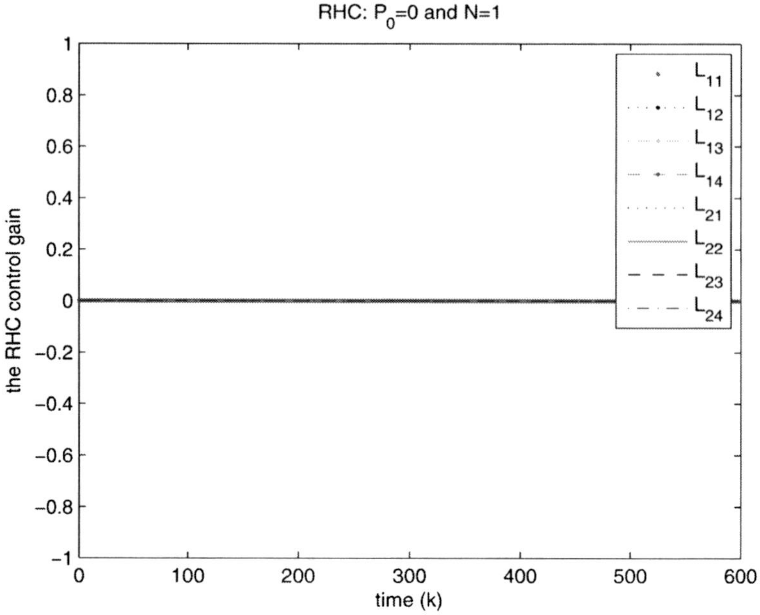

Fig. 14.13 Control gain of RHC algorithm for $N = 1$.

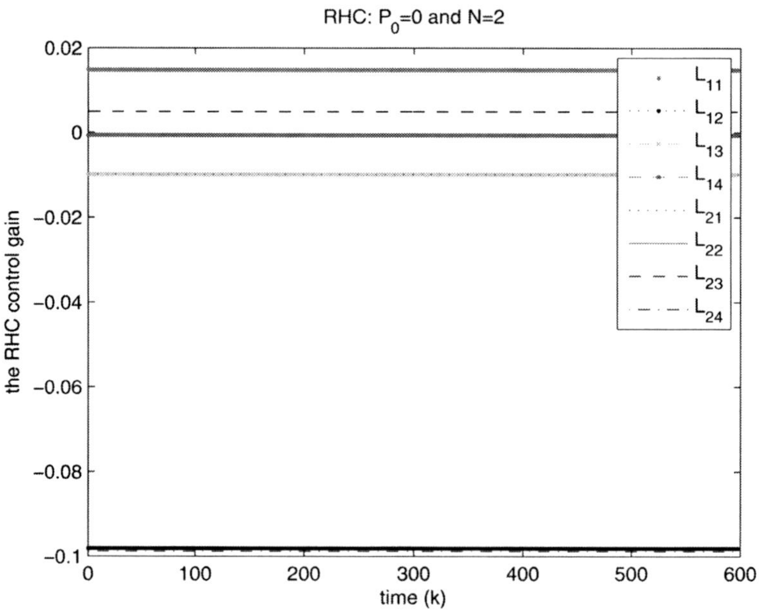

Fig. 14.14 Control gain of RHC algorithm for $N = 2$.

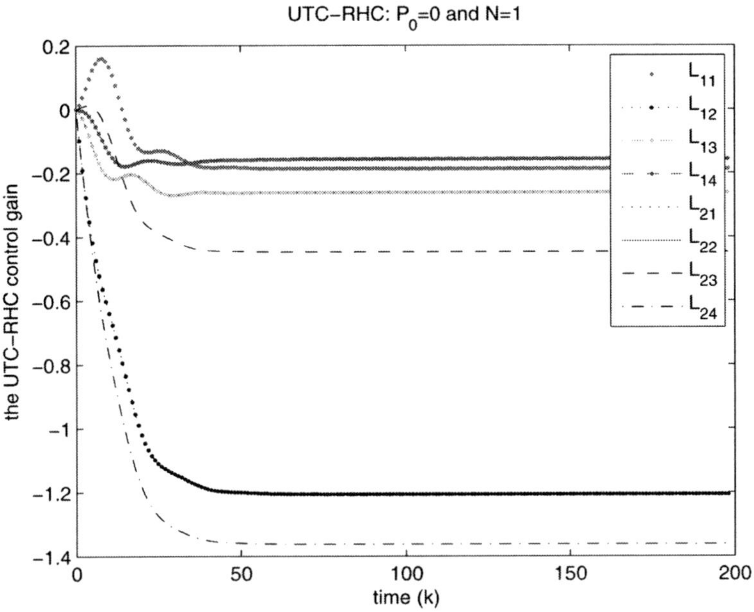

Fig. 14.15 Control gain of UTC-RHC algorithm for $N = 1$.

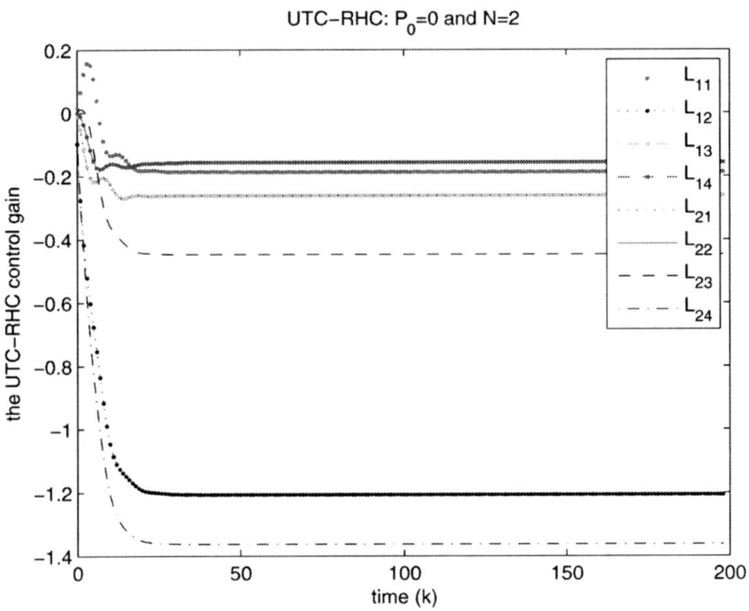

Fig. 14.16 Control gain of UTC-RHC algorithm for $N = 2$.

References

1. A. Al-Tamimi, F.L. Lewis and M. Abu-Khalaf, Discrete-time nonlinear HJB solution using approximate dynamic programming: Convergence proof, *IEEE Trans. on Systems Man and Cybernetics, Part B: Cybernetics* **38**, 943–949, 2008.
2. A.G. Barto, R.S. Sutton and C.W. Anderson, Neuronlike elements that can solve difficult learning control problems, *IEEE Trans. on Systems Man and Cybernetics* **SMC-13**, 835–846, 1983.
3. D.P. Bertsekas, Dynamic programming and suboptimal control: A survey from ADP to MPC, Laboratory for Information and Decision Systems Report 2632, MIT, 2005.
4. D.P. Bertsekas and J.N. Tsitsiklis, *Neuro-Dynamic Programming*, Athena Scientific, Belmont, MA, 1996.
5. R.R. Bitmead, M. Gever and V. Wertz, *Adaptive Optimal Control: The Thinking Man's GPC*, Prentice Hall, New York, 1990.
6. R.R. Bitmead and M. Gever, Riccati difference and differential equations: convergence, monotonicity and stability, in *The Riccati Equation*, S. Bittani, A.J. Laub and J.C. Willems (Eds.), Springer-Verlag, New York, pp 263–292, 1991.
7. R.R. Bitmead, M. Gever, I.R. Petersen and R.J. Kaye, Monotonicity and stabilizability properties of solutions of the Riccati difference equation: Propositions, lemmas, theorems, fallacious conjectures and counterexamples, *Systems and Control Letters* **5**, 309–315, 1985.
8. C.T. Chen, *Linear System Theory and Design*, Holt, Rinehart and Winston, New York, 1984.
9. H. Chen and F. Allgower, A quasi-infinite horizon nonlinear model predictive control scheme with guaranteed stability, *Automatica* **34**, 1205–1217, 1998.
10. L. Chisci, A. Lombardi and E. Mosca, Dual receding horizon control of constrained discrete time systems, *European Journal of Control* **2**, 278–285, 1996.
11. G. De Nicolao, L. Magni and R. Scattolini, Stabilizing receding horizon control of nonlinear time-varying systems, *IEEE Trans. on Automatic Control* **43**, 1030–1036, 1998.
12. G. Grimm, M.J. Messina, S.E. Tuna and A.R. Teel, Model predictive control: for want of a local control Lyapunov function, all is not lost, *IEEE Trans. on Automatic Control* **50**, 546–558, 2005.
13. R.A. Howard, *Dynamic Programming and Markov Processes*, MIT Press, Cambridge, MA, 1960.
14. A. Jadbabaie and J. Hauser, On the stability of receding horizon control with a general terminal cost, *IEEE Trans. on Automatic Control* **50**, 674–678, 2005.
15. A. Jadbabaie, J. Yu and J. Hauser, Unconstrained receding horizon control of nonlinear systems, *IEEE Trans. on Automatic Control* **46**, 776–783, 2001.
16. S.S. Keerthi and E.G. Gilbert, Optimal infinite-horizon feedback laws for a great class of constrained discrete-time systems: Stability and moving horizon approximations, *Journal of Optimization Theory and Applications* **57**, 65–293, 1988.
17. W.H. Kwon and S. Han, *Receding Horizon Control: Model Predictive Control for State Models*, Springer-Verlag, London, 2005.
18. W.H. Kwon and A.E. Pearson, A modified quadratic cost problem and feedback stabilization of a linear system, *IEEE Trans. on Automatic Control* **22**, 838–842, 1977.
19. W.H. Kwon and A.E. Pearson, On feedback stabilization of time-varying discrete linear systems, *IEEE Trans. on Automatic Control* **23**, 479–481, 1978.
20. T. Landelius, Reinforcement learning and distributed local model synthesis, PhD Thesis, Linkoping University, Sweden, 1997.
21. J.W. Lee, W.H. Kwon and J.H. Choi, On stability of constrained receding horizon control with finite terminal weighting matrix, *Automatica* **34**, 1607–1612, 1998.
22. F.L. Lewis and V.L. Syrmos, *Optimal Control*, 2nd edn. John Wiley and Sons, New York, 1995.
23. D.Q. Mayne, J.B. Rawlings, C.V. Rao and P.O.M. Scokaert, Constrained model predictive control: stability and optimality, *Automatica* **36**, 789–814, 2000.

24. H. Michalska and D.Q. Mayne, Robust receding horizon control of constrained nonlinear systems, *IEEE Trans. on Automatic Control* **38**, 1623–1633, 1993.
25. M. Morari and J.H. Lee, Model predictive control: Past, present and future, *Computers and Chemical Engineering* **23**, 667–682, 1999.
26. J.A. Primbs and V. Nevistic, Feasibility and stability of constrained finite receding horizon control, *Automatica* **36**, 965–971, 2000.
27. J.A. Primbs, V. Nevistic and J. Doyle, Nonlinear optimal control: A control Lyapunov function and receding horizon perspective, *Asian Journal of Control* **1**, 14–24, 1999.
28. S.J. Qin and T.A. Badgwell, A survey of industrial model predictive control technology, *Control Engineering Practice* **11**, 733–764, 2003.
29. Z. Quan, S. Han and W.H. Kwon, Stability-guaranteed horizon size for receding horizon control, *IEICE Trans. on Fundamentals of Electronics, Communications and Computer Sciences* **E90-A**, 523–525, 2007.
30. W.J. Rugh, *Linear System Theory*, 2nd edn. Prentice Hall, Upper Saddle River, NJ, 1996.
31. P.O.M. Scokaert, D.W. Mayne and J.B. Rawlings, Suboptimal model predictive control (feasibility implies stability), *IEEE Trans. on Automatic Control* **44**, 648–654, 1999.
32. J. Si, A.G. Barto, W. Powell and D.C. Wunsch, *Handbook of Learning and Approximate Dynamic Programming*, Wiley-Interscience, Hoboken, NJ, 2004.
33. R.S. Sutton and A.G. Barto, *Reinforcement Learning: An Introduction*, MIT Press, Cambridge, MA, 1998.
34. D. Vrabie, M. Abu-Khalaf, F.L. Lewis and Y. Wang, Continuous time ADP for linear systems with partially unknown dynamics, in *Proceedings of IEEE International Symposium on Approximate Dynamic Programming and Reinforcement Learning*, pp. 247–253, 2007.
35. C. Watkins, Learning from delayed rewards, PhD Thesis, Cambridge University, Cambridge, England, 1989.
36. P.J. Werbos, A menu of designs for reinforcement learning over time, in *Neural Networks for Control*, W.T. Miller, R.S. Sutton and P.J. Werbos (Eds.), MIT Press, Cambridge, MA, pp. 67–95, 1990.
37. R.L. Williams II and D.A. Lawrence, *Linear State-Space Control Systems*, John Wiley and Sons, Hoboken, NJ, 2007.
38. H.W. Zhang, J. Huang and F.L. Lewis, Algorithm and stability of ATC receding horizon control, in *Proceedings of IEEE International Symposium on Approximate Dynamic Programming and Reinforcement Learning*, pp. 28–35, 2009.

Chapter 15
Identifier-Based Discovery in Large-Scale Networks
An Economic Perspective

Joud Khoury and Chaouki T. Abdallah

Abstract The design of any network mechanism that requires collaboration among selfish agents could only benefit from accounting for the complex social and economic interactions and incentives of the agents using the design. This chapter presents a broad treatment of the main economic issues that arise in the context of identifier-based discovery on large scale networks, particularly on the Internet. An "identified" object (such as a node or service), referred to as a player, demands to be discoverable by the rest of the network on its "identifier". A discovery scheme provides such a service to the players and incurs a cost for doing so. Providing such a service while accounting for the cost and making sure that the incentives of the players are aligned is the general economic problem that we address in this work. After introducing the identifier-based discovery problem, we present a taxonomy of discovery schemes and proposals based on their business model and we pose several questions that are becoming increasingly important as we proceed to design the inter-network of the future. An incentive model for distributed discovery in the context of the Border Gateway Protocol (BGP) and path-vector protocols in general is then presented. We model BGP route distribution and computation using a game in which a BGP speaker advertises its prefix to its direct neighbors promising them a reward for further distributing the route deeper into the network. The neighbors do the same thing with their direct neighbors, and so on. The result of the this cascaded route distribution is a globally advertised prefix and hence discoverability. We present initial results on the existence of equilibria in the game and we motivate our ongoing work.

Joud Khoury · Chaouki T. Abdallah
Department of Electrical & Computer Engineering, University of New Mexico, Albuquerque, NM 87131, USA; e-mail: {jkhoury, chaouki}@ece.unm.edu

K.P. Valavanis (ed.), Applications of Intelligent Control to Engineering Systems, 395–425.

15.1 Introduction

Traditionally, the design process in the context of the Internet has focused on sources of value as they relate to performance, robustness, resilience, reliability, etc. with less emphasis on the socio-economical dynamics that underly the latter. The value of any new design in the new era does not solely depend on performance and must take into account the complex social and economic interactions and incentives of the agents using the design if success is to be reached [26, 38]. Check [38] for an interesting overview of several tools that are important in bridging computer science and economics to better understand the complex socio-economic interactions in the context of the Internet, and [26] for an interesting overview of several of the problems and applications arising at the interface between information and networks.

Almost every networking application relies on discovery and naming (alternatively referred to as identification throughout the discussion since we ignore name semantics) services. An identifier (or alternatively a name) in this context refers to an address that is independent of the network topology but that could nevertheless be routable. Identifier-based discovery (simply referred to as discovery hereafter) is a core network service aimed at discovering a network path to an identified object. Discovery is usually the first step in communication, before a path to the destination object is established. Given an identifier of some object on the network, discovering a path to the object could either utilize mapping/resolution where the identifier is mapped to some locator (see for example [18, 33, 35], and the Domain Name System (DNS)), or it could utilize routing-on-identifiers (see [5, 7, 12, 28] etc.). In either case however, an underlying routing scheme that routes on locators typically exists and is utilized after a path has been discovered for efficient communication. Note that the terms identifier and locator are both addresses at different layers of abstraction. We differentiate the two terms only after we fix an upper layer: an identifier at the upper layer maps into a locator which is an address relative to the upper layer. The locator itself is a path identifier at a lower layer.

In this discussion we assume that a naming or identification system for a large scale network, the Internet mainly, is required given the network's mobile and ubiquitous usage models. For example, on the Internet, this translates into either designing a new system or enhancing the current ones (for example DNS). While there is a rich literature on applying game theory and economics models to Internet games, we find in the networking literature a number of proposals for Internet discovery schemes (and id routing) requiring significant coordination among selfish users while ignoring the economic aspects that may possibly render them infeasible or inefficient (and we shall give several examples of such system or proposals later in Section 15.3). In a future Internet in which domains or Autonomous Systems (ASes) are selfish agents trying to maximize their local utilities, the design of any identifier based discovery scheme could benefit from establishing the right economic models. The problem on the Internet specifically is exacerbated as there are multiple layers of identification managed by different systems, mainly DNS at the application and the Border Gateway Protocol (BGP) [39] at the network layer. Note that in the case of the latter, the Internet Protocol (IP) address space has been aggressively de-aggregated for reasons that we shall discuss shortly.[1] In this sense, the IP address space has been transformed into an identifier space in which the address is almost independent of the topological location especially in the case of Provider Independent (PI) addressing. In this work, we take a first step at designing discovery schemes for a future Internet (check NSF's FIND [3], and the GENI [1] initiatives for current efforts to redesign and implement the future Internet). We introduce two interrelated design goals that we have identified as missing in the current design process, *service differentiation* and *incentives*, and we elaborate on the latter.

Simply stated, a named object (such as a node or service), referred to as a player, demands to be discoverable by the rest of the network. A discovery scheme provides such service to the players. We define the *discovery level* to be a measure of "how discoverable" a player is by the rest of the network (this is "how easy" it is for the network to discover the player not the opposite). The performance of discovery, or the discovery level, could significantly affect the player's business model especially in time-sensitive application contexts. If discovering an object takes a significant time relative to the object's download time, the requesting user's experience suffers. As an example, when no caching is involved, the DNS resolution latency comprises a significant part of the total latency to download a webpage (10–30%) [9, 23]. This overhead becomes more obvious in Content Distribution Networks (CDNs), where content objects are extensively replicated throughout the network closer to the user and the discovery (or resolution) could potentially become the bottleneck. Traditionally, the design of discovery schemes has assumed that all players have the same discovery performance requirements, thus resulting in homogeneous demand. In such a setting, the discovery schemes deliver a discovery service that is oblivious to the actual, possibly heterogeneous, discovery requirements – and valuations – of the different players. In reality however, the CNN site will likely value a higher discovery level more than a generic residential site. An interesting question to ask is therefore the following: should the design of discovery mechanisms account for discovery service differentiation? To further motivate the need for differentiation, we note that on the current Internet, Akamai provides such an expedited resolution service [4]. However, the service which is based on DNS suffers from the same pitfalls of the latter (expensive first lookup and critical dependence on caching) and tightly couples the content distribution provider with the resolution service provider. In a recent work [25], we have presented a first attempt to answer this question by introducing the *multi-level discovery* framework which is concerned with the design of discovery schemes that can provide different service levels to different sets of nodes.

Obviously, there is a cost associated with being discoverable. This could be the cost of distributing and maintaining information (state) about the identifiers. Accounting for and sharing the cost of discovery is an interesting problem whose absence in current path discovery schemes has led to critical economic and scalability concerns. As an example, the Internet's BGP [39] control plane functionality is oblivious to cost. More clearly, a node (BGP speaker) that advertises a

[1] IP is the network layer protocol in the current Internet that allows interconnecting disparate Internet domains or ASes.

provider-independent prefix (identifier) does not pay for the cost of being discoverable. Such a cost may be large given that the prefix is maintained at every node in the Default Free Zone (DFZ)[2] (the rest of the network pays!). Such incentive mismatch in the current BGP workings is problematic and is further exacerbated by provider-independent addressing, multi-homing, and traffic engineering practices [34]. Notice here that BGP with its control and forwarding planes represents a discovery scheme on prefixes which are technically flat identifiers in a largely de-aggregated namespace. Hereafter, we refer to this form of BGP as BGP-DA for De-Aggregation. Hence, *we conjecture that a discovery scheme should be aware of incentives and cost necessitating that players/nodes pay for the cost of getting the service*.

The rest of the chapter is organized as follows: first we motivate the notion of strategic interactions on networks by presenting three games in Section 15.2 that we shall refer to throughout the discussion. Section 15.3 presents a taxonomy of discovery schemes based on their business models as well as our initial thoughts on suitable economic models for the different discovery models. An incentive model for distributed discovery in the context of BGP and path-vector protocols in general is then presented in Section 15.4 before concluding.

15.2 Networks and Strategic Behavior

Game theory is a fundamental mathematical tool for understanding the strategic interactions among selfish network agents, particularly on the Internet over which autonomous agents (e.g. ASes) interact. The theory provides several solution concepts to help study games that arise in different situations and that have specific requirements and varying underlying assumptions [37]. We overview some basic ones here and we provide examples to illustrate each. The most central and widely applicable solution concept is the *pure strategy* Nash equilibrium (PSNE or NE) which could be simply thought of as a set of strategies that forms a stable solution of the game. A set of strategies for the players is termed a *strategy profile*. Under NE strategy profile, no player can move profitably (i.e. increase her payoff) by deviating from her strategy given every other player's strategy. Despite its wide applicability, the NE solution has several shortcomings in that it may not exist (and hence might require mixing), there could be multiple equilibria, and there is no clear way of how to get to it. In this sense, the *mixed strategy* solution concept was developed by Nash to guarantee that an equilibrium will always exist in the game by mixing the player's strategies (introducing probability distributions over the pure strategies and hence rendering the strategy space a convex set). A more stringent solution concept is the *dominant strategy* solution. Unlike the pure strategy solution, a dominant strategy yields a player the highest payoff independent of the strategies of the rest of the players. Dominant strategies are very attractive solutions when they exist, and when

[2] The DFZ refers to the set of BGP routers in the Internet that do not have any default route as part of their routing table i.e. any such router keeps state about every advertised prefix/destination.

they don't exist game designers might try to design for them. For example, when a player's strategy is to declare some private information that is necessary to the social welfare of the game, an attractive solution would be to make the truthful revelation a dominant strategy hence making sure that the player will never have an incentive to lie. The mechanism design framework [32] provides exactly this solution allowing the mechanism "designer" to achieve a dominant strategy solution (in addition to other design goals). A extension to mechanism design, Algorithmic Mechanism Design (AMD) [36], deals with the computation complexity of the solution and Distributed AMD [16] further considers the "network complexity" in distributed settings. Several other solution concepts exist; however, we will only overview one more which is the *subgame perfect* Nash equilibrium (SPNE) which extends the one-shot NE concept to settings in which players take turns playing (e.g. player 1 plays first, then player 2 plays). In such setting, the subgame perfect NE becomes more "natural" as it captures the order of decision taking. Briefly, a subgame perfect NE is a NE in every subgame of the original game where a subgame could be informally defined as a portion of the game that can be independently analyzed. Note that by the formal definition of a subgame, every game is a subgame of itself and hence every SPNE is necessarily a NE. For formal definitions of the solution concepts and a comprehensive treatment of the topic, we refer the reader to [17].

How does strategy factor into networking problems? To motivate the importance of strategic behavior, we hereby present three networking applications that employ different solution concepts and that we shall refer to throughout the discussion. Our hope is that the games highlight some of the basic economic issues that are of interest to network settings and the tools that are useful in studying these settings. Note that the games we present here might not be straightforward for the unexperienced reader who we refer to [17, 37] for introductory material on the subject. The first application we present is that of "query incentive networks" and is due to Kleinberg and Raghavan [27]. The second application is that of "trading networks with price setting agents" due to Blume et al. [10]. The common aspect of the first two games is that price setting is a strategic behavior of the players which is not the case with the third application we present, "Incentive-compatible interdomain routing" due to Feigenbaum et al. [15]. Additionally, while the first two games are solely interested in studying the equilibria, the third presents a distributed mechanism that achieves the solution.

15.2.1 Nash Equilibria and Query Incentive Networks Game [27]

Query incentives are motivated in peer-to-peer and in social networks where some root node issues a query seeking a piece of information or a service on the network. The seeker does not know which nodes on the network have the answer (neither does any other node) and hence the only way to find the answer is to propagate the query deeper into the network until a node with an answer is reached. In order to do so, every node needs to incentivize its direct children to propagate the query deeper

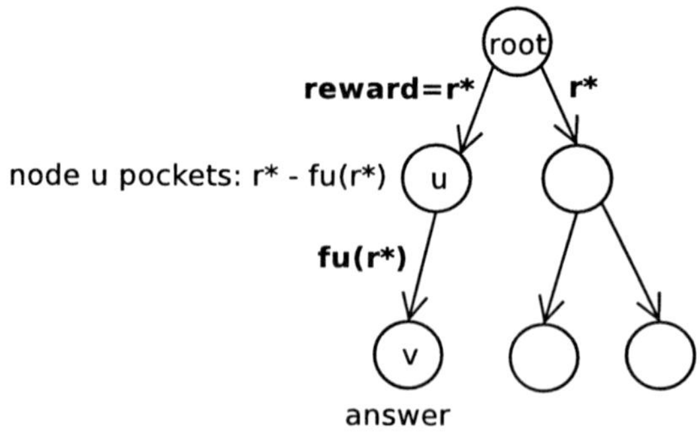

Fig. 15.1 Query Incentive Game: node v has an answer to the query.

where hopefully a destination node with an answer will be reached. Propagation is assumed to occur on a tree and incentives are provided by each parent on the tree to its children in the form of rewards. A node that gets offered a reward will itself offer a smaller reward to its children if its does not posses the answer hence pocketing some reward if an answer to the query is found under the node's subtree. We shall refer to this game hereafter as the QUERY-GAME and we note that this game is based on a similar game initially introduced by Li et al. [31].

Formally, each node (player) u receives a reward r from its parent and offers the same reward $f_u(r) < r$ to its children if it does not have the answer. Otherwise, if u has the answer to the query it responds to its parent with the answer. Each node holds the answer with probability $1 - p$ and on average one in every n nodes holds the answer (n is referred to as the rarity of the answer). The node's strategy is hence $f_u(r)$ which is assumed to be integer-valued and the payoff is simply $(r - f_u(r))\alpha_u(\mathbf{f})$ where $\alpha_u(\mathbf{f})$ is the probability that an answer is found in the subtree rooted at u given that node u has played f_u and every other node's strategy is given by $\mathbf{f} = \{f_v, \forall v\}$ ($\mathbf{f}$ is a *strategy profile*). Figure 15.1 depicts a sample game on a tree.

There are several questions that arise in such a game: how will a node act strategically to tradeoff its payoff and the probability that an answer is found in its subtree knowing that a higher promised reward potentially means higher probability of finding an answer but less payoff? how much initial investment r^* is required (as a function of the tree structure and the rarity of the answer n) in order to find an answer with high probability? The authors answer these questions in [27] by modeling a general class of branching processes parametrized on the branching factor b, where the latter is the mean number of active offsprings (or children) per node in the tree constructed using a random branching process [27] (when $b < 1$, the tree is almost surely finite while it is infinite when $b > 1$ with positive probability). When looking for the equilibria, one important point to notice in this game is the interdependency of the players' strategies as given by the tree structure - the strategy of a player will depend on the strategies of its children and so on. The authors show that the Nash equilibrium exists (and is unique with some caveats) by constructing a set of functions **g** (a strategy profile) inductively and showing that the resulting strategy profile is indeed an equilibrium. This result simply says that there exists a stable solution to the game such that if the nodes play the strategies **g** then no node will be able to move profitably given the strategy profile of the rest of the nodes. However, the model does not provide a recipe to get to the solution. Knowing that a solution exists, the next step is to study the breakpoint structure of rewards to be able to say something about the initial investment required (check [27] for results there). In summary, the goal of this game (and the one in [31]) is to provide incentives for query propagation in decentralized networks with complete uncertainty about the destination of the answer knowing that such a process could incur cost that must be paid for by someone to keep the incentives aligned. In the next game, we shall discuss a game the uses the SPNE solution.

15.2.2 Subgame Perfect Nash Equilibria and Trading Networks Game [10]

The next game we present is that of trading networks which despite being more motivated from a markets angle will provide several insights into networking games that involve competition. A set of sellers S wish to sell their goods to a set of buyers B indirectly through a set of traders T. While [10] studies both cases where the goods are distinguishable or not, in this brief overview we shall only focus on indistinguishable goods i.e. a single type of good where all copies are identical. Each seller holds exactly one copy of the good initially and each seller is only interested in buying one copy of the good as well. Trade between the buyers and the sellers can only happen through a set of traders T as specified by a graph G. G specifies how sellers and buyers are connected to the traders where each edge in G connects a node in $B \cup S$ to a node in T. Sellers are assumed to have zero value for the good while each buyer j has a value θ_j for the good. Figure 15.2 depicts such a setting where the indices i, j, t are used to refer to the sellers S, the buyers B, and the traders T, respectively.

We shall refer to this game as the TRADE-GAME. The game aims at studying the process of strategic price setting in markets with intermediaries, and proceeds as follows: first each trader offers a bid price β_{ti} to each seller i to which it is connected, and an ask price α_{tj} to each buyer j to which it is connected. The vector of bid/ask prices is the strategy profile of the traders. Then buyers and sellers choose among the offers they got, the traders pay the sellers the bid price and get the ask price from the buyers. If a trader gets more buyer offers than the seller offers it has, the trader will have to pay a large penalty. This is so that such a scenario will never happen at equilibrium. The payoffs of the different players are as follows: a player that does not participate in a trade gets no payoff. A buyer that participates in a trade through some trader t gets a payoff of $\theta_j - \alpha_{tj}$, while a seller i that participates in a trade with trader t gets a payoff of β_{ti} (again here assuming the seller has no value for keeping the good). Finally, a trader that participates in trade with a set of buyers

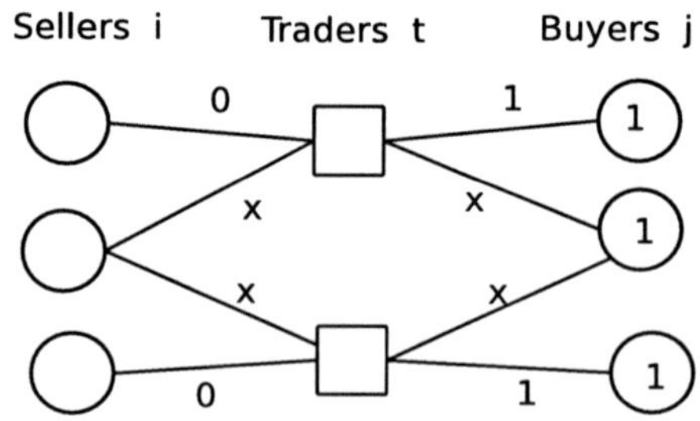

Fig. 15.2 Trading Network Game: sellers S to the left (circles) connect to traders T (squares) who in turn connect to buyers B to the right (circles). The buyers' values are indicated inside the circles (1 in this case). Equilibrium bid and ask prices are shown above the links.

and sellers gets a payoff of $\sum_r(\alpha_{tj_r} - \beta_{ti_r})$ minus a penalty if more buyers than sellers accept its offer (where the index r runs for each distinct buyer, seller combination that have accepted t's offer). It is important to notice that price setting in this game is strategic. Hence, as in the previous game, the first question to ask is how will the traders act strategically to set the market prices knowing that multiple traders could be competing for the same business? and what solution concept is most suitable to studying this game? The solution concept used in this game is the subgame perfect NE which is suitable in such a two stage game where traders play first and then buyers and sellers react. With this in mind, the next step to understanding the strategic behavior of the players (or equivalently the price setting dynamics) is to ask whether a solution (equilibrium) exists and to understand the structure of any such solution. In Figure 15.2, the equilibrium strategies are shown above the links. Two interesting equilibrium phenomena in this game are the effects of monopoly and perfect competition. Both traders in this example make a maximum profit (of 1) from the single monopolized buyer/seller pairs that have access to one trader, while the traders make zero profit when competing for the business of the middle seller and buyer. This must be the case at equilibrium. It turns out as shown by the authors that the equilibrium always exists and that every equilibrium is welfare maximizing (where the welfare of an outcome is simply the difference between the values of the buyers and those of the sellers). These results are shown by resorting to the primal/dual solutions of a welfare maximization linear program. In any solution, no trader will be able to make any profit unless the latter is essential for the social welfare of the game (this result captures the case where traders could have different costs and hence only the cheaper ones will be part of the equilibrium). The game (with distinguishable goods) could be directly extended to account for trading costs i.e. where traders incur costs to perform the trade and the same results hold i.e. a trader will be able to make profit only when the trader is crucial to the social welfare.

15.2.3 Mechanism Design and Interdomain Routing Game [15]

The third game we present in this section is that of interdomain routing incentives, particularly for BGP. First, we briefly overview how BGP operates after which we proceed to describe the incentive mechanism. The Internet is mainly composed of independent Autonomous Systems (ASes), or administrative domains, that must coordinate to implement a distributed routing algorithm that allows packets to be routed between the domains to reach their intended destinations. BGP is a policy-based path vector protocol and is the de-facto protocol for Internet interdomain routing. The protocol's specification [39] was initially intended to empower domains with control over *route selection* (which path or route to pick among multiple advertised routes to a destination), and *route propagation* (who to export the route to among an AS's direct neighbors). The commercialization of the Internet quickly transformed ASes into economic entities that act selfishly when implementing their internal policies and particularly the decisions that relate to route selection and propagation [13]. Intuitively, selfishness and the lack of coordination could potentially lead to instabilities in the outcome of the protocol, as is actually the case with BGP. Griffin et al. have studied this problem and the authors provided the most widely accepted formulation, the stable paths problem, with sufficient conditions under which the protocol converges to a stable solution, the *no dispute wheel* condition [21]. In addition to the algorithmic side of BGP which deals with convergence and stability, recent work has focused on the economic side particularly studying the equilibria of a BGP game and trying to align the incentives of the players (check [15, 30] and references therein).

The interdomain routing incentive game of [15], hereby referred to as ROUTING-GAME, aims to study the policies (strategies) under which BGP is welfare maximizing (i.e. it maximizes the social welfare), and incentive-compatible (i.e. no player has an incentive to deviate from telling the truth where the player's action is to declare private information), and to design a distributed mechanism to provide these attractive properties. Formally, in this game we are given a graph $G = (N, L)$ that represents the AS level topology (nodes N are the ASes and L the set of links between them). The route computation problem is studied for a single destination d and may be directly extended to all destinations assuming route computation is performed independently per destination. Hence, there exists a set of n players indexed by i, and the destination d. Each player has a valuation function $v_i : P^i \rightarrow \mathbb{R}$ which assigns a real number to every permitted route to d, P^i being the set of all permitted routes from i to d. Note that a route is permitted if it is not dropped by i and its neighbors. No two paths are assumed to have the same valuation. Social welfare of a particular outcome, an allocation of routes R_i, $\forall i$ that forms a tree T_d, is defined to be $W_{T_d} = \sum_{i=1}^{n} v_i(R_i)$. Clearly, the concept of internal policy is captured with the strict valuation or preference function v_i over the different routes to d which is private information given that the nodes are autonomous. In this sense, and as mentioned earlier, the goal of this problem is to design a mechanism that can maximize the social welfare despite the fact that its components, the v_i functions, are unknown or private. The mechanism design framework and particularly the Vickery–Clark–Groves (VCG) mechanism provides the solution [37]. To do so, a central bank is assumed to exist whose sole task is to allocate a payment $p_i(T_d)$ to each node i based on the outcome. More clearly, a player may either truthfully reveal her valuation to the mechanism (by always picking the best valued routes to d) or not, hoping to manipulate the outcome to her advantage. Based on the players' actions and hence on the outcome tree T_d, a payment $p_i(T_d)$ will be made by the central bank to each player. The utility of each player from an outcome will then be $u_i(T_d) = v_i(R_i) + p_i(T_d)$. The VCG payment scheme is intentionally designed to make the truthful action a *dominant strategy* for all players, hence no player has an incentive to lie about her valuation. To achieve this, AS i will be compensated an amount p_i proportional to the decrease in the value of all upstream ASes that have picked their best route to d through i when the latter does not participate. This is exactly the impact on the social welfare when i is not playing [37]. From a game standpoint, the solution concept that was targeted is the dominant strategy solution - playing truthfully is a dominant strategy and achieving such an attractive solution comes at the expense of assuming a central bank that regulates payments. The authors show that BGP augmented with a VCG payment scheme is incentive-compatible and welfare maximizing in several well studied settings (assumptions on policies or valuation functions).

In the above problem, and generally in problems involving mechanism design, the common scenario is an allocation mechanism that distributes some resource to a set of participating players. In order for a mechanism to implement the Social Choice Function (SCF), for example maximizing the social welfare of all players, the mechanism needs to know the real private information (such as true valuations for example) of the players. This is the case because players might be able to strategically manipulate the output of the mechanism by lying about their private information or strategies. Hence, *dominant strategies* is a desired solution concept that the mechanism would value and would aim towards in such a way so as to implement the SCF.

15.3 Identifier-Based Discovery

We now proceed to formalize the general discovery problem. We start by modeling the network as a graph $G = (V, E)$ with a set of nodes V, $|V| = n$, where each node $u \in V$ can host at most a single default object. The objects residing on the nodes are the players that demand to be discoverable by the rest of the network. The default object is meant to capture the case when the node itself is the object. An object has a unique identifier. Hereafter, we shall refer to the objects by index $i = 1, \ldots, n$ and it should be clear that anytime we refer to node i or player i, we are actually referring to the default object hosted on the node.

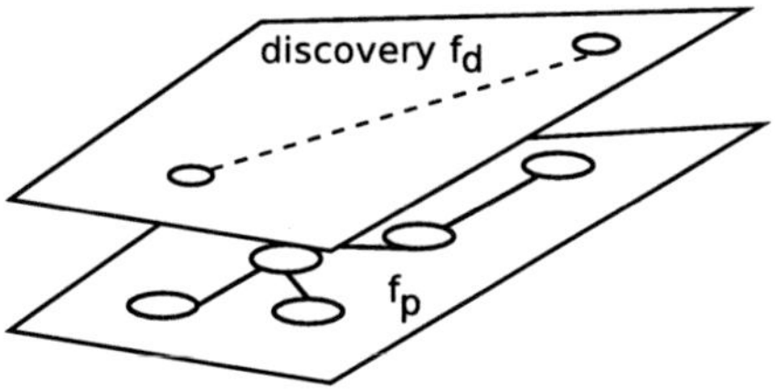

Fig. 15.3 Two layer model.

15.3.1 What Is Discovery?

Recall that an identifier represents an object's identity and remains unchanged when location (e.g. topology) information changes. A locator identifies the location of an object and must change when the object's location information changes.

The notion of discovery throughout this chapter refers to *path discovery based on identifiers*. A discovery scheme f_d generally operates on top of a routing scheme f_p that routes based on locators. We refer to this model as the two-layer discovery model as depicted in Figure 15.3. Whenever f_p exists, all that remains to be discovered by f_d is the identifier-to-locator mapping (e.g. DNS name to IP address mapping). When f_p is not available, then path discovery needs to be performed by f_d as well (e.g. BGP-DA on provider-independent prefixes). The two-layer model may be applied recursively i.e. a new discovery function f_d^* may operate on top of f_d where the latter is virtualized as the locator routing function. This chapter is particularly concerned with the design of mechanisms that implement the discovery function f_d.

Generally, the process of identifier-based path discovery involves a search or discovery query that is forwarded based on a series of calls "forward to *next node* that should have more ($\geq$) information about the named destination" starting at a source node. Discovery schemes in large-scale networks require maintaining distributed state about the identifier space in the upper plane of the two-layer model. Note here that by considering path discovery that involves distributed in-network state, we are clearly restricting the discussion to stateful routing (proactive) schemes which seem to be more common in large-scale networks. Reactive or on-demand discovery schemes generally involve flooding which renders them less efficient to implement at large scales.

From an algorithmic standpoint, a generalized discovery scheme provides the following operations:

- Discovery operations: encapsulate the interface that the players P use to communicate with the mechanism and include two operations:
  - *join(i, level)*: allows player i to request a discovery service possibly expressing a desired service level (and potentially her valuation of some service level).
  - *discover(i, j)*: allows player i to discover player j.

- Service operations: are implemented on the service nodes and dictate a set of rules for maintaining state about the namespace and for handling the above queries.

The main reason that we separate the two routing functions f_d and f_p is because there are instances where the two functions are managed by different entities that can minimally collaborate to jointly optimize the two functions. For example, with the current Internet where BGP implements some form of f_p (routing on IP addresses), discovery schemes are being introduced in a separate plane that is not necessarily provisioned by ISPs (players) but rather by other economic entities (as in DNS, and recently [18]). On the other hand, in name-independent compact routing design [7], it is assumed that the two functions are being jointly optimized to achieve a single global goal of efficient communication/discovery. This requirement has motivated us to study discovery mechanisms separately and to deviate from the pure algorithmic treatment of the topic towards *solution concepts* that are based in economics.

15.3.2 Discovery versus Search: Why Receiver-Based Discovery?

In order to frame our work, we introduce the notions of *advertisers* and *seekers*. In identifier-based discovery, advertisers are the entities that wish to be discoverable by the rest of the network using their identifiers. They utilize the *join(i, level)* interface to express their wish to the mechanism. Seekers, who could be advertisers as well, wish to locate the advertisers and they utilize the *discover(i, j)* operation to do so. In our model players are advertisers who may simultaneously be seekers (think of a node in a Distributed Hash Table (DHT) for example as in [41]).

It is important to distinguish two different classes of problems that relate to discovery and that have been considered in the literature. The first, distributed information retrieval, is that of locating information without prior knowledge of the providers or the location of the information (information could be located anywhere in the network). This problem is generally referred to as unstructured search (as in Gnutella, Freenet P2P networks, social networks, etc.). One key idea here is that in order for the requester to find the requested information, she must search for it and be willing to invest in the search. The provider either can not or is not willing to do so. A prominent work in this vein that addresses cost and incentive structures includes the work by Kleinberg [27].

The second class of problems, which we are more interested in and which we refer to as identifier based discovery, aim at discovering a path to a uniquely identified entity assuming the seeker is given the identifier(s) of the destination beforehand. This problem is common in service centric networks where there generally exists many competing providers for the same service. Within this class of problems, we distinguish two subclasses based on the cost model employed. The first subclass deals with routing problems and focuses on the transit or forwarding cost which is to be bore by the seeker. Several proposals fall under this subclass and

Table 15.1 Identifier-based discovery schemes.

Model	Representative Schemes
Model I	DNS, DONA [28], eFIT [33], LIS ([18], etc.)
Model II	DHTs (Chord [41], etc.)
Model III	NICR ([5,7], etc.), BGP-DA, ROFL [12]

many utilize economic tools based in mechanism design [14,36,42]. We distinguish another flavor of the problem by noticing that in service-centric network environments, the seeker gets no utility from the discovery part but rather gets the utility from consuming the service itself. In this sense, the utility of discovery is mainly to the provider or the advertiser: the provider wishes to sell the service and can efficiently do so only when the service is "discoverable". This is the main point that distinguishes our work from the literature on routing and forwarding incentives. The players may be thought of as providers that receive a utility from being discoverable by the rest of the network, the utility of being famous, the latter being inevitably related to the player's business. Hence, in the receiver-based business model, the player does not care about whether other players are discoverable or not, whereas with general P2P resource sharing applications the player's utility is to share the resources of other players and hence to be able to discover the rest of the network (originator-based).

15.3.3 A Taxonomy of Discovery Schemes

Figure 15.4 shows some classic models used by current discovery schemes (and proposals) following the two-layer model. Big circles (light and dark) represent nodes used by f_p at the lower layer (nodes V). At the upper layer, big dark circles represent a subset of those nodes that maintains state about the virtual namespace (service nodes V_D where $V_D \subseteq V$); small dark circles are the objects that wish to be discovered or the players (players P). Figure 15.4 tries to illuminate the relationship between the players P (who receive the discovery service), and the nodes V_D (who provide the service and incur the cost). This relationship is important in an economic setting, such as when studying pricing schemes and when devising a strategic model (and solution concept) for the problem at hand. For example, service nodes in model I (described shortly) may be generally considered to be obedient (i.e. to follow the protocol) as they belong to the same administrative entity (or to multiple competing entities each providing the same service). In models II and III however one needs to consider strategic service nodes in addition to the strategic agents where the two sets could be the same. Some of the representative schemes in the literature that follow these service models are listed in Table 15.1.

In model (I) $[V_D \neq P]$, there is a dedicated set of nodes V_D (possibly infrastructure) that keep the state information about the virtual namespace while the players

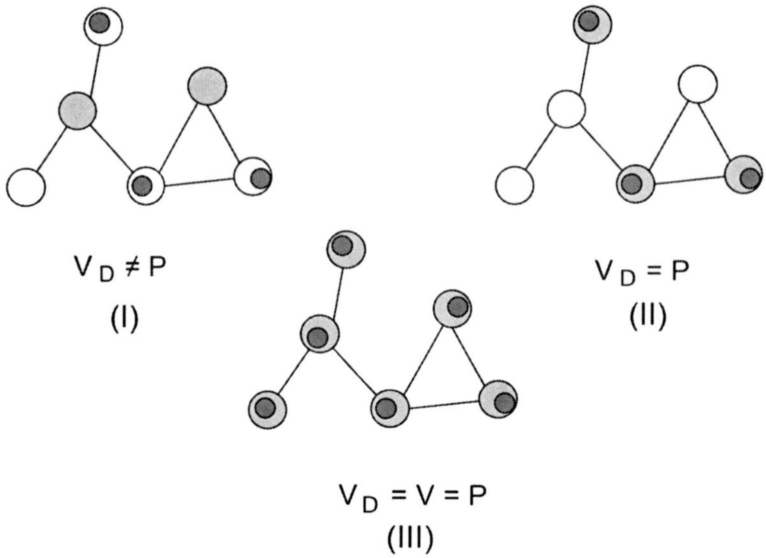

Fig. 15.4 Representation of some common models for discovery.

P reside on different nodes. DNS is one example of a centralized scheme that follows this model. In DNS, V_D is the set of root/gTLD (for global Top Level Domain) servers and the players are domain servers that keep zone files. Another scheme that uses this model and that is distributed is the recent DONA proposal [28] where V_D is the set of resolution handlers, and the players are generally objects on edge nodes. Another set of proposals that fits under this model is embodied by the Locator-ID-Split (LIS) work which aims at providing discoverability to edge sites (e.g. [18]) or nodes (e.g. [35]) in the Internet.

In model (II) $[V_D = P]$, the state is kept on the same set of nodes that the players reside on. In such a model, the players themselves have a common interest in implementing the discovery scheme f_d. The typical example here is Distributed Hash Tables (DHT).

In model (III) $[V_D = V = P]$, the state is maintained on all the nodes V and the players are all the nodes. This model is common to proposals that perform native routing on flat identifiers. One class of schemes that fits under this model is represented by the Name Independent Compact Routing (NICR) [7]. In NICR, the upper and lower layer functions are jointly designed and closely related. Another class of schemes that belong to this model does not utilize an underlying f_p i.e. f_d is basically a simultaneous discovery and forwarding scheme. BGP-DA is the representative scheme here where the players or nodes are the ASes advertising the prefixes and where it is necessary for all nodes V to keep the state in order for prefix path discovery (i.e. routing in this case) to succeed. Another recent scheme is the DHT-based ROFL [12], in which the routers are the nodes (if we ignore objects here) that are themselves the players identified by flat identifiers (hashes). Note that models (II) and (III) are the same for our purposes and we shall not make the distinction between the two hereafter.

It is worth noting that each of the schemes in Table 15.1 is designed to satisfy a set of requirements and is based on a set of assumptions about the two-layer functions. Some of the common requirements observed in the literature include *efficiency*, *scalability*, *trust*, *user-control*, *robustness*, *economic requirements*, etc. Some of assumptions address the underlying graph structure (e.g. *scale-free*, or *small-world*) assumptions, or more specific structural assumptions of underlying metric embeddings.

15.3.4 Incentives and Pricing

Having introduced the discovery problem and overviewed different discovery models used in the literature, we now proceed to motivate the need for incentives in discovery. Recall that in order for a node to be discoverable, a cost must be incurred by the set of service nodes V_D generally for maintaining *state* about the node's identifier. The term *state* in this context refers to the information stored on the service nodes to allow the players to be discoverable. The per-node state may be thought of as simply the node's routing table which is generally comprised of mappings from identifier to location information. The question that arises then is who pays for maintaining the state, and what incentive models are suitable for the different discovery models. In this section, we present some solution concepts that are applicable to each of the discovery models, and set the stage for the BGP incentive model which will be discussed in the next section.

15.3.4.1 Model I: $V_D \neq P$

Recall that in this setting, the players P are requesting a discovery service from a set of infrastructure service nodes V_D. When $V_D \neq P$, mechanism design and particularly Distributed Algorithmic Mechanism Design (DAMD) [16] in addition to general cost-sharing models [37] seem to be intuitive frameworks for modeling incentives and pricing. Different situations may arise based on whether the service nodes are obedient or not (obedient service nodes will not try to manipulate the protocol), belong to multiple competing economic entities or not, and on whether the mechanism is subsidized or not. Note that when the mechanism is subsidized, the designer of the mechanism does not have to worry about budget-balance where the latter means that the total payments made by the players must offset the total cost of providing the service.

Assume the service nodes to be obedient and no competition dynamics present, and consider the following DAMD model: each player has a valuation of being discoverable, which she presents to the mechanism. The mechanism logically controls (and operates on) the service nodes collecting all the players' valuations, the demand, and allocating payments back to the players to achieve a mix of goals. These goals could potentially include incentive-compatibility (or strategy-proofness), welfare maximization (or efficiency), and/or budget-balance. When the mechanism is subsidized, the goal of the mechanism is to maximize the social welfare (instead of budget-balance) under the constraint that a cost is associated with providing the service. In this sense, truthful valuations of the service need to be declared by the players, and hence the goal of incentive-compatibility (especially when the mechanism is able to provide different levels of the service). We have presented such a DAMD model that accounts for service differentiation in a recent work [25]. A one-shot VCG variant [37] is a natural solution here that could achieve efficiency and incentive-compatibility again assuming that the mechanism could be subsidized in other ways. The VCG pricing scheme is a cost-sharing scheme i.e. it shares the total cost of providing the service among the participating players. The mechanism will always maximize the social welfare of all the players and will pick prices (cost shares) such that a player i pays an amount equal to the difference in the total welfare of the other players with and without player i's participation - the damage caused by player i's participation.[3] Note that the budget-balance requirement becomes essential when the subsidization assumption does not hold since the total cost must be collected so that service nodes are paid for participating. For example, if a node j is not compensated for the cost of keeping state about the rest of the network, the node will have no incentive to participate. It has been proven by Laffont and Green and later by Satterwaite impossibility theorems [37] that cost-sharing mechanisms can be either strategy-proof and efficient, or strategy-proof and budget-balanced but not both.

When competition among the service providers is present, then the one-shot mechanism design framework seems less practical. This case is more representative of model (I) than the no-competition case. The main idea here is that multiple competing Discovery Service Providers (DSPs) offer the service to the players. Each DSP is assumed to be owned and operated by an autonomous economic entity and DSPs compete for service or market share. Dynamic pricing is more suitable in such a model and a realistic strategic model for this setting based on repeated games was introduced by Afergan [6]. The model discusses price strategies at Internet interchange locations, such as multiple ISPs providing service to a customer (e.g. a CDN). The same model may potentially apply to the discovery mechanism pricing where multiple competing DSPs compete for market share.

15.3.4.2 Models II, III: $V_D = P$

When the set of service nodes $V_D = P$, players incur a cost due to participation of other players and the issue of incentive and pricing becomes even more challenging. In this distributed setting, the traditional game theoretic and economic tools seem to be more applicable, since the centralized designer and the obedient service nodes

[3] This VCG pricing scheme is referred to as the Clark Pivot rule [37].

assumptions inherent to the mechanism design framework no longer hold. Consider BGP for example where every node that wishes to be discoverable introduces state about its identifier on every other node in the DFZ. NICR [5, 7] schemes on the other hand are less costly as they try to optimize the tradeoff between state and *stretch* (stretch is defined in the context of routing as the ratio between the cost of the path taken by the routing scheme, to the minimum cost path where cost could be defined differently based on the setting (e.g. hops or delay); the maximum of the ratio for all source destination pairs is generally referred to as the stretch [7]). In this sense, a node that wishes to be discoverable must introduce state on a subset of other nodes in the network. In both examples above, one can directly recognize the incentive mismatch issue and the challenges inherent to the design of incentive and pricing models that are suitable for this setting. In the next section, we present one such model for BGP.

15.4 An Incentive Model for Route Distribution and Discovery in Path Vector Protocols

The main motivator for devising a model to account for the cost of distribution in BGP is the recent attention in the research community to the incentive mismatch when it comes to the cost of discovery in BGP. Herrin has analyzed in [22] the non-trivial cost of maintaining a BGP route and has highlighted the inherent incentive mismatch in the current BGP system where the rest of the network pays for a node's route advertisement. This reality is exacerbated by the fact that number of BGP prefixes in the global routing table (Routing Information Base – RIB) is constantly increasing at a rate of roughly 100,000 entries every 2 years and is expected to reach a total of 388,000 entries in 2011 [2].

BGP is intrinsically about distributing route information to destinations (which are IP prefixes) to establish paths in the network (route distribution and route computation). Path discovery is the outcome of route distribution and route computation. A large body of work has focused on putting the right incentives in place knowing that ASes are economic agents that act selfishly in order to maximize their utilities. In dealing with the incentive problem, previous work has ignored the control plane incentives (route distribution) and focused on the forwarding plane incentives (e.g. transit costs) when trying to compute routes. One possible explanation for this situation is based on the following assumption: a node will have an incentive to distribute routes to destinations since the node will get paid for transiting traffic to these destinations and hence route distribution becomes an artifact of the transit process and is ignored. The majority of previous work that tries to introduce the required incentive models do so by introducing per-packet transit costs. Nodes declare these costs to a mechanism and receive payments from the latter. The mechanism design framework is generally employed here and the mechanism is generally assumed to be subsidized (hence budget-balance is not a design goal). In this work, we conjecture that forwarding is an artifact of route distribution (and definitely computation) where the latter happens first in the process and hence our main focus is on incentivizing nodes to distribute route information. Clearly, we separate the BGP distribution game from the forwarding game and we focus solely on the former. Whether the two games can be combined and studied simultaneously is an open question at this point.

In this section, we synthesize many of the ideas and results from [14,15,20,27,31] into a coherent model for studying BGP route distribution incentives. The model we employ is influenced by the query propagation model studied by Kleinberg [27] in the context of social networks.

15.4.1 A Simple Distribution Model

A destination d advertises its prefix and wishes to invest some initial amount of money r_d in order to be globally discoverable (or so that the information about d be globally distributed). Since d can distribute its information to its direct neighbors only, d needs to provide incentives to get the information to propagate deeper into the network. d wants to incentivize its neighbors to be distributors of its route who then incentivize their neighbors to be distributors and so on. A transit node i will be rewarded based on the role it plays in the outcome routing tree to d, T_d (whether the outcome is a tree should become clear later in the discussion). The utility of the transit node i from distributing d's route, as we shall describe shortly, increases with the number of nodes that route to d through i – hence the incentive to distribute.

The model seems to correctly capture many of the details behind how policy-based BGP (and in general path-vector protocols) works and the inherent incentives required. Additionally, the model is consistent with the simple path vector formulation introduced by Griffin in [20]. More clearly, it is widely accepted that each AS participating in BGP has as part of its decision space, the following decisions to make:

- import policy: a decision on which routes to d to consider,
- route selection: a decision on what route to d to pick among the multiple possible routes,
- export policy: a decision on who to forward the advertisement to among its direct neighbors.

All three policies are captured in the game model we describe next.

There are two main properties of interest when it comes to the BGP game model: *convergence*, and *incentives*. The BGP inter-domain routing protocol handles complex interactions between autonomous, competing economic entities that can express local preferences over the different routes. Given the asynchronous interactions among the ASs and the partial information, convergence of BGP to a stable solution becomes an essential property to aim for when studying policies. Griffin et al. [20] defined the stable paths problem which is widely accepted as the general problem that BGP is solving. The authors formulated a general sufficient condition under which the protocol converges to an equilibrium state, mainly the "no dispute wheels" condition. A game theoretic model was recently developed by Levin et al. [30] that enhances the stable paths formalization and studies the incentive-compatibility question. In addition to convergence, incentive issues are crucial to the success and stability of BGP mainly since nodes are assumed to be selfish entities that will act strategically to optimize their utility. In this sense, any distribution and route computation mechanism or policy can only benefit from aligning the incentives of the players to achieve the mechanism's goals [14, 15, 30, 37].

15.4.2 Related Work

The Simple Path Vector Protocol (SPVP) formalism [20] develops sufficient conditions for the outcome of a path vector protocol to be stable. The two main components of the formalism are *permitted paths* and local strict *preference* relations over alternate paths to some destination. A respective game-theoretic model was developed by Levin [30] that captures these conditions in addition to incentives in a game theoretic setting. Other traditional BGP inventive models have not accounted for distribution or discovery costs and incentives and have assumed that every BGP speaker has value in knowing about all destinations and is hence willing to tolerate the cost of such assumption. Our work is fundamentally different than previous models particularly in regard to the incentive structure. The aim in our model is for a destination d to become discoverable by the rest of the network.

Feigenbaum et al. study incentive issues in BGP by considering least cost path (LCP) policies [14] and more general poilicies [15]. The ROUTING-GAME presented in Section 15.2 describes [15]. Our model is fundamentally different from [14] (and other works based in mechanism design) in that the prices are strategic, and it does not assume the existence of a bank (or a central authority) that allocates payments to the players but is rather completely distributed as in real markets. The main element of [15] is payments made by the bank to nodes. The model assumes that the route to d is of value to a source node where the latter will strategically pick among the multiple routes to d. A bank is required to make sure payments are correctly allocated to nodes based on their contribution to the outcome. The bank assumption is troublesome in a distributed setting such as the Internet, and an important question posed in [15] is whether the bank can be eliminated and replaced by direct payments by the nodes.

Li et al. [31] study an incentive model for query relaying in peer-to-peer (p2p) networks based on rewards, on which Kleinberg et al. [27] build to model a more general class of trees. We have introduced the latter model in Section 15.2 with the QUERY-GAME. Both of these models do not account for competition. Similar to the problem setting of [31], an advertiser does not know in advance the full topology neither the resulting outcome of route distribution. Designing payment schemes for such settings generally requires revelation of "non-private value" information such as topology information [40] which might not be available to the players neither to the mechanism designer. The dynamic pricing scheme introduced in [31] avoids such revelation by pricing only based on local information. While we borrow the basic idea from [31] and [27], we address a totally different problem which is that of route distribution versus information seeking.

In addition, our work relates to price determination in network markets with intermediaries (refer to the work by Blume et al. [10] and the references therein). We have introduced the TRADE-GAME of [10] as well in Section 15.2. A main differentiator of this class of work from other work on market pricing is its consideration of intermediaries and the emergence of prices as a result of strategic behavior rather than competitive analysis or truthful mechanisms. Our work specifically involves cascading of traders (or distributors) on complex network structures mainly the Internet.

15.4.3 The General Game

We focus in this work on path-vector protocols and we reuse notation from [15, 30, 31]. We are given a graph $G = (V, E)$ where V consists of a set of n nodes (alternatively termed players, or agents) each identified by a unique index $i = \{1, \ldots, n\}$, and a destination d, and E is the set of edges or links. Without loss of generality, we will study the BGP discovery/route distribution problem for some fixed destination AS with prefix d (as in [15, 20, 30] and [27, 31]). The model is extendable to all possible destinations (BGP speakers) by noticing that route distribution and computation is performed independently per prefix. The destination d is referred to as the *advertiser* and the set of players in the network are termed *seekers* in the discovery model. Seekers may be distributors who actively participate in distributing d's route information to other seeker nodes while retailers simply consume the route (leaf nodes in the outcome distribution tree). For each seeker node j, Let $P(j)$ be the set of all routes to d that are known to j through advertisements, $P(j) \subseteq \mathcal{P}(j)$, the latter being the set of all simple routes from j. The empty route $\phi \in \mathcal{P}(j)$. Denote by $R_j \in P(j)$ a simple route from j to the destination d with $R_j = \phi$ when no route exists at j, and let $(k, j)R_j$ be the route formed by concatenating link (k, j) with R_j, where $(k, j) \in E$. Denote by $B(i)$ the set of direct neighbors of node i and let $\text{next}(R_i)$ be the next hop node on the route R_i from i to d. Define node j to be an *upstream* node relative to node i, $j = \text{next}(R_i)$. The opposite holds for *downstream* node. Finally, let D_i denote the degree of node i, $D_i \in \mathbb{N}$. For example in Figure 15.5, $\text{next}(R_5) = 3$ and 3 is the upstream node relative to 5.

The general discovery game is simple: destination d will first export its prefix (identifier) information to its neighbors promising them a reward r_d ($r_d = 10$ in Figure 15.5) which directly depends on d's utility of being discoverable. A distributor node j (a player) in turn strategizes by selecting a route among the possibly multiple advertised routes to d, and deciding on a reward $r_{jl} < r_{ij}$ to send to each *candidate* neighbor $l \in B(j)$ that it has not received a competing offer from (i.e. s.t. $r_{lj} < r_{jl}$ where $r_{lj} = 0$ means that j did not receive an offer from neighbor

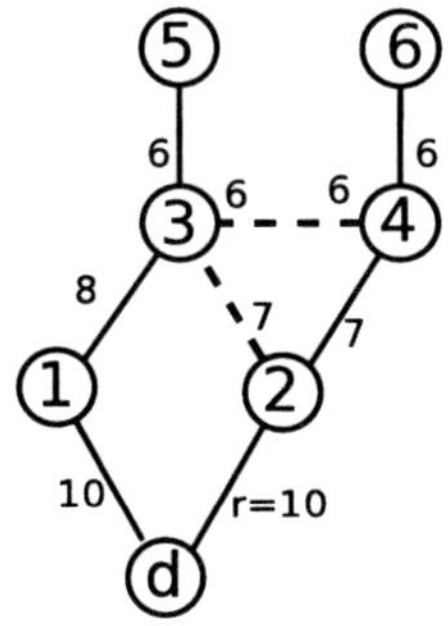

Fig. 15.5 Sample network (not at equilibrium): Solid lines indicate an outcome tree T_d under the advertised rewards.

l) pocketing the difference $r_{ij} - r_{jl}$. The process repeats up to some depth that is directly dependent on the initial investment r_d as well as on the strategies of the nodes. A reward r_{ij} that a node i promises to some direct neighbor $j \in B(i)$ is a contract stating that i will pay j an amount that is a function of r_{ij} and of the set of downstream nodes k that decide to route to d through j (i.e. $j \in R_k$ and $R_j = (j, i)R_i$). Note that such a reward model requires that the downstream nodes k notify j of their best route so that the latter can claim its reward from its upstream parent. We intentionally keep this reward model abstract at this point and we shall revisit it later in the discussion when we define more specific utility functions. For example, in Figure 15.5, node d promises $\{1, 2\}$ a reward $r_d = 10$. Node 1 exports route $(1, d)$ to its neighbor promising a reward $r_{13} = 8$. Similarly node 2 exports the route $(2, d)$ to its neighbor set $\{3, 4\}$ with $r_{23} = r_{24} = 7$ and so on. Clearly in this model, we assume that a player can strategize per neighbor, presenting different rewards to different neighbors. We take such assumption based on the autonomous nature of the nodes and the current practice in BGP where policies may differ significantly across neighbors (as with the widely accepted Gao–Rexford policies [19] for example).

15.4.3.1 Assumptions

To keep our model tractable, we take several simplifying assumptions. In particular, we assume that:

1. the graph is at steady state for the duration of the game i.e. we do not consider topology dynamics;
2. the destination d (the advertiser) does not differentiate among the different ASes in the network;
3. the advertised rewards are integers i.e. $r_{ij} \in \mathbb{Z}^+$ and that $r_{ij} < r_{\text{next}(R_i)}$, where the notation $r_{\text{next}(R_i)}$ refers to the reward that the upstream node from i on R_i

offers to i. A similar assumption was taken in [27] and is important to avoid the degenerate case of never running out of rewards, referred to as "Zeno's Paradox";

4. a node that does not participate will have a utility of zero; additionally, when the best strategy of a node results in a utility of zero, we assume that the node will prefer to participate than to default as this could lead to future business opportunities for the node;
5. finally, we only study the game for a class of policies which we refer to as the Highest Reward Policy (HRP) and we accordingly define a utility function for the players. As the name suggests, HRP policies incentivize players to choose the path that promises the highest reward. Such class of policies may be defined general enough to account for complex cost structures as part of the decision space. Despite the fact that the distribution model we devise is general, we assume for the scope of this work that transit costs are extraneous to the model and we refer to resulting preference function as *homogeneous preferences*. This is a restrictive assumption at this point given that BGP allows for arbitrary and complex policies among the players. Such policies are generally modeled with a valuation or preference function that assigns strict preference to the different routes to d. Transit cost is one form of such functions [15], and more complex ones (for example next-hop preferences or metric based preferences) have been studied and modeled [15, 20]. In BGP, such preferences are reflected in contractual agreements between the ASes.

15.4.3.2 Strategy Space, Cost, and Utility

Strategy Space

We now proceed to define the strategy space. Given a set of advertised routes $P(i)$ where each route $R_i \in P(i)$ is associated with a promised reward $r_{\text{next}(R_i)} \in \mathbb{Z}^+$, the strategy $s_i \in S_i$ of an autonomous node i comprises two decisions:

- After receiving update messages from neighboring nodes, pick a single best route $R_i \in P(i)$;
- Pick a reward vector $r_i = [r_{ij}]$ promising a reward r_{ij} to each candidate neighbor j (and export route and respective reward to *all* candidate neighbors).

A strategy profile $\mathbf{s} = (s_1, \ldots, s_n)$ defines an outcome of the game, and a utility function $u_i(\mathbf{s})$ associates every outcome with a real value in $\mathbb{R}$. We shall show shortly that for a certain class of utility functions, every outcome uniquely determines a set of paths to destination d given by $O_d = (R_1, \ldots, R_n)$ and that O_d is always a tree T_d. We use the notation s_{-i} to refer to the strategy profile of all players excluding i. For a given utility function, the Nash equilibrium is defined as follows:

Definition 1 *A Nash Equilibrium is a strategy profile* $\mathbf{s}^* = (s_1^*, \ldots, s_n^*)$ *such that no player can move profitably by changing her strategy, i.e. for each player* i, $u_i(s_i^*, s_{-i}^*) \geq u_i(s_i, s_{-i}^*)$, $\forall s_i \in S_i$.

Cost

There are two classes of cost within the distribution model. The first class defines the cost of participation while the second defines the "per-sale" costs. More clearly, cost of participation is local to the node and includes for example the cost associated with the effort that a node spends in maintaining and distributing the route information to its neighbors. The participation cost may additionally include the amortized per-route operational cost of the hardware (the DFZ router). Herrin [22] estimates this cost to be \$0.04 per route/router/year for a total cost of at least \$6,200 per year for each advertised route assuming there are around 150,000 DFZ routers that need to be updated. Per-sale costs, on the other hand, are incurred by the node for each sale that it makes and is generally proportional to the number of its downstream nodes in the outcome O_d. As mentioned earlier in the assumptions, we ignore this class of cost in the current model leaving it as part of our future work. Hence, while in general the cost function may be more complicated, we simply assume that the distribution cost c_i is composed of two components: c_i^{dist} representing the distribution cost and is only incurred by the distributors, and c_i^{state} represents the cost of maintaining the state and is incurred by all participating players.

Utility

Earlier in the discussion, we briefly alluded to a rewarding model in which node i rewards a neighbor node j based on some function of r_{ij} and of the set of downstream nodes of j (the latter corresponding to the number of sales node j made). Defining a concrete rewarding function (and hence utility function) for the players is a questions that the game modeler is left with. Specifically, we seek to identify the classes of utility functions and the underlying network structure for which equilibria exist. As a first step, we experiment with a simple class of functions which rewards a node linearly based on the number of sales that the node makes. This first model incentivizes distribution and potentially requires a large initial investment from d. More clearly, define the set $N_i(\mathbf{s}) = \{j \in N \backslash \{i\} | i \in R_j\}$ to be the set of nodes that pick their best route to d going through i (nodes downstream of i) and let $\delta_i(\mathbf{s}) = |N_i(\mathbf{s})|$. Additionally, let $I(x)$ denote the indicator function which evaluates to 1 when $x > 0$ and to 0 otherwise. Thus, $I(\delta_i(\mathbf{s}))$ indicates whether i is a distributor or not. We are now ready to define the utility of a node i from an outcome or strategy profile $\mathbf{s}$ as follows:

$$u_i(\mathbf{s}) = (r_{\text{next}(R_i)} - c_i^{\text{state}}) - c_i^{\text{dist}} I(\delta_i(\mathbf{s})) + \sum_{\{j:i=\text{next}(R_j)\}} (r_{\text{next}(R_i)} - r_{ij})(\delta_j(\mathbf{s}) + 1). \tag{15.1}$$

The first term in the utility function $(r_{\text{next}(R_i)} - c_i^{\text{state}})$ is incurred by every participating node and is the one unit of reward from the upstream parent on the chosen best path minus the local state cost. The second and third terms are only incurred by distributors. The second term $c_i^{\text{dist}} I(\delta_i(\mathbf{s}))$ denotes the distribution cost while the last term given by the summation is the total profit made by i where $(r_{\text{next}(R_i)} - r_{ij})(\delta_j(\mathbf{s}) + 1)$ is i's profit from the sale to j (which depends on the size of j's subtree given by δ_j). We assume here that node i gets no utility from an oscillating route and gets positive utility when R_i is stable. A rational selfish node will always try to maximize its utility by picking $s_i = (R_i, [r_{ij}])$. Equation (15.1) indicates that a node increases its utility linearly in the number of downstream seekers it can recruit, given by the summation. However, to increase the third term given by the summation, node i should carefully pick its rewards r_{ij} given that there might be other nodes competing with i for the route. There is an inherent tradeoff between $(r_{\text{next}(R_i)} - r_{ij})$ and $(\delta_j(\mathbf{s}))$ s.t. $i = next(R_j)$ when trying to maximize the utility in Equation (15.1) in the face of competition as shall become clear shortly. A lower promised reward allows the node to compete but will cut the profit margin. Finally, we assume implicitly that the destination node d gets a fixed incremental utility of r_d for each distinct player that maintains a route to d – the incremental value of being discoverable by any seeker.

15.5 HRP: Convergence, and Equilibria

Before discussing BGP convergence and equilibria under our assumptions and the utility function defined in Equation (15.1), we first prove the following lemma which results in the Highest Reward Path (HRP) policy:

Lemma 1. *In order to maximize its utility, node i must always pick the route R_i with the highest promised reward i.e.* $r_{\text{next}(R_i)} \geq r_{\text{next}(R_l)}, \forall R_l \in P(i)$.

The proof of Lemma 1 is given in Appendix A. The lemma implies that a player could perform her two actions sequentially, by first choosing the highest reward route R_i, then deciding on the reward vector r_{ij} to export to its neighbors. When the rewards are equal however, we assume that a node breaks ties uniformly.

15.5.1 Convergence

A standard model for studying the convergence of BGP protocol dynamics was introduced by Griffin et al. [20] (and later studied by Levin et al. [30]), and assumes BGP is an infinite round game in which a *scheduler* entity decides on which players participate at each round (the *schedule*). Any schedule must be fair allowing each player to play indefinitely and to participate in an infinite number of rounds. Convergence here refers to the convergence of BGP protocol dynamics to a unique outcome tree T_d under some strategy profile $\mathbf{s}$. The "no dispute wheels" condition, introduced by Griffin et al. [20], is the most general condition known to guarantee convergence of possibly "conflicting" BGP policies to a unique stable solution (tree). From Lemma 1, it may easily be shown that "no dispute wheels" exist under

HRP policy i.e. when the nodes choose highest reward path breaking ties uniformly. No dispute wheel can exist under HRP policy simply because any dispute wheel violates the assumption of strictly decreasing rewards on the reward structure induced by the wheel. Hence, the BGP outcome converges to a unique tree T_d [20] under any strategy profile **s**. The tree is stable given **s** which itself is only stable at equilibrium. Note that this is true for every strategy profile (i.e. independent of how the nodes pick their rewards) as long as the strictly decreasing rewards assumption holds, $r_{ij} < r_{\text{next}(R_i)}, \forall i, j$.

Lemma 2 ([20]). *The equilibrium outcome O_d under* $\mathbf{s}^*$ *is a stable routing tree T_d.*

Having said that, the next set of questions is targeted at finding the equilibrium profile $\mathbf{s}^*$. Particularly, does such an equilibrium exist and under what conditions? Is it unique? And how hard is it to find? In this work, we study the existence of equilibria on special network topologies leaving the other questions for future work.

15.5.2 Equilibria

In the game defined thus far, notice first that every outcome (including the equilibrium) depends on the initial reward/utility r_d of the advertiser as well as on the tie-breaking preferences of the nodes, where both of these are defining properties of the game. Studying the equilibria of the general game for different classes of utility functions and for different underlying graph structures is not an easy problem due to the complexity of the strategic dependencies and the competition dynamics. We are not aware of general equilibria existence results that apply to this game. Hence, we start by studying the game on the simplest possible class of graphs with and without competition. Particularly, we present existence results for the simplest two graphs: 1) the line which involves no competition, and 2) the ring which involves competition. We additionally assume in this discussion that the costs are constant with $c_i^{\text{dist}} = c_i^{\text{state}} = 1$.

Before trying to understand the equilibria of the game on these simple graphs, there is an inherent order of play to capture in the model in order to apply the right solution concept. Recall that the advertiser d starts by advertising itself and promising a reward r_d. The game starts at stage 1 where the direct neighbors of d, i.e. the nodes at distance 1 from d, observe r_d and play simultaneously by picking their rewards while the rest of the nodes "do-nothing". At stage 2, nodes at distance 2 from d observe the stage 1 strategies and then play simultaneously by picking their rewards. At stage 3, nodes at distance 3 from d observe the stage 1 and stage 2 strategies and then play simultaneously and so on. Stages in this multi-stage game with observed actions [17] have no temporal semantics. Rather, they identify the network positions which have strategic significance. The closer a node is to the advertiser, the more power such a node has due to the strictly decreasing rewards assumption. The multi-stage game model seems intuitive based on the assumptions of strictly decreasing rewards and the ability of the node to strategize independently

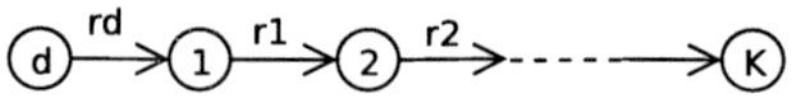

Fig. 15.6 Line netowk: a node's index is the stage at which the node plays; d plays at stage 0.

on each of the downstream links. We resort to the multi-stage model on these simple graphs simply because any equilibrium in the multi-stage game is a stable outcome in BGP no matter how the scheduler schedules the nodes as long as the schedule is fair and infinite.

Formally, each node or player i plays only once at stage $k > 0$ where the latter is the distance from i to d in number of hops; at every other stage the node plays the "do nothing" action. The set of player actions at stage k is the stage-k action profile, denoted by $a^k = (a_1^k, \ldots, a_n^k)$. Further, denote by $h^{k+1} = (a^1, \ldots, a^k)$, the *history* at the end of stage k which is simply the sequence of actions at all previous stages. We let $h^1 = (r_d)$. Finally, $h^{k+1} \subset H^{k+1}$ the latter being the set of all possible stage-k histories. When the game has a finite number of stages, say $K + 1$, then a terminal history h^{K+1} is equivalent to an outcome of the game (which is a tree T_d) and the set of all outcomes is H^{K+1}.

The strategy of node i who plays at stage $k > 0$ is then $s_i : H^k \rightarrow \mathbb{R}^{m_i}$ where m_i is the number of node i's direct neighbors at stage $k+1$. It is important to notice in this multi-stage setting that the node's strategy is explicitly defined to account for the order of play given by the graph structure (i.e. a pure strategy of a player is a function of the history). Starting with r_d (which is h^1), it is clear how the game produces actions at every later stage given by the node strategies resulting in a terminal action profile or outcome. Hence an outcome in H^{K+1} may be associated with every strategy profile $\mathbf{s}$. We now proceed to study the equilibria on the line and the ring network topologies.

15.5.2.1 No Competition: The Line

In the same spirit as [27] we inductively construct the NE for the line network of Figure 15.6 given the utility function of Equation (15.1). We present the result for the line network which may be directly extended to general trees. Before proceeding with the construction, notice that for the line network, $m_i = 1$ for all nodes except the leaf node since each of those nodes has a single downstream child. In addition, $\delta_i(\mathbf{s}) = \delta_j(\mathbf{s})+1, \forall i, j$ where j is i's child ($\delta_i = 0$ when i is a leaf). We shall refer to both the node and the stage using the same index since our intention should be clear from the context. For example, the child of node i is $i+1$ and its parent is $i-1$ where node i is the node at stage i. Additionally, without loosing any information, we simply represent the history $h^{k+1} = (r_k)$ for $k > 0$ where r_k is the reward promised by node k (node k's action), and hence the strategy of node k is $s_k(h^k) = s_k(r_{k-1})$.

We construct the equilibrium strategy s^* as follows: first, for all players i, let $s_i^*(x) = 0$ when $x \leq c_i^{\text{state}}$ (where c_i^{state} is assumed to be 1). Then assume that $s_i^*(x)$

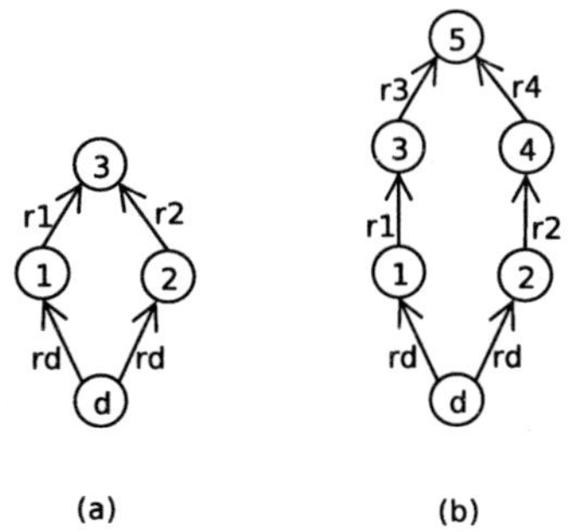

Fig. 15.7 Ring network with even number of nodes: (a) 2-stage game, (b) 3-stage game.

is defined for all $x < r$ and for all i. Obviously, with this information, every node i can compute $\delta_i(x, s^*_{-i})$ for all $x < r$. This is simply due to the fact that δ_i depends on the downstream nodes from i who must play an action or reward strictly less than r. Finally, for all i we let $s_i^*(r) = \arg\max_x (r_{i-r} - x)\delta_i(x, s^*_{-i})$ where $x < r$.

Claim 1 *The strategy profile s^* is a Nash equilibrium.*

Proof. The proof of is straightforward: given the utility function defined in Equation (15.1), no node can move profitably. Notice that in general when $r_{\text{next}(R_i)} \leq c^{\text{state}}$, propagation of the reward will stop simply because at equilibrium no node will accept to make negative utility and will prefer to not participate instead (the case with the leaf node). □

Clearly, the NE is not unique since different strategies could result in the same utility. This occurs on the line particularly when a node could get the same utility from being a distributor or not due to the incurred distribution cost. If we assume that a node will always prefer to distribute when the utility is the same, then it can be shown that the NE is unique.

Notice that in the line network, the NE exists for all values of r_d and in any equilibrium node 1 will always be able to make the maximum profit given r_d due to its strategic network position as the first and only distributor.

15.5.2.2 Competition: The Ring

Contrary to the line network, we will present a negative result in this section for the ring network. In a ring, each node has a degree of 2 and $m_i = 1$ again for all nodes except the leaf node. We will consider rings with an even number of nodes due to the direct competition dynamics. In the multi-stage game, after observing r_d, nodes 1 and 2 play simultaneously at stage 1 promising rewards r_1 and r_2 respectively to their downstream children, and so on. Figure 15.7 shows the 2-stage and the 3-stage versions of the game.

For the 2-stage game in Figure 15.7a, it is easy to show that an equilibrium always exists in which $s_1(r_d) = s_2(r_d) = (r_d - 1)$ when $r_d > 1$ and 0 otherwise. This means that node 3 enjoys the benefits of *perfect competition* due to the Bertrand-style competition [17] between nodes 1 and 2 which drives their profit to the minimum possible profit (and hence drives 3's reward to the maximum possible reward). The equilibrium in this game is independent of 3's preference for breaking ties which is not the case with the 3-stage game as we shall show next.

We now present the following negative result when the utility function is given by Equation (15.1):

Claim 2 *For the multi-stage game induced by the ring network, a Nash equilibrium does not always exist, i.e. there exists a value of r_d for which no equilibrium exists.*

Proof. The proof makes use of a counterexample. Particularly, the 3-stage game of Figure 15.7b does not have an equilibrium for $r_d > 5$. This is mainly due to oscillation of the best response dynamics and may be shown by examining the strategic form game, in matrix form, between players 1 and 2. We leave this as an exercise for the interested reader. Briefly, and assuming that node 5 breaks ties by picking route $(5, 3)R_3$, $r_d > 5$ signifies the breaking point of equilibrium or the reward at which node 2, when maximizing its utility $(r_d - r_2)\delta_2$, will always oscillate between competing for 5 (large r_2) or not (small r_2). □

Note that under the linear utility given in Equation (15.1), the NE is not guaranteed to exist on the simple ring network. This result is an artifact of the utility function. Finding the classes of utility functions for which equilibria always exist, for the ring network initially and for more general topologies as well, is part of our ongoing work.

15.6 Conclusions and Ongoing Work

In this chapter we have presented a general treatment of the incentive issues that might arise in the context of identifier-based discovery. The BGP incentive model presented has several advantages, mainly providing a dynamic, distributed pricing scheme for route distribution that is partially immune to manipulation and does not require a centralized bank. However, several forms of manipulation may occur in the rewards model. For example, node i may declare a reward $r_i > r_{\text{next}(R_i)}$ when competing on a route to possibly increase its utility, or nodes may lie about the real values of δ when declaring these values to their upstream nodes. Such forms of manipulation may be avoided by route verification and secure cryptographic mechanisms (check secure BGP for example [11]). We have not considered such issues in this chapter. The cascaded rewarding model is similar in spirit to network marketing in economics. One of the main pitfalls of network marketing (alternatively referred to as Multi-Level Marketing or MLM) is that it may put more incentives on recruiting distributors rather than on making a sale. In our model, recruiting a downstream distributor is equivalent to making a sale since a downstream node to i must route to d through i. In addition, the assumption $r_i < r_{\text{next}(R_i)}$ eliminates the pyramid effect common to network marketing schemes.

We are currently working on establishing equilibria results for general classes of utility functions and for general graph structures. The natural next step after that would be to study distributed algorithms that converge to the equilibria, particularly focusing on scalable extensions to BGP [24]. Additionally, we plan to quantify the cost of being discoverable, or in other words the initial investment r_d required by d to guarantee *global reachability* – as a function of the network structure. Interestingly here, for the Internet AS level topology, it was shown by Krioukov et al. [29] that the average distance between any two nodes is small (around 3.5 hops). This property lends itself to the *small world* phenomenon in complex networks [8].

Appendix A: Proof of Lemma 1

Proof. The proof is straightforward. The case for $|B(i)| = 1$ is trivial. The case for $|B(i)| = 2$ is trivial as well since i will not be able to make a sale to the higher reward neighbor by picking the lower reward offer. Assume that node i has more than 2 neighbors and that any two neighbors, say k, l advertise routes $R_k, R_l \in P(i)$ s.t. $k = next(R_k), l = next(R_l)$ and $r_{ki} < r_{li}$, and assume that i's utility for choosing route R_k over R_l either increases or remains the same i.e. $u_i^{R_k} \geq u_i^{R_l}$. We will show by contradiction that neither of these two scenarios could happen.

Scenario 1: $u_i^{R_k} > u_i^{R_l}$

From Equation (15.1), it must be the case that either (case 1) node i was able to make at least one more sale to some neighbor j who would otherwise not buy, or (case 2) some neighbor j who picks $(j, i)R_i$ can strictly increase her $\delta_j(\mathbf{s})$ when i chooses the lower reward path R_k. For case 1, and assuming that r_{ij} is the same when i chooses either route, it is simple to show that we arrive at a contradiction in the case when $j \in \{k, l\}$ (mainly due to the strictly decreasing reward assumption i.e. $r_i < r_{\text{next}(.)}$); and in the case when $j \notin \{k, l\}$, it must be the case that j's utility increases with i's route choice i.e. $u_j^{(j,i)R_k} > u_j^{(j,i)R_l}$. This contradicts with Equation (15.1) since w.r.t. j, both routes have the same next hop node i. The same analogy holds for case 2.

Scenario 2: $u_i^{R_k} = u_i^{R_l}$

Using the same analogy of scenario 1, there must exist at least one neighbor j of i that would buy i's offer only when the latter picks R_k, or otherwise node i will be able to strictly increase its utility by picking R_l pocketing more profit. □

References

1. GENI: Global environment for network innovations. http://www.geni.org/
2. G. Huston, BGP in 2008, http://www.potaroo.net/ispcol/2009-03/bgp2008.html, March 2008.
3. NSF Nets FIND initiative. http://www.nets-find.net/
4. Akamai Technologies. http://www.akamai.com, 2008.
5. I. Abraham, C. Gavoille, D. Malkhi, N. Nisan and M. Thorup, Compact name-independent routing with minimum stretch, in *Proceedings of ACM SPAA '04*, ACM Press, New York, pp. 20–24, 2004.
6. M. Afergan, Using repeated games to design incentive-based routing systems, in *Proceedings of 25th IEEE International Conference on Computer Communications, INFOCOM 2006*, pp. 1–13, 2006.
7. B. Awerbuch, A. Bar-Noy, N. Linial and D. Peleg, Compact distributed data structures for adaptive routing, in *Proceedings of ACM STOC '89*, ACM, New York, pp. 479–489, 1989.
8. A.L. Barabasi, *Linked*, Perseus Publishing, 2002.
9. L. Bent and G.M. Voelker, Whole page performance, Workshop on Web Content Caching and Distribution, Boulder CO, 2002.
10. L. Blume, D. Easley, J. Kleinberg and E. Tardos, Trading networks with price-setting agents, in *EC '07, Proceedings of the 8th ACM Conference on Electronic Commerce*, ACM, New York, pp. 143–151, 2007.
11. K. Butler, T. Farley, P. McDaniel and J. Rexford, A survey of bgp security issues and solutions, Tech. Report, AT&T Labs, 2004.
12. M. Caesar, T. Condie, J. Kannan, K. Lakshminarayanan and I. Stoica, Rofl: routing on flat labels, in *Proceedings of SIGCOMM 2006*, ACM Press, New York, pp. 363–374, 2006.
13. M. Caesar and J. Rexford, Bgp routing policies in isp networks, *IEEE Network* **19**(6), 5–11, 2005.
14. J. Feigenbaum, C. Papadimitriou, R. Sami and S. Shenker, A bgp-based mechanism for lowest-cost routing, *Distrib. Comput.* **18**(1), 61–72, 2005.
15. J. Feigenbaum, V. Ramachandran and M. Schapira, Incentive-compatible interdomain routing, in *EC '06: Proceedings of the 7th ACM conference on Electronic Commerce*, ACM, New York, pp. 130–139, 2006.
16. J. Feigenbaum and S. Shenker, Distributed algorithmic mechanism design: Recent results and future directions, in *Proceedings of the 6th International Workshop on Discrete Algorithms and Methods for Mobile Computing and Communications*, ACM Press, pp. 1–13, 2002.
17. D. Fudenberg and J. Tirole, *Game Theory*, MIT Press, 1991.
18. V. Fuller, D. Meyer and D. Farinacci, Lisp alternative topology (lisp+alt), http://tools.ietf.org/html/draft-fuller-lisp-alt-03.txt.
19. L. Gao, On inferring autonomous system relationships in the internet, *IEEE/ACM Trans. Netw.* **9**(6), 733–745, 2001.
20. T.G. Griffin, F.B. Shepherd and G. Wilfong, Policy disputes in path-vector protocols, in *ICNP '99: Proceedings of the Seventh Annual International Conference on Network Protocols*, IEEE Computer Society, Washington, DC, p. 21, 1999.
21. T.G. Griffin, F.B. Shepherd and G. Wilfong, The stable paths problem and interdomain routing, *IEEE/ACM Trans. Netw.* **10**(2), 232–243, 2002.
22. W. Herrin, What does a bgp route cost?, http://bill.herrin.us/network/bgpcost.html, 2008.
23. C. Huitema and S. Weerahandi, Internet measurements: The rising tide and the dns snag, in *Proceedings of the 13th ITC Specialist Seminar on IP Traffic Measurement Modeling and Management*, IPseminar, ITC, Monterrey, CA, 2000.
24. J. Khoury, Discovery in large-scale networks, Ph.D. Thesis, University of New Mexico, 2009 (to appear).
25. J. Khoury and C.T. Abdallah, Identifier-based discovery mechanism design in large-scale networks, in *Future-Net'09: Proceedings of the International Workshop on the Network of the Future, with IEEE ICC'09*, Dresden, Germany, 2009.

26. J. Kleinberg, The convergence of social and technological networks, *Commun. ACM* **51**(11), 66–72, 2008.
27. J. Kleinberg and P. Raghavan, Query incentive networks, in *FOCS '05: Proceedings of the 46th Annual IEEE Symposium on Foundations of Computer Science*, IEEE Computer Society, Washington, DC, pp. 132–141, 2005.
28. T. Koponen, M. Chawla, B.G. Chun, A. Ermolinskiy, K.H. Kim, S. Shenker and I. Stoica, A data-oriented network architecture, in *Proceedings of SIGCOMM'07*, ACM, Kyoto, Japan, 2007.
29. D. Krioukov, k.c. claffy, K. Fall and A. Brady, On compact routing for the internet, *SIGCOMM Comput. Commun. Rev.* **37**(3), 41–52, 2007.
30. H. Levin, M. Schapira and A. Zohar, Interdomain routing and games, in *STOC '08: Proceedings of the 40th Annual ACM Symposium on Theory of Computing*, ACM, New York, NY, pp. 57–66, 2008.
31. C. Li, B. Yu and K. Sycara, An incentive mechanism for message relaying in unstructured peer-to-peer systems, in *AAMAS '07: Proceedings of the 6th International Joint Conference on Autonomous Agents and Multiagent Systems*, ACM, New York, pp. 1–8, 2007.
32. A. Mas-Colell, M.D. Whinston and J.R. Green, *Microeconomic Theory*, Oxford University Press, 1995.
33. D. Massey, L. Wang, B. Zhang and L. Zhang, A scalable routing system design for future internet, in *ACM SIGCOMM Workshop on IPv6 and the Future of the Internet*, ACM Press, New York, 2007.
34. D. Meyer, L. Zhang and K. Fall, Report from the IAB Workshop on Routing and Addressing, Internet RFC 4984, Septenber 2007.
35. R. Moskowitz, P. Nikander and P. Jokela, Host identity protocol architecture, RFC 4423, 2006.
36. N. Nisan and A. Ronen, Algorithmic mechanism design, in *Proceedings of the 31st ACM Symposium on Theory of Computing*, Atlanta, GA, 1999.
37. N. Nisan, T. Roughgarden, E. Tardos and V.V. Vazirani, *Algorithmic Game Theory*, Cambridge University Press, New York, 2007.
38. Papadimitriou, C.: Algorithms, games, and the internet, in *STOC '01: Proceedings of the Thirty-Third Annual ACM Symposium on Theory of Computing*, ACM, New York, pp. 749–753, 2001.
39. Y. Rekhter, T. Li and S. Hares, RFC 4271: A border gateway protocol 4 (bgp-4), 2006.
40. J. Shneidman, D.C. Parkes and M. Seltzer, Overcoming rational manipulation in distributed mechanism implementations, Tech. Rep., Harvard University, 2003.
41. I. Stoica, R. Morris, D. Liben-Nowell, D.R. Karger, M.F. Kaashoek, F. Dabek and H. Balakrishnan, Chord: A scalable peer-to-peer lookup protocol for internet applications, *IEEE/ACM Trans. Netw.* **11**(1), 17–32, 2003.
42. S. Yuen and B. Li, Strategyproof mechanisms towards dynamic topology formation in autonomous networks, *Mob. Netw. Appl.* **10**(6), 961–970, 2005.

Printed in the United States
149013LV00003B/64/P

9 789048 130177